Consolidated Bioprocessing for Sustainable Fuel Production

Consolidated Bioprocessing for Sustainable Fuel Production

Edited by

Neha Srivastava
IIT (BHU)
Varanasi, India

WILEY

Registered Offices
John Wiley & Sons, Inc., 111 River Street, Hoboken, NJ 07030, USA
John Wiley & Sons Ltd., New Era House, 8 Oldlands Way, Bognor Regis, West Sussex, PO22 9NQ

For details of our global editorial offices, customer services, and more information about Wiley products visit us at www.wiley.com.

The manufacturer's authorized representative according to the EU General Product Safety Regulation is Wiley-VCH GmbH, Boschstr. 12, 69469 Weinheim, Germany, e-mail: Product_Safety@wiley.com.

Wiley also publishes its books in a variety of electronic formats and by print-on-demand. Some content that appears in standard print versions of this book may not be available in other formats.

Library of Congress Cataloging-in-Publication Data Applied for:

Hardback ISBN: 9781394381159

Cover Design: Wiley
Cover Image: Pixabay

Set in 9.5/12.5pt STIXTwoText by Straive, Chennai, India

Contents

List of Contributors *xv*
Preface *xix*
Acknowledgment *xx*

1 **Overview of Green Industrial Energy** *1*
Sheetal Thakur, Subhadra Rajpoot, Shagun Sharma, and Ankita Sharma
1.1 Introduction *1*
1.2 Current Market Scenario of Green Industrial Energy *3*
1.2.1 Market Size and Growth *3*
1.2.2 Regional Trends *4*
1.2.3 Policy and Market Drivers *6*
1.3 Technology Highlights of Green Industry Energy *7*
1.3.1 Advanced Solar Technologies *7*
1.3.2 Wind Energy Innovations *8*
1.3.3 Green Hydrogen Production *8*
1.3.4 Artificial Intelligence and Digitalization *9*
1.3.5 Carbon Capture and Sustainable Fuels *10*
1.4 Green Fuel Currently Present in the Market *10*
1.4.1 Biofuels *11*
1.4.2 Other Green Fuels *13*
1.5 Advantages of Green Industrial Energy *13*
1.5.1 Significant Reduction in Green House Gas Emission *13*
1.5.2 Long-Term Cost Efficiency and Operational Savings *14*
1.5.3 Job Creation and Economic Growth *14*
1.5.4 Enhanced Energy Security and Independence *15*
1.5.5 Decarbonization of Hard-to-Abate Sectors *15*
1.5.6 Innovation, Technological Advancement, and Competitiveness *15*
1.5.7 Public Health and Environmental Benefits *16*
1.5.8 Grid Stability and Flexibility *16*
1.5.9 Support for Circular Economy and Resource Efficiency *17*
1.5.10 Regulatory and Brand Advantages *17*
1.6 Challenges and Market Gaps *18*
1.6.1 Supply–Demand Mismatch *18*

1.6.2 Energy Storage and Grid Reliability *18*
1.6.3 Work Force and Skill Shortage *19*
1.6.4 Infrastructure and Supply Chain Constrains *19*
1.6.5 Policy and Regulation Uncertainty *19*
1.6.6 Geographic and Temporal Mismatches *20*
1.6.7 High Initial Capital Cost *20*
1.7 Integration of Emerging Technologies *21*
1.8 Conclusion *21*
References *21*

2 Fundamentals of Bioprocesses in Green Energy Production *29*
Avanish Kumar, Girdhari Lal Devnani, and Dan Bahadur Pal
2.1 Introduction *29*
2.2 Basic Concepts of Bioprocess Engineering *30*
2.3 Feedstock for Bioprocesses *32*
2.3.1 First-Generation Feedstock *33*
2.3.2 Second-Generation Feedstock *33*
2.3.3 Third-Generation Feedstock *33*
2.3.4 Fourth-Generation Feedstock *33*
2.4 Microorganisms and Enzymes in Bioprocesses *34*
2.5 Types of Bioprocesses *35*
2.5.1 Hydrolysis *35*
2.5.2 Acidogenisis *35*
2.5.3 Acetogenesis *35*
2.5.4 Methanogens *35*
2.6 Production of Bioprocess Products *36*
2.6.1 Biogas Production *36*
2.6.2 Fermentation *37*
2.6.2.1 Ethanol Production *37*
2.6.2.2 Transesterification (for Biodiesel Production) *39*
2.6.2.3 Biohydrogen Production *40*
2.7 Bioreactors and Process Design *42*
2.7.1 Types of Bioreactors *42*
2.7.1.1 Batch Bioreactor *42*
2.7.1.2 Fed-Batch Bioreactor *42*
2.7.1.3 Continuous Bioreactor (Chemostat) *43*
2.7.1.4 Anaerobic Bioreactor *43*
2.7.1.5 Photo Bioreactor (PBR) *43*
2.7.1.6 Airlift Bioreactor *43*
2.7.1.7 Packed Bed and Fluidized Bed Bioreactors *43*
2.7.1.8 Key Design and Operational Parameters *43*
2.8 Process Integration Techniques *44*
2.8.1 Simultaneous Scarification and Fermentation (SSF) *44*
2.8.2 Consolidated Bioprocessing (CBP) *45*
2.8.3 Aspen Plus and Similar Process Simulation Tools *45*

2.9 Environmental and Economic Evaluations *45*
2.9.1 Energy Balance *45*
2.9.2 Life Cycle Assessment (LCA) *46*
2.9.3 Cost–Benefit Analysis (CBA) *46*
2.9.4 Sustainability and the Circular Economy *46*
2.10 Challenges and Future Directions in Bioenergy Bioprocesses *47*
2.10.1 Technical Barriers *47*
2.10.2 Innovations in Synthetic Biology and Metabolic Engineering *47*
2.10.3 Policy, Funding, and Market Considerations *47*
2.10.4 Emerging Trends *48*
2.11 Conclusion *48*
References *48*

3 Biochemical Processes to Enhance Bioenergy Production *57*
Madhavi Tiwari
3.1 Introduction *57*
3.1.1 Advanced Microbial Strain Development *57*
3.1.2 Enzyme Engineering and Catalytic Enhancement *59*
3.1.3 Co-Culture and Consortium-Based Strategies *59*
3.1.4 Process Intensification Techniques *60*
3.1.5 Omics-Driven Process Optimization *60*
3.1.6 Commercialization and Scale-Up Challenges *61*
3.1.7 Future Prospects and Emerging Technologies *61*
3.2 Two Microbial Biofactories of the Future for the Production of Energy *61*
3.2.1 Metabolic Engineering: Redirecting Cellular Machinery *62*
3.2.2 Genome Editing: Precision with CRISPR/Cas9 *62*
3.2.3 Synthetic Biology: Designing Novel Biofuel Pathways *63*
3.3 Tailoring Enzymes for Efficient Bioenergy Conversion *63*
3.3.1 Directed Evolution: Mimicking Nature's Selection *64*
3.3.2 Rational Design: Engineering with Structure in Mind *64*
3.3.3 Enzyme Immobilization: Stability and Reusability *65*
3.3.4 Cocktail Optimization: Synergistic Enzyme Blends *65*
3.4 Co-Culture and Microbial Consortium Strategies for Bioenergy *65*
3.4.1 Synergistic Microbial Interactions *65*
3.4.2 Division of Metabolic Labor *66*
3.4.3 Community Stability and Resilience *66*
3.5 Advanced Process Integration for Enhanced Biofuel Production *67*
3.5.1 High-Solid Fermentation (HSF) *67*
3.5.2 Simultaneous Saccharification and Fermentation (SSF) *67*
3.5.3 Fed-Batch and Continuous Fermentation *68*
3.5.4 In Situ Product Recovery (ISPR) *68*
3.6 Integrative Omics Approaches for Optimizing Bioenergy Yields *68*
3.6.1 Genomics *68*
3.6.2 Transcriptomics *68*
3.6.3 Proteomics *69*

3.6.4 Metabolomics *69*
3.7 Commercialization and Scale-Up Challenges in Bioenergy Production *70*
3.7.1 Technical Barriers *70*
3.7.2 Economic Constraints *71*
3.7.3 Regulatory and Policy Challenges *71*
3.7.4 Infrastructure and Logistics *71*
3.7.5 Social Acceptance and Public Perception *71*
3.8 Conclusion *72*
References *72*

4 Bioprocess Technology: Commercial Challenges for Application *75*
Muhammad Islam, Muhammad Zubair Ali, Yuliya Kulikova, Olga Babich, Afsana Huseynova Anvar, Hafiz Abdullah Shakir, Muhammad Khan, Titin Haryati, Marcelo Franco, and Muhammad Irfan
4.1 Introduction *75*
4.2 Biomass Conversion Strategies *78*
4.2.1 Biochemical Approach *78*
4.2.2 Thermochemical Approach *79*
4.3 Green Biofuels Produced via Bioprocess Technology *80*
4.3.1 First-Generation Biofuels *80*
4.3.2 Second-Generation Biofuels *81*
4.3.3 Third-Generation Biofuels *82*
4.3.4 Fourth-Generation Biofuels *82*
4.4 Green Biofuels Examples *82*
4.4.1 Gaseous Biofuels *82*
4.4.2 Liquid Biofuels *83*
4.4.2.1 Recent Advancements in Bioprocessing Technology *84*
4.5 Challenges *85*
4.5.1 Sustainability and Availability of Feedstock *85*
4.5.2 Diversity of Feedstock *85*
4.5.3 Efficiency of Conversion *85*
4.5.4 Developments in Technology *86*
4.5.5 High Production Costs and Initial Investment *86*
4.6 Frameworks of Regulations and Policies for the Development of Biofuels *87*
4.6.1 Blending Goals and Renewable Fuel Standards (RFS) *87*
4.6.2 Regulations for Feedstock *87*
4.6.3 Monetary Rewards *87*
4.6.4 Support for Research and Development *87*
4.6.5 Policies for Imports and Exports *87*
4.6.6 Rules Pertaining to the Environment *88*
4.6.7 Support for Infrastructure *88*
4.6.8 Public Education and Awareness *88*
4.7 Conclusion *88*
References *89*

5 Consolidated Bioprocess Technology: Principles and Applications *95*
Gursharan Singh and Shailendra Kumar Arya
5.1 Introduction *95*
5.2 Important Aspects of CBP *96*
5.2.1 CBP, A Single Step Technology and Economical Process *96*
5.2.2 Importance of the Baker's Yeast (*Saccharomyces cerevisiae*) for CBP *97*
5.2.3 Strain Development/Improvement for the Successful Operation of CBP *98*
5.2.4 CBP in the Presence of More Than One Microbe or Consortium *100*
5.2.5 CBP at High Temperatures *101*
5.3 Advantages and Disadvantages of CBP Technology *102*
5.4 Conclusions *103*
References *103*

6 Microbial Consortia in Consolidated Bioprocess Technology *107*
Tehreem Maryam, Maryam Kainat, Maryam Maqsood, Yuliya Kulikova, Olga Babich, Afsana Huseynova Anvar, Hafiz Abdullah Shakir, Muhammad Khan, Titin Haryati, Marcelo Franco, and Muhammad Irfan
6.1 Introduction *107*
6.2 Microbial Consortia *109*
6.2.1 Ecological Basis of Microbial Consortia *110*
6.2.2 Types of Microbial Consortia *111*
6.3 Consolidated Bioprocessing *113*
6.4 Integration of Microbial Consortia in CBP *114*
6.5 Future Perspectives *116*
6.6 Conclusion *117*
References *118*

7 Challenges and Solutions of Consolidated Bioprocessing: A Case Study of Biogas Production *125*
Muhammad Anas, Waseem Ahmed Khattak, Erdogan E. Hakki, Javed Iqbal, Banzeer Ahsan Abbasi, Ziaur Rehman, and Akhtar Munir
7.1 Introduction *125*
7.1.1 Overview of Consolidated Bioprocessing (CBP) *125*
7.1.2 Importance in Biogas Production *126*
7.1.3 Challenges in Consolidated Bioprocessing *127*
7.1.3.1 Limitations of Microbial Strain *127*
7.1.3.2 Metabolic Limitations and Genetic Problems *128*
7.1.3.3 Trade-Offs of Enzyme Production *128*
7.1.3.4 Tolerance Problems and Stability Problems *128*
7.2 Substrate Complexity and Pretreatment Needs *128*
7.2.1 Lignocellulosic Heterogeneity *128*
7.2.1.1 Pretreatment Is Nonetheless Required *128*
7.2.1.2 Degradation By-Product Inhibition *128*
7.2.2 Process Efficiency and Yield Optimization *129*
7.3 Case Study: Biogas Production via CBP *129*

7.3.1 System Design and Operational Parameters *129*
7.3.2 Key Performance Metrics *130*
7.4 Solutions and Innovations *131*
7.4.1 Genetic and Metabolic Engineering of Microbial Consortia *131*
7.4.2 Advanced Pretreatment Techniques *131*
7.4.3 Process Integration and Intensification *132*
7.5 Future Perspectives *133*
7.5.1 Technological Advancements *133*
7.5.2 Commercial Viability *134*
7.5.3 Sustainability Outlook *134*
7.6 Conclusion *135*
References *136*

8 Technological Assessment of Bioprocess and Consolidated Process: A Way Forward *141*
Waseem Ahmed Khattak, Erdogan E. Hakki, Muhammad Anas, Javed Iqbal, and Banzeer Ahsan Abbasi
8.1 Introduction *141*
8.1.1 Overview of Bioprocess Technologies *141*
8.1.2 Importance of Technological Assessment in Bioprocessing *142*
8.1.3 Scope and Objectives of the Chapter *143*
8.2 Fundamentals of Bioprocessing *144*
8.2.1 Key Components and Stages of Bioprocesses *144*
8.2.2 Classification of Bioprocesses: Batch, Fed-Batch, and Continuous *145*
8.2.2.1 Batch Bioprocessing *145*
8.2.2.2 Fed-Batch Bioprocessing *145*
8.2.2.3 Continuous Bioprocessing *145*
8.2.3 Role of Microorganisms and Enzymes *145*
8.2.3.1 Microorganisms in Bioprocessing *146*
8.2.3.2 Enzymes in Bioprocessing *146*
8.3 Consolidated Bioprocessing (CBP): Concept and Evolution *146*
8.3.1 Definition and Core Principles *146*
8.3.2 Historical Development of CBP *147*
8.3.3 Comparison with SHF and SSF *148*
8.4 Technological Advancements in Bioprocess and CBP *149*
8.4.1 Synthetic Biology and Metabolic Engineering *149*
8.4.2 Enzyme Engineering and Cocktail Optimization *149*
8.4.3 Reactor Design and Process Integration *150*
8.4.4 Process Monitoring and Automation *151*
8.5 Assessment Parameters for Bioprocess Technologies *151*
8.5.1 Technical Feasibility and Efficiency Metrics *151*
8.5.2 Economic Evaluation and Cost–Benefit Analysis *152*
8.5.3 Environmental Impact and Life Cycle Assessment (LCA) *152*
8.5.4 Scalability and Industrial Applicability *153*
8.6 Challenges and Bottlenecks *154*

8.6.1 Substrate Limitations and Pretreatment Issues *154*
8.6.2 Product Inhibition and Yield Losses *155*
8.6.3 Enzyme Cost and Recycling *155*
8.6.4 Regulatory and Market Constraints *156*
8.7 Future Trends and Opportunities *156*
8.7.1 Integration with Artificial Intelligence and Digital Twins *156*
8.7.2 Circular Bioeconomy and Sustainable Development Goals *157*
8.7.3 Policy Support and Public–Private Partnerships *157*
8.8 Conclusion *158*
References *159*

9 Reactor Design and Consolidated Bioprocess Technology: Practical Use and Impact *165*
Navodita Maurice
Abbreviations *165*
9.1 Introduction *167*
9.2 Bioreactor Design *170*
9.2.1 Scale Up *174*
9.2.2 Mixing *174*
9.2.3 Mass Transfer *175*
9.2.4 Shear *177*
9.2.5 Pressure *179*
9.2.6 Scale-Up Criteria *179*
9.2.7 Scale Down *180*
9.3 Types of Bioreactors *180*
9.3.1 Stirred Tank Reactors (STRs) *184*
9.3.1.1 Continuous Stirred Tank Reactors (CSTRs) *185*
9.3.1.2 Features of the STRs *186*
9.3.2 Bubble Column Bioreactors (BCRs) *187*
9.3.3 Airlift Bioreactors (ARLs) *188*
9.3.4 Packed Bed Bioreactors (PBRs) *188*
9.3.5 Fluidized Bed (FB) Reactors *189*
9.3.6 Membrane Bioreactors (MBs) *191*
9.3.7 Perforated Trays *191*
9.3.8 Fixed Bed Bioreactors *192*
9.3.9 Drum Bioreactor *192*
9.3.10 Wave Bioreactors *193*
9.3.11 Disposable Bioreactors *194*
9.3.12 Photobioreactors *194*
9.4 Recent Advances in Bioreactor Applications *196*
9.4.1 Production of Biopharmaceuticals *196*
9.4.1.1 Bioartificial Liver Devices *197*
9.4.1.2 Blood-Brain Barrier (BBB) Models *199*
9.4.1.3 Bioartificial Kidney (BAK) Devices *200*
9.4.2 Production of Biofuels *202*

9.4.3 Bioremediation *206*
9.4.4 Cell Culture and Tissue Engineering *207*
9.4.5 Food Production *209*
9.5 Conclusion and Future Perspectives of Bioreactors *210*
Acknowledgments *210*
References *210*

10 Economic Analysis of Consolidated Bioprocess Technology in Green Fuel Production *227*
Susheel Kumar Singh, Pradeep Kumar Mishra, and Ravi Prakash Jaiswal
10.1 Introduction *227*
10.2 Overview of CBP *228*
10.2.1 Conventional Bioconversion vs. CBP *228*
10.2.2 Engineering Strategies and Synthetic Biology in CBP *229*
10.3 Economic Components of Biofuel Production *230*
10.3.1 Feedstock Costs *230*
10.3.2 Capital Expenditures (CAPEX) *231*
10.3.3 Operating Expenditures (OPEX) *232*
10.3.4 Product Yields and Conversion Efficiency *232*
10.4 Integration with Circular Bioeconomy Models *233*
10.5 Scalability and Deployment Potential *234*
10.6 Factors Influencing CBP in Scaling Up *234*
10.6.1 Market Viability in the Low-Carbon Economy *234*
10.6.2 Innovation and Strain Development Trajectory *236*
10.7 Future Outlook and Strategic Recommendations *236*
10.8 Conclusion *237*
References *238*

11 Frequency Adaptation of Consolidated Bioprocess Technology in Green Fuel Production *243*
Susheel Kumar Singh, Pradeep Kumar Mishra, and Ravi Prakash Jaiswal
11.1 Introduction *243*
11.2 Technological Basis of Consolidated Bioprocessing (CBP) *244*
11.2.1 Core Components of CBP *244*
11.2.2 Microorganisms in CBP: Native to Engineered *244*
11.2.3 Feedstock Compatibility and Pretreatment Considerations *245*
11.2.4 Advanced CBP Strategies: Thermophiles, Consortia, and Hybrid Processes *245*
11.3 Drivers of CBP Adoption *245*
11.3.1 Economic Drivers *245*
11.3.2 Technological and Innovation Drivers *246*
11.3.3 Policy and Regulatory Drivers *246*
11.4 Technological Enhancements and Digestibility Improvements *246*
11.4.1 Engineered *Clostridium thermocellum* for Hemicellulose Co-Utilization *247*
11.4.2 Discovery of Novel Thermophiles for Hydrogen Production *247*
11.4.3 Thermophile-Based CBP Platforms and Synthetic Biology *247*

11.4.4 Co-Culture and Semi-Fed-Batch Strategies for Enhanced Sugar Release *247*
11.5 Challenges in the Adoption of CBP Technology *248*
11.5.1 Genetic Instability and Limited Host Diversity *248*
11.5.2 Enzyme Activity and Product Yield *249*
11.5.3 Feedstock Variability and Pretreatment Dependence *249*
11.5.4 Lack of Scalable Bioreactor Designs *249*
11.5.5 Regulatory and Market Uncertainty *249*
11.6 Case Studies and Industrial Applications of CBP *249*
11.6.1 Industrial Actors Leading CBP Implementation *249*
11.6.2 Case Studies and Pilot Facilities *250*
11.6.2.1 Case Studies on the Mascoma Corporation Pilot Plant and Industrial Applications of CBP *250*
11.6.3 Geographic and Sectoral Trends *250*
11.7 Emerging Trends in CBP Technology *251*
11.8 Integration with Circular Bioeconomy *251*
11.9 Future Prospects and Policy Recommendations in Sustainable Biofuel Production *251*
11.9.1 Vision for the Future *252*
11.9.2 Strategic Policy Recommendations *252*
11.9.3 A Path Toward a Carbon-Neutral Future *253*
11.10 Conclusion *253*
References *254*

12 Cost Analysis of Consolidated Bioprocess Technology for Green Energy Production and Environmental Sustainability Impact *259*
Vikram Singh, Neeraj Gupta, Deepti Singh, Gyanendra Tripathi, Ashish, and Fredrick Kayusi
12.1 Introduction *259*
12.1.1 Scope and Objectives of the Chapter *260*
12.2 Fundamentals of Consolidated Bioprocessing (CBP) *261*
12.2.1 Definition and Evolution of CBP *261*
12.2.2 Microorganisms Used in CBP (Native and Engineered Strains) *261*
12.2.2.1 Native CBP Microorganisms *261*
12.2.2.2 Engineered CBP Microorganisms *262*
12.2.3 Substrates Used: Lignocellulosic Biomass and Agricultural Waste *263*
12.3 Process Flow and System Design of Consolidated Bioprocessing (CBP) *263*
12.3.1 One-Pot Conversion: Hydrolysis and Fermentation *263*
12.3.2 Role of Enzyme Production *In Situ* *264*
12.3.3 Integrated Bioreactor Design *264*
12.3.4 Process Monitoring and Control *265*
12.4 Cost Components of CBP *265*
12.4.1 Feedstock Collection and Pretreatment *265*
12.4.2 Microbial Strain Development and Inoculum Preparation *266*
12.4.3 Capital Costs: Equipment, Bioreactor, and Utilities *267*
12.4.4 Operating Costs: Enzymes, Media, Energy, and Labor *267*

12.4.5 Downstream Processing and Product Recovery *267*
12.5 Economic Feasibility and Techno-Economic Assessment *268*
12.5.1 Case Study: Cost–Benefit Analysis of a CBP Facility *268*
12.5.1.1 Cost Considerations *268*
12.5.1.2 Benefit Analysis *269*
12.5.2 Sensitivity Analysis: Effect of Feedstock Price and Yield *269*
12.5.3 Comparison with Traditional Bioprocessing and Other Green Technologies *269*
12.6 Challenges and Opportunities *270*
12.6.1 Technical Barriers in Microbial Engineering and Scale-Up *270*
12.6.2 Economic Viability for Developing Countries *270*
12.6.3 Opportunities in Synthetic Biology and AI-Driven Process Control *270*
12.6.4 Future Directions for Sustainable and Circular CBP Models *270*
References *271*

13 Biochemical Enzymes to Influence the Bioprocess of Lignocellulosic Biomass for Bioethanol Production Process and Cost: An Analysis *275*
Nihan Arabacı
13.1 Introduction *275*
13.2 Bioethanol Production *276*
13.3 Biomass Feedstocks in the Manufacturing of Bioethanol *277*
13.3.1 First-Generation Feedstocks *277*
13.3.2 Second-Generation Feedstocks *277*
13.3.3 Third-Generation Feedstocks *278*
13.4 Lignocellulosic Biomass in the Bioethanol Process *278*
13.5 Microbial Cellulases in Hydrolyzing Lignocellulose: Biological Pretreatment *282*
13.5.1 Endoglucanase (Endo β-1,4-D-Glucan Glucanohydrolase, EC 3.2.1.4) *286*
13.5.2 Exoglucanase or Cellobiohydrolase (CBH) (β-1,4-D-Glucan Cellobiohydrolase, EC 3.2.1.91) *286*
13.5.3 β-Glucosidase or Cellobiase (β-D-Glucoside Glucohydrolase, EC 3.2.1.21) *286*
13.6 Conclusion *287*
References *287*

Index *293*

List of Contributors

Banzeer Ahsan Abbasi
Department of Botany
Rawalpindi Women University
Rawalpindi
Pakistan

Muhammad Zubair Ali
Department of Biotechnology
University of Sargodha
Sargodha
Pakistan

Muhammad Anas
Department of Soil Science & Plant Nutrition
Selçuk University
Konya
Türkiye

Afsana Huseynova Anvar
Department of Life Science
Khazar University
Baku
Azerbaijan

and

Department of Molecular Biology and Biotechnology
Baku State University
Baku
Azerbaijan

Nihan Arabacı
Department of Biology
Faculty of Arts and Sciences
University of Cukurova
Adana
Türkiye

Shailendra Kumar Arya
UIET
Biotechnology Branch
Panjab University
Chandigarh
India

Ashish
Department of Bioengineering
Integral University
Lucknow
Uttar Pradesh
India

Olga Babich
Research and Education Center "Industrial Biotechnologies"
Immanuel Kant Baltic Federal University
Kaliningrad
Russia

Girdhari Lal Devnani
Department of Chemical Engineering
Harcourt Butler Technical University
Kanpur
Uttar Pradesh
India

Marcelo Franco
Department of Exact Science
State University of Santa Cruz
Ilheus
Brazil

Neeraj Gupta
Institute of Biosciences and Technology
Shri Ramswaroop Memorial University
Lucknow
Uttar Pradesh
India

Erdogan E. Hakki
Department of Soil Science & Plant Nutrition
Selçuk University
Konya
Türkiye

Titin Haryati
Research Center for Applied Microbiology
National Research and Innovation Agency
Bogor
West Java
Indonesia

Javed Iqbal
Department of Botany
Bacha Khan University
Charsadda
Pakistan

Muhammad Irfan
Department of Biotechnology
University of Sargodha
Sargodha
Pakistan

Muhammad Islam
Department of Biotechnology
University of Sargodha
Sargodha
Pakistan

Ravi Prakash Jaiswal
Department of Chemical Engineering & Technology
IIT (BHU)
Varanasi
Uttar Pradesh
India

Maryam Kainat
Department of Biotechnology
University of Sargodha
Sargodha
Pakistan

Fredrick Kayusi
Department of Environmental Studies
Geography and Planning
Maasai Mara University
Narok
Kenya

and

Department of Environmental and Earth Sciences
Pwani University
Kilifi
Kenya

Muhammad Khan
Institute of Zoology
University of the Punjab
Lahore
Pakistan

Waseem Ahmed Khattak
Department of Plant Sciences
Faculty of Biological Sciences
Quaid-I-Azam University
Islamabad
Pakistan

Yuliya Kulikova
Research and Education Center "Industrial Biotechnologies"
Immanuel Kant Baltic Federal University
Kaliningrad
Russia

Avanish Kumar
Department of Chemical Engineering
Harcourt Butler Technical University
Kanpur
Uttar Pradesh
India

Maryam Maqsood
Department of Biotechnology
University of Sargodha
Sargodha
Pakistan

Tehreem Maryam
Department of Biotechnology
University of Sargodha
Sargodha
Pakistan

Navodita Maurice
Prophyl Ltd.
Bár
Hungary

Pradeep Kumar Mishra
Department of Chemical Engineering & Technology
IIT (BHU)
Varanasi
Uttar Pradesh
India

Akhtar Munir
Department of Chemistry
Quaid-i-Azam University
Islamabad
Pakistan

Dan Bahadur Pal
Department of Chemical Engineering
Harcourt Butler Technical University
Kanpur
Uttar Pradesh
India

Subhadra Rajpoot
Department of Applied Sciences
Amity University
Greater Noida
Uttar Pradesh
India

Ziaur Rehman
Department of Chemistry
Quaid-i-Azam University
Islamabad
Pakistan

Hafiz Abdullah Shakir
Institute of Zoology
University of the Punjab
Lahore
Pakistan

Ankita Sharma
School of Legal Studies
CGC University
Jhanjeri
Punjab
India

Shagun Sharma
Department of Biotechnology
Chandigarh University
Gharuan
Punjab
India

Deepti Singh
Institute of Biosciences and Technology
Shri Ramswaroop Memorial University
Lucknow
Uttar Pradesh
India

Gursharan Singh
School of Allied Medical Sciences
Lovely Professional University
Phagwara
India

Susheel Kumar Singh
Department of Chemical Engineering & Technology
IIT (BHU)
Varanasi
Uttar Pradesh
India

Vikram Singh
Department of Bioengineering
Integral University
Lucknow
Uttar Pradesh
India

Sheetal Thakur
Department of Biotechnology
University Centre for Research & Development
Chandigarh University
Gharuan
Punjab
India

Madhavi Tiwari
Department of Applied Sciences
Shri Rawatpura Sarkar University
Raipur
Chhattisgarh
India

Gyanendra Tripathi
Department of Bioengineering
Integral University
Lucknow
Uttar Pradesh
India

Preface

The present book entitled, *Consolidated Bioprocessing for Sustainable Fuel Production*, provides a way forward towards sustainable bioenergy production via consolidated bioprocessing (CBP). It is a strategy where all different steps of bioenergy production happen in a single reactor via single microbial consortium which is directly responsible for pretreatment, bioconversion and fermentation of the final product. Single process platform of CBP can save all: cost involved in multistage steps of bioenergy production, time and energy consumption. Moreover, due to single-stage processing, this is among one of the most sustainable and green process for green fuel production at commercial stage.

Therefore, focusing on these facts, this proposed book aims to explore and describe the current situation of green energy production via applying sustainable bioprocess and CBP approach, a strategy that is aimed at the best option for the sustainable production of green energy. This book is divided in total 13 chapters, where Chapters 1 and 2 provide current status update on industrial bioenergy production and fundamental principles of bioprocess in green energy production. Chapters 3 and 4 are focused on details about biochemical process involved in bioenergy production and about existing challenges in commercial bioprocess technology for bioenergy production. Furthermore, Chapters 5 and 6 are based on delivering information about principles and application of CBP process and significance of microbial population in this process to improve the sustainability of the overall process. Whereas, Chapters 7 and 8 explain about existing challenges in CBP process with the feasible solution and technological assessment of this process for further improvement. Furthermore, Chapters 9 and 10 are specific on reactor design specificity of CBP process and economic challenges in the area of CBP with possible future suggestions; while Chapters 11 and 12 focuses on case study trials of CBP-inspired green energy production at a commercial scale and general overview cost analysis of CBP process. At last, Chapter 13 provides new insights about the microbial enzymes which are the main players of CBP process to provide feasibility in pretreatment, bioconversion and bioenergy production. This book certainly sets a potential milestone to overcome the serious concern associated with the sustainability of the CBP process and bioenergy production via CBP on a commercial scale.

Acknowledgment

The editor is thankful to all the academicians and scientists whose contributions have enriched this volume. I also express our deep sense of gratitude to our parents whose blessings have always prompted us to pursue academic activities deeply. It is quite possible that in a work of this nature, some mistakes might have crept in text inadvertently and for these we owe undiluted responsibility. I am grateful to all authors for their contribution to present book. I am deeply grateful to Sakeena Qureshi and Wiley for giving this opportunity to editor and Department of Chemical Engineering & Technology, IIT (BHU) Varanasi, Uttar Pradesh, India, for all technical support. I thank them from the core of our heart.

1

Overview of Green Industrial Energy

Sheetal Thakur[1], Subhadra Rajpoot[2], Shagun Sharma[3], and Ankita Sharma[4]

[1] *Department of Biotechnology, University Centre for Research & Development, Chandigarh University, Gharuan, Punjab, India*
[2] *Department of Applied Sciences, Amity University, Greater Noida, Uttar Pradesh, India*
[3] *Department of Biotechnology, Chandigarh University, Gharuan, Punjab, India*
[4] *School of Legal Studies, CGC University, Jhanjeri, Punjab, India*

1.1 Introduction

Green industrial energy refers to the usage of renewable and low-carbon energy sources/resources and energy-efficient operation principles that contribute to the overall sustainability of the environment. Although different types of energy resources are available, green energy focuses on energy derived from natural mechanisms. Renewable energy is often used interchangeably with green energy despite clear differences. While the former focuses on the source's renewal, the latter relies on energy sustainability and a low carbon footprint. The statement "All green energy is renewable, but not all renewable energy is green" can be supported by a comparison of biomass that may be renewable, but its burning emits harmful gases and is limited for the same reason. Green industrial energy refers to the use of green energy sources, including solar, rivers, wind, and biomass, among others, in industrial activities instead of conventional fossil fuels, thus minimizing carbon emissions into the atmosphere. Societies are anxious about the increased techniques' implications on global warming, increasing the melting ice cap, and related environmental consequences, as well as the rapid exhaustion of natural resources and excessive waste production and related pollution. The global industrial space is currently undergoing radical reforms, which are largely due to the extensive need to address climate change, tackle resource depletion, and curb the severe environmental challenges. In this context, industries, which have been the largest consumers of fossil fuels and high-energy-consuming filling processes, have shown impeccable leadership in embracing sustainable principles. These principles are represented by green industrial energy that focuses on the use of renewable energy sources (RESs) and energy-efficient technologies in industries to lessen the negative implications of environmental abuse induced by production activities.

Green industrial energy can be understood through three key dimensions, which include the use of green energy, products, and processes in business management. Green energy refers to the generation, as well as the use, of power from sources that are renewable,

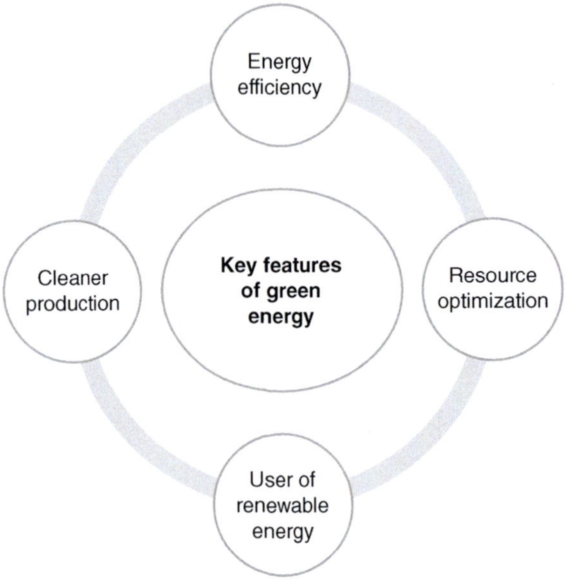

Figure 1.1 Key features of green energy.

including solar, wind, hydro, and biomass sources (Figure 1.1). When compared to fossil fuels, green energy sources considerably decrease the carbon emissions and greenhouse gas footprints. The most significant beneficiary of the move will be the manufacturing sector, which is accountable for a large portion of global energy use and greenhouse gas emissions. In turn, the transition not only mitigates environmental degradation but also enhances power security and business productivity and saves a lot of money in the long term. Promising economic opportunities for governments and businesses are expanding as the global role of green industrial strategies rises. As a result, policymakers in developing countries face a challenge when helping capture ongoing transformation opportunities because they struggle to integrate rapid change to make people more robust and economically viable. Consequently, it is necessary to combine industrial and environmental strategies, including diplomacy to better understand the environmental context while also examining industrial practices and expectations. Industrial contribution to job creation remains a bright spot coupled with rapid technological advancements that provide excellence in the acceleration of the transition; new solar panels, biomass boilers, and grid-scale battery technologies increase the reliability, adaptability, and affordability of RESs for industrial use. Moreover, artificial intelligence (AI) tools that refine energy management, as well as new materials and digital systems, are raising the efficiency of industrial systems (Figure 1.1).

There are several benefits to the adoption of green industrial energy, but it remains a challenge in one way or the other. However, the pros are multiple: lowering environmental harm, increasing people's well-being and state of protection, improving brand value, and building resilience against ever-changing prices of fossil fuels. Nonetheless, green industrial energy would necessitate the adoption of coordinated policy frameworks, strategic funding, and cooperation from governments, key industrial interests, and research establishments. This chapter focuses on the current market view of green industrial energy from the market size and growth element. This chapter explains the certainty that transition to green industrial energy is a crucial move toward more sustainable economic growth and environmental health. It will also be considering the green energy technologies, kinds of green fuels already on the market, significant obstacles, and existing market failures in the green industrial energy systems

1.2 Current Market Scenario of Green Industrial Energy

It can generally be seen that the shortage of non-renewable resources requires society to create multi-determined strategic plans. To tackle the shortage of nature's resources, human society is developing green energy, which has been mainly responsible for the revolution and rapid transformation, strengthening sustainable economic growth, and environmental accessibility. Consequently, governments of all countries have been defining policy frameworks and policy implementation that target clean and green energy and have been developing both new innovations and strong sustainability goals. Accordingly, the green industrial energy market is currently spreading rapidly, based on technology innovation, policy support, and global growth in demand for clean energy. This chapter displays an outline of the current market that has been developed based on the new proposed methods confirmed by new data and surveys.

1.2.1 Market Size and Growth

As per the recent report published by the Green Energy Market Report (2025), the global renewable energy market is valued at approximately USD 1.74 trillion in 2025, with projections to reach USD 7.28 trillion by 2034, reflecting a compound annual growth rate (CAGR) of 17.23% (StartUs Insights, 2025). Other estimates place the 2025 market size at USD 128.53 billion, with a projected rise to USD 379.23 billion by 2037 (CAGR > 9.2%). The green power segment specifically is forecasted to grow from USD 56.95 million in 2025 to USD 146.13 million by 2033, at a CAGR of 12.5%.

Solar and wind power are still driving most of the new energy capacity being added. In 2024, renewables made up over 90% of all new global power, with 585 GW added—a record-breaking 15.1% growth rate for renewable energy. In the United States, 98% of the new power-generating capacity in the first quarter of 2025 was for solar and wind; in this, solar alone tends to hold over two-thirds of that growth. Other major markets are also aligned with this trend, in particular, India added a record 25 GW of the renewable energy capacity in the FY 2024–25, led by the solar power.

The renewable energy market is divided into three main segments:

- By nature: solar, wind, bioenergy, hydro energy, and geothermal energy.
- By capacity: 0–1000, 1100–5000, and > 5000 kWh.
- By consumer: domestic, commercial, industrial, utility, and others.

Furthermore, there are some of the subsegments:

- In the category of the solar energy: photovoltaic (PV) systems, solar water heating, concentrated solar power (CSP).
- In the category of wind energy: onshore and offshore wind energy.

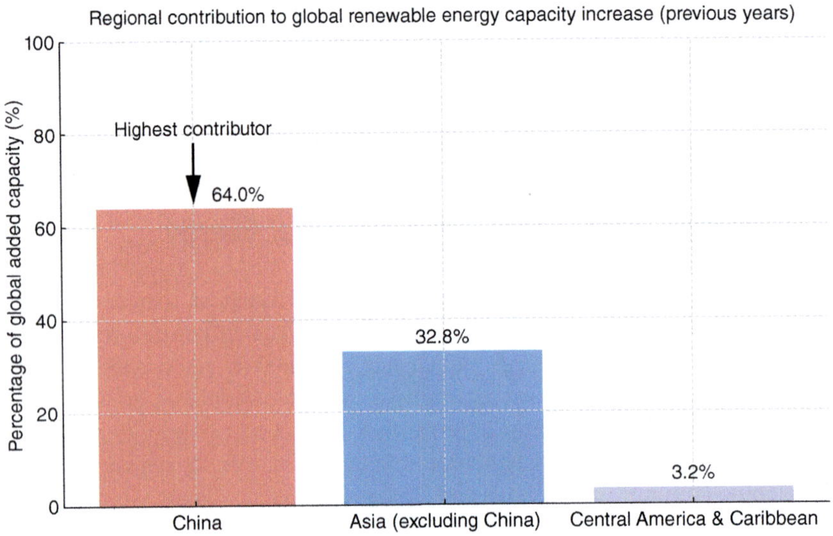

Figure 1.2 Bar graph showing region-wise contribution to global renewable energy.

- In the category of bioenergy: biomass, biogas, biofuels (ethanol, biodiesel).
- In the category of hydro energy: small and large hydropower, run-of-river hydropower.
- In the category of geothermal energy: flash steam and dry steam plants, binary cycle power plants.

Renewables have the power to revitalize economies. However, the transition to clean energy needs to happen faster and more fairly, ensuring that every country can fully access affordable, clean renewable power.

The bar chart (Figure 1.2) shows regional contributions to global renewable energy capacity increases in previous years, with China highlighted as the top contributor.

Geographic disparities are again highlighted while looking at the progress in capacity growth. In the past few years, the major increase in global renewable energy has been observed to occur in the Asian region, where China alone is contributing nearly 64% of the global capacity. However, Central America and the Caribbean had the negligible share of almost 3.2% only.

> The renewable energy market is projected to grow significantly, reaching \$1.73 trillion by 2029, with a steady CAGR of 9.8%.

1.2.2 Regional Trends

The United States, Germany, the United Kingdom, India, and China are the major market centers, with cities such as London, New York, Melbourne, Sydney, and Singapore as activity epicenters. However, Asia Pacific, despite existing problems, including China and India, is the fastest-growing market due to rapid population growth, industrialization, and huge investments in green power projects. The percentages of electricity generation from renewable energy are shown in Table 1.1. It has already become more economically

Table 1.1 Region-Wise Trends of Green Industrial Energy.

Region	Key Trends and Highlights	References
United Kingdom	• Renewables now generate more electricity than fossil fuels, with 42.3% of energy production from renewables in 2025. • Offshore and onshore wind are rapidly expanding, supported by government investment and policy changes, including lifting the onshore wind ban and targeting 40 GW offshore wind by 2030. • Solar capacity is set to triple, with major new solar farms approved and under construction. • The Clean Power 2030 plan aims for 95% clean energy by 2030, requiring major grid upgrades and storage solutions. • New technologies such as sustainable aviation fuels, carbon capture, and microgrids are entering the market.	RSMUK (2025) Uswitch (2025) Apertus Group (2025) BDO UK (2025)
United States	• Export capacity for renewables is knocked to grow by 86%, with substantial new projects coming online in early 2025. • The United States is investing heavily in wind, solar, and battery storage, with a focus on grid modernization and resilience. • Industrial green hydrogen projects are scaling up, and policy incentives are accelerating clean energy adoption in manufacturing and heavy industry.	Deloitte (2025)
Asia (India and China)	• Asia Pacific is the rapidly expanding marketplace for green industrial energy, driven by fast automation, expansion, and strong government targets. • China leads global wind and solar capacity additions, with aggressive investments in green hydrogen and battery storage. • India is ramping up solar and wind installations, aiming for 500 GW of non-fossil capacity by 2030, and is addressing grid and workforce challenges to meet surging industrial demand.	Gopinath (2016) Swain (2014) Li and Lin (2016)
Europe (excluding the United Kingdom)	• The European Union is advancing large-scale renewable integration, with a focus on cross-border grid interconnections, offshore wind in the North Sea, and green hydrogen corridors. • Policy frameworks such as Fit for 55 and the Green Deal Industrial Plan are driving industrial decarbonization and innovation.	Mao et al. (2019) Tagliapietra and Veugelers (2023)
Other regions	• Australia is scaling up solar and wind for industrial mining and export-oriented green hydrogen projects. • Middle Eastern countries are investing in solar and green ammonia for export, leveraging abundant renewable resources. • Africa is expanding decentralized renewables for industrial clusters, focusing on energy access and local manufacturing.	Curtis and McLellan (2023) Baruah and Enweremadu (2019)

beneficial for electricity generation. Simultaneously, the main factor in a green economy is renewable energy, and European countries strive to attain 100% climate neutrality by the year 2050, which requires 2% of Gross Domestic Product (GDP) to collect green funds. Electricity provided more than 80% of green generation in several countries. Furthermore, 47–8% was hydropower and 8–6% the solar-wind power generation. Thus, the most positive dynamics of electricity generation positively affect RESs are China, India, Turkey, Russia, Iran, and Nigeria. In 2019, RESs increased by more than 200 GW of installed capacity. Furthermore, the total capacity of RESs was 47%, which was considered the most ecologically clean industry of generation by scientists. Positive changes of such a scale are supported by investment activity dynamics. The ratio of the first half of 2020 indicates a 5% increase in investment and renewables. In 2019, a total of $333.6 billion was allocated to clean energy. Asia Pacific's project countries accounted for the majority of them and amounted to 49%. In the solar-wind industry, the first six months of 2020 amounted to $35 billion, which amounted 319% more than in the whole of 2019. The second significant trend is a slow growth in power consumption: 1–3% in 2019. Renewables and natural gas were responsible for growth by 70%, and the leaders in gas suppliers were China, India, and Indonesia. The decline in consumption was most significant in the United States over the past decades, followed by Germany.

1.2.3 Policy and Market Drivers

Green industrial policies are at the core of national specials in the United Kingdom, the United States, the European Union (EU), China, and India. They result in reducing carbon emissions, increasing competitiveness, driving energy self-sufficiency by promoting national clean technology manufacturing, R&D, and the whole of the supply. To be most specific, the US Inflation Reduction Act has led to substantial investments in renewables and green hydrogen, while the EU's Green Deal Industrial Plan and subsequent carbon border adjustment mechanisms stimulate industrial decarbonization. Due to the good performance of carbon markets and carbon pricing drives, industrial players are incentivized to reduce emissions and invest in cleaner technologies. Carbon trading has become more sufficient and favorable as a result of less shrinkage in aligned carbon markets, the implementation of tax incentives for trading, and the spread of global standards. Policies supporting workforce reskilling, especially in the cases of workers transitioning from fossil fuel sectors, are critical budget allocations, challenging talent shortages in the green energy sector and securing a just and inclusive transition. Policies supporting the expansion and sophistication of electric grid infrastructure, including regulatory power, energy storage, permitting, and coordination of the integration, are critical in the longer term to spread more renewables through industrial processes. This is essential to manage power demand growth from new sectors, such as cleantech manufacturing and data centers. Industrial use of green energy is quickly growing due to rising demand from domains such as cleantech manufacturing, data centers, AI, and carbon management. By 2030, these industries will need a further 57 GW, beyond even surpassing current renewable capacity. The plummeting costs and rising reliability and performance of solar, wind, and battery technologies make smart energy increasingly attractive to industry. There is a growing trend of private sector investment in renewables, instigated by financial innovation, including novel instruments like green bonds, electricity derivatives, contracts for difference, and virtual power purchase agreements. These contraptions reduce returning volatility and stimulate the flow of investment to green portfolios. The evacuation of global supply chains and the resurgence of local energy autonomy are encouraging the development of—mostly solar and wind—production capabilities, particularly in the United States, Europe, and India. Corporate decarbonization and net-zero commitment targets are another powerful driver, encouraging industry players to adopt green energy production, increase energy efficiency, and reduce direct as well as top of in-line carbon, driven by customer expectations and investors' Environmental, Social, Governance (ESG) compliance demand. These multiple drivers foster process innovation and support the global shift toward a low-carbon economy.

1.3 Technology Highlights of Green Industry Energy

1.3.1 Advanced Solar Technologies

Largely, the energy consumption is higher in the industrial sector when compared to any other segment. Nonetheless, the rampant thermal demand for several processes can only be fulfilled by solar energy. A solar energy system can work as an alternative source in developing nations, as there are several other important sub-sectors viz., textile, agriculture, pharmaceutical, food processing, and brick industries. Solar energy is used for several processes in these sectors such as boiling, drying, dying, washing, sterilization, and pasteurization (Hasanuzzaman et al., 2012; Schmitt, 2016). In addition, studies reveal that the energy utilization in glass, steel, and ceramic industries is divided into different temperature range requirements, where only 13% of the process implementation requires thermal energy at a low temperature of up to 100°C, whereas 27% of processes require up to 200°C temperature and the rest of the procedures need very high temperatures (Farjana et al., 2018a). In European industries, the majority of the energy demand, i.e., around 27%, is consumed in the form of heat (Heß & Oliva, 2010; Kylili et al., 2018). Furthermore, the sustainable solution for the future industry and the process applications is solar energy systems, as fossil fuels usage for the heat production in industry is the primary reason for the rising source of greenhouse gas (GHG) emissions and climate change (Farjana et al., 2018b). Therefore, considering the benefits of harvesting solar energy as a source of heat production in industrial processes, it reduces the reliance on fossil fuels and hence the decline in GHG emissions. Conversely, there are various challenges on commercial scale for integrating this solar heat into the different industrial processes along with the assortment of augmented solar thermal collectors. Research studies in this aspect have identified a suitable industrial process where solar heat can be utilized either globally or regionally (Kumar et al., 2019).

Derakhshandeh et al. (2021) discussed various solar panel cleaning techniques, such as passive coating, electrostatic and mechanical self-cleaning, and surface acoustic waves. In oppressive circumstances, water spraying on solar panels can enhance the equipment proficiency as well as utilization. Perovskite solar cells have emerged as a game-changer, achieving conversion efficiencies above 25% in 2024 and offering lower production expenditure and more flexibility compared to that of the traditional silicon-based panels. Research at the University of Oxford has pushed these efficiencies beyond 28%, signaling imminent commercial viability and the potential to surpass silicon panels in industrial applications (FirstIgnite, 2024). Flexible solar panels and high-efficiency PV cells are being deployed on diverse industrial surfaces, maximizing space utilization and enabling more widespread solar adoption (Green City Times, 2025).

1.3.2 Wind Energy Innovations

Energy production from the airstream has become another valuable choice as an RES as it is advantageous in the reduction of carbon footprint and the partial fulfilment of expanding power supply demand. This sustainable energy source works with the help of the most prevalent and commonly used method of electromagnetic generators (EMGs)-based wind turbines, which transform wind energy into electricity. Nevertheless, EMGs have disadvantages, such as high maintenance costs and limited voltage output, which necessitate the development of alternative energy conversion technologies (Ellabban et al., 2014). Researchers have developed nanogenerators to serve as self-powered wind speed sensors or to power commercial solar cells (Zhao, Wu, et al., 2022). Triboelectric nanogenerators (TENGs) are the alternative options, emerging as promising devices capable of capturing a range of mechanical energies, including wind energy. TENGs offer benefits such as high voltage, low weight, and simple production, but their output is still modest (Qian & Jing, 2018). Novel hybrid nanogenerators have also been developed that combine TENG and EMG technologies for self-powered wind speed sensing and fueling commercial solar cells (Guo et al., 2019).

Recent innovations, such as a pinwheel that generates electricity continuously by combining triboelectrification and electromagnetic induction, demonstrate the potential to capture small quantities of wind energy (Guo et al., 2019). Further research and expansion in the utilization of wind energy may help in the formulation of more proficient and low-cost wind energy harvesting procedures.

1.3.3 Green Hydrogen Production

In the current scenario, green hydrogen is another effective solution on the global platform, with the capacity of contributing to decarbonization and sustainable energy solutions, offering a carbon-neutral future (Chakraborty et al., 2022). Several methods have been designed for the production of hydrogen, and each method varies depending upon its energy source and environmental influence. Several research reports have proposed hydrogen generation techniques by classifying them on the basis of their source, designing a color-coded scheme to differentiate between the numerous methods (Moreno-Brieva et al., 2023; Scovell & Walton, 2024). As an example, green hydrogen is produced via the electrolysis of water, generating electricity sourced from other renewable sources, namely, solar, wind, hydro, and biomass, and is characterized by its zero-carbon emission manner. Hence, the integration of green hydrogen with other green technologies, like green buildings, can produce an interdependent ecosystem where renewable hydrogen powers energy-efficient infrastructures developed with a motive of enhancing overall sustainability and reducing carbon footprints (Massarweh et al., 2023).

Electrolysis with RESs and biomass gasification are measured as the most efficient and economical techniques for the production of green hydrogen. Along with this, green hydrogen production can be achieved by several other methods, namely, encompassing oil distillation, coal mining, brackish water desalination, and natural gas conversion. However, the accurate utilization of such methods demands thorough planning, technology optimization, infrastructure building, and potential markets for green hydrogen applications (Marouani et al., 2023). The use and application of green hydrogen as an energy source are diverse in major industry setups, including transportation and heating–cooling systems.

Currently, green hydrogen production has become very popular worldwide as it is a cleaner and more environment-friendly energy option compared to blue and yellow hydrogen derived from natural gas and coal. Major technology advanced countries such as Japan, China, the EU, the United States, and South Korea are designing and incorporating the plans of expanding green hydrogen production in the near future. These countries aspire to harness hydrogen as a clean and RES, actively striving to advance the requisite technology to achieve higher efficiency and reduced costs in green hydrogen production (Marouani et al., 2023).

1.3.4 Artificial Intelligence and Digitalization

AI has emerged as a transformative force in the renewable energy sector, potentially revolutionizing solar and wind energy solutions by optimizing production, increasing environmental awareness, and enhancing management and distribution (Belu, 2013; Jha et al., 2017). AI applications have proven their ability to improve wind farm management, decision-making, and offshore wind energy production (Mitchell et al., 2022). Machine learning (ML), a part of AI, has been gradually implemented to solar and wind energy setups with a motive of optimizing their efficacy, decreasing the costs, and improving their predictive capacities. ML methods have become the advanced routes playing a crucial role in the enhancement of the use and implementation of RESs, making them increasingly viable for addressing global energy demands. In solar energy systems, ML algorithms improve solar panel performance, enhance energy forecasting, and boost energy storage solutions. For instance, ML techniques have been used to detect and localize solar panel faults, drastically reducing the time required to identify and rectify faulty cells (Ahan et al., 2021). In addition, ML-based solar energy forecasting systems have demonstrated high accuracy (Partheeban et al., 2022). Similarly, ML has shown promise in wind energy systems, where it can enhance wind energy forecasting and wind speed predictions (Zhao, Sun, et al., 2022). Moreover, ML algorithms have optimized wind farm planning, resulting in more efficient design and implementation (Wu et al., 2014). Overall, the extrapolative and optimization capacities of ML can elevate RES with more effective management that can further contribute to the attainment of global energy demands and sustainable development goals (SDGs).

By incorporating ML and AI, smart grids can intelligently respond to sudden shifts in power supply and demand and forecast energy needs and potential equipment malfunctions (Azad et al., 2019). In urban areas, the traditional fuels and batteries are not capable of storing sufficient electricity, which makes the load forecasting unfavorable. However, ML algorithms, such as ensemble learning, have shown promising results, frequently outperforming conventional forecasting models (Rangel-Martinez et al., 2021). The work by

Naganathan et al. (2016) further supports the versatility of ML, deploying semi-supervised learning to model energy consumption for buildings and at an industrial scale, respectively. They effectively tackled the challenge of insufficiently labeled data, a common issue in ML. Other ML applications include semi-supervised learning for modeling energy consumption in buildings and dynamic security assessment (DSA) for evaluating the system resilience.

1.3.5 Carbon Capture and Sustainable Fuels

Carbon capture and storage (CCS) technologies are being piloted in energy-intensive industries like cement and steel, enabling substantial emissions reductions where direct electrification is challenging. Sustainable bioenergy and biowaste utilization offer additional pathways for decarbonizing industrial heat and power generation.

Carbon capture, utilization, and storage (CCUS) technologies are potential elucidations of converting the energy-demanding functions, facilitating the conversion to a low-carbon economy (Metz et al., 2005). Generally, the CC from industrial developments can be achieved in three modes, i.e., pre-, post-, and oxy-combustion capture. Post-combustion capture is the most reliable (end of the pipe) approach in which the flue gases stemming from the polluting route are treated for CO_2 capture. The major problem with this process stems from the fact that in flue gases, the partial pressure of CO_2 is low which denotes the elevated energy utilization for the capture process (Fernandez et al., 2014). Pre-combustion capture comprises a partial fuel oxidation (e.g., catalytic reforming, gasification) for syngas production which further, subject to CC, produces a hydrogen-abundant gas. The distinctive benefit of pre-combustion capture over post-combustion capture is that the partial pressure of CO_2 in the shifted syngas is high, resulting in a significant lowering of energy consumption for the CC process (Theo et al., 2016). Also, the hydrogen production occurs during the pre-combustion route, which can be utilized for the manufacturing of synthetic fuels/chemicals (Budzianowski, 2012). Oxy-combustion route uses an oxygen-rich burning environment, which avoids contamination with nitrogen and, therefore, streamlines the carbon capture process. However, there are certain important parameters which need to be carefully considered, such as the ancillary energy utilization for cryogenic air separation and the redesigning of the boilers to survive the oxy-fuel combustion (Cormos, 2016).

1.4 Green Fuel Currently Present in the Market

The extensive use of fossil fuels in the market led to their exhaustion stage to arrive early than the projected time, hence, green fuel has become a replacement factor. Green fuel is a type of renewable as well as clean energy that is specifically engineered for lowering the environmental impact, such as causing depletion in the carbon emissions. Green fuels are rapidly expanding in global markets, driven by regulatory mandates, corporate sustainability goals, and technological innovation. Green fuel's key segment is green hydrogen, which holds a great market value of about USD 12.31 billion in 2025 and is foreseen to reach USD 199.22 billion by 2034. From the production front and deployment, Asia Pacific is envisioned to surpass the market.

1.4.1 Biofuels

Biofuels are other sources of renewable energy, which are derived from organic materials like plants, algae, agri-waste, and animal fats. Presently, biofuels have been categorized as first to fourth generation, depending on the raw material and/or biosynthetic pathway (i.e., genetic engineering).

First-Generation Biofuels: First-generation biofuels are generally classified as biodiesel and bioethanol. The production of bioethanol is based upon the microbial fermentation of food-grade feedstocks rich in sucrose and starch, like wheat, corn, and sugarcane. Commercial strains utilized for the same include, but are not limited to, *Saccharomyces cerevisiae*, *Scheffersomyces stipites*, and *and Schizosaccharomyces pombe*. Depending upon the type of strain and feedstock used, bioethanol can also be categorized as second- and third-generation biofuels (Azhar et al., 2017; Favaro et al., 2019; Jacobus et al., 2021). Biodiesel is majorly produced from edible soy, rapeseed, or palm oil. In comparison with bioethanol, only partial biosynthesis of biodiesel is achieved, as its making involves chemically catalyzed procedures like transesterification of lipids with alcohols. Another method is enzymatic catalysis, which is currently applicable only at the laboratory scale (Aarthy et al., 2014; Fattah et al., 2020). Further, the production of biobutanol is also achievable by sugar fermentation using corn, cane, wheat, and other food crops. However, lower yield and productivity, product inhibition, and high costs are the major limitations in biobutanol production (Jiang et al., 2019; Karthick & Nanthagopal, 2021).

Second-Generation (Advanced) Biofuels: In order to overcome the drawbacks encountered in first-generation biofuels, second-generation biofuels were designed by using lignocellulosic biomass from agri- and wood residues and other waste streams like wheat bran/animal fat from food industry or leftover of cooking/frying oil. Certain non-food plants like the drought-enduring *Jatropha curcas*, capable of growing in wastelands, may act as other promising sources for second-generation biofuels (Kumar & Sharma, 2008). Hence, second-generation biofuels evade the requirement of agricultural land usage and do not compete with food resources. However, the feedstocks used in second-generation fuel production show more complexity in comparison to that of sugarcane or palm oil as this waste stream contains components like lignin, which reduce the fermentation efficiency. This leads to the requirement of pre-treatment procedures, which ultimately increase the processing time as well as cost (Joshi et al., 2017; Mischko et al., 2018). Catalytic cracking and hydrogenation methods are also used for sourcing second-generation biodiesel from waste oils. However, the incomplete conversion and cake formation are the major drawbacks of this process, which lead to the catalyst deactivation (Yano et al., 2015; Zhang et al., 2022).

One of the ancient, most commonly used, and long-established fermentative processes for butanol production is based upon the *Clostridia* fermentation which utilizes lignocellulose biomass and other waste streams. Several *Clostridia* are recognized as natural butanol producers, capable of metabolizing a wide range of substrates. However, similar to its first-generation precursors, this process is having the limitations of low butanol titers and product inhibition (Jiang et al., 2019; Karthick & Nanthagopal, 2021). Scientific data estimates that second-generation biofuels have the potential to meet up to 30% of the world's transportation energy demand (Malode et al., 2021). However, in spite of being highly capable of valorizing waste streams, second-generation biofuels alone cannot meet the huge current energy supply demand worldwide. Similarly, the biomass generated from food crops for production of first-generation biofuels is obtainable in restricted amount. Hence, second-generation biofuels must be integrated with other technologies to ensure sufficient supply of fuels, which prompted the research on third-generation biofuels.

Third-Generation Biofuels: Third-generation biofuels are primarily sourced from the biomass of microalgae and cyanobacteria, which naturally generate alcohols and lipids to further convert into biodiesel or any other high-energy fuel products. The capability of algae of exhibiting twofold to fourfold higher photosynthesis rates than terrestrial plants results in faster biomass production (Li et al., 2008). Several cultures can be cultivated using waste water, brackish or salt water, as algae cultivation does not necessitate the availability of fresh water, which is low cost and avoids interference with agricultural activity (Melcher et al., 2021). Most importantly, direct CO_2 supply is required for the efficient algae cultivation, which can be attained from industrial emitters or by atmospheric CC. In conventional cultivation setups, nearly 70% of supplied CO_2 is utilized for photosynthesis and hence, for biomass production (Doucha & Lívanský, 2006). Therefore, algae-based biofuels could have a negative carbon footprint as microalgae directly sequester GHG within their biomass. One of the most pronounced third-generation processes involved the production of biodiesel or other high-energy-density biofuels, like biokerosene, by using oleaginous microalgae (Lee et al., 2015).

Fourth-Generation Biofuels: Created using synthetic biology and genetically engineered microorganisms, these fuels aim for even greater sustainability and lower emissions, though they are still emerging at commercial scale. The most recent category is termed fourth-generation biofuels, which embodies the use of genetic engineering to enhance the desirable traits of organisms utilized in biofuel production. This encompasses various traits such as the ability to utilize multiple sugars like pentoses and hexoses, improved lipid synthesis, and enhanced photosynthesis and carbon fixation. In model organisms like *Escherichia coli* and *S. cerevisiae*, extensive genetic engineering tools exist to modify endogenous pathways or introduce new methods. Currently, two major strategies are employed: one focuses on optimizing native producers by improving growth rates, broadening substrate utilization, redirecting metabolic flux toward biofuel synthesis, and increasing yields; the other involves reconstructing biofuel-producing pathways from native organisms into genetically tractable models. A range of microorganisms, including bacteria, yeast, and algae, serve as heterologous hosts, affecting their metabolic flexibility to convert diverse substrates into various biofuels.

Various industries, including automobiles, shipping, and aerospace, frequently use green fuels, such as biofuels, synthetic fuels, and hydrogen. This pattern can be seen in the automotive sector alone, which accounts for over 33% of the e-fuel market share in 2024, highlighting the regular and widespread use in transportation. The agricultural sector also has a great usage of biofuels in heavy-duty machinery, such as tractors, irrigation pumps, etc., all of which require fuel for working. There is a lower impact on the environment, all with the lower costs of the fuel in order to keep the machinery running for longer periods of time. This helps in the improvement of energy self-sufficiency in rural areas; furthermore, it promotes the circular economy by utilizing agricultural waste. Industrial sectors and transportation are the primary drivers of bioenergy adoption, but power generation, shipping, chemicals, and buildings are making significant strides, supported by technological advances and strong policy frameworks.

1.4.2 Other Green Fuels

Green hydrogen is steadily gaining momentum as a clean fuel option for industrial use, fuel cell vehicles, and energy storage. Significant government initiatives and supportive policies are helping drive large-scale production and its integration into current energy systems. Generated by electrolysis powered by renewable energy, green hydrogen is gaining real market traction, especially for use in industry (e.g., steel, chemicals), heavy transport, and as a feedstock for other green fuels. Sustainable Aviation Fuel (SAF) production is growing quickly, with next-generation fuels doubling output to one million tons in 2024, marking a significant step toward dropping carbon emissions in the aviation sector. SAF is generated from a range of feedstocks, including waste oils, fats, and advanced bio-based materials. Airlines are locking in multi-year supply agreements, and SAF is central to decarbonizing aviation. E-fuels, i.e., synthetic fuels made using renewable electricity and captured carbon, are being scaled up as promising alternatives for hard-to-electrify sectors like shipping, aviation, and heavy industry. E-fuels are manufactured by combining green hydrogen with captured CO_2, creating synthetic hydrocarbons that can be utilized in existing engines and infrastructure. These are gaining attention for their potential in sectors where direct electrification is challenging, such as aviation and shipping. Renewable Natural Gas (RNG), also known as biomethane, is produced from organic waste via anaerobic digestion or gasification. It is used in transportation, electricity generation, and industrial applications, and can be injected into existing natural gas grids. Advanced biofuels (second-generation) are derived from non-food feedstocks, such as agri-waste, municipal solid waste, or lignocellulosic biomass. They address concerns around food security and land use associated with first-generation biofuels. Green methanol and green ammonia, produced using renewable energy and sustainable feedstocks, are emerging as alternative fuels for shipping, power generation, and chemical industries. They are especially important for decarbonizing maritime transport and industrial processes.

1.5 Advantages of Green Industrial Energy

1.5.1 Significant Reduction in Green House Gas Emission

The green industrial energy is extensively being uplifted due to the foremost reason of having very low or zero carbon emissions that makes it more environmentally friendly and also the need of the hour. This reduces the load on the environment and also turns to be cost-effective, as its sources are natural like wind, sun, ocean, and geothermal energy that are present in the environment free of cost. They also make the countries self-sufficient contrary to the case in non-renewable sources and moreover these non-renewable ones are not good for the environment as well as the health. Climate change is also prevented along with the biodiversity (Sharma & Dubey, 2022). The EU has set ambitious climate goals as part of its commitment to achieving climate neutrality by 2050. The European Climate Law, adopted in 2021, establishes a legally binding target to reduce net GHG emissions by at least 55% by 2030, compared to 1990 levels. Additionally, the EU aims to increase the share of energy from renewable sources to a minimum of 42.5% by 2030, with an aspiration to reach 45%. These measures are integral to the EU's strategy to become the first climate-neutral continent by 2050 (Androniceanu & Sabie, 2022).

1.5.2 Long-Term Cost Efficiency and Operational Savings

The evolution to sustainable energy and environmentally friendly economic growth has highlighted the key role of green finance in supporting this shift. For all this, China has put a dual-carbon goal in the Paris Agreement, focusing on the efforts for the reduction in consumption of energy intensity and the utilization efficiency also need to be enhanced. To achieve these goals, China must continue to accelerate its green transformation by guiding and restructuring the operations and industrial frameworks of businesses toward environmentally sustainable practices. This involves integrating green development principles across all sectors, including industry, agriculture, and services, to safeguard energy, lessen emissions, and enhance efficiency. By promoting advancements in technology, models, and standards, China aims to foster a comprehensive green transformation of traditional industries. Encouraging enterprises to adopt green, low-carbon, and eco-friendly processes and equipment is crucial in this endeavor. Additionally, optimizing the spatial layout of industries and infrastructure in industrial parks, and promoting the circular combination of industries and circular production in enterprises, are essential steps toward a sustainable future (Wang & Xu, 2025). The completion of the SDGs by 2030 needs developing a green economy as a key factor. For achieving a positive influence in the free economic efficiency nationwide, an advanced industrial structure is needed, whereas the energy utilization efficiency has partial positive intermediary effect in the influence path. The rationalization of industrial structure is strongly and positively associated with nationwide green economic efficiency, with energy utilization serving as a significant positive mediator in this relationship (Jiang et al., 2024).

1.5.3 Job Creation and Economic Growth

Concerning job creation, federal and state government policy initiatives focused on the United States as the point of focus. Analysis reveals that whereas a $1 million investment in the green energy sector could generate approximately 17 jobs, fossil fuel sectors yielded, on average, about 5 jobs for the same amount of investment. Furthermore, these jobs spread across varying levels of skills, making green job creation an all-inclusive growth strategy. After analysis, research demonstrates that policy-driven demand for green energy skilled labor facilitated the green economy at play. There were cases of Ontario's Green Energy Act in Canada, which supercharged and fueled demand for skilled labor in clean technology, made exploitative green economy. Another G-7 country that emerged as a global leader in renewable energy and green jobs is Germany. This achievement has been propelled by the Energiewende policy. The International Labour Organization (ILO) report highlights that energy efficiency, waste management, and public transportation sectors are playing an equal role in this achievement: there seem thousands of jobs within the green energy sector in the United Kingdom; there were 10,000 jobs in the solar and wind energy sector in France. In the world, Japan had the highest-paid job market in solar energy, which facilitated one of their power national policies to increase investment in these energy sectors. Italy, in the renewable energy sector, has adopted an energy system which covers green jobs in the circular economy sector. According to the International Renewable Energy Agency (IRENA) report, solar and wind power are spearheading green energy in the G-7 countries.

1.5.4 Enhanced Energy Security and Independence

Non-renewables have started to be replaced by renewables, and these are also replacing the energy imports, thereby helping in the enhancement of energy independence and security (Gökgöz & Güvercin, 2018). Energy security is very important to every country, and it is determined by the diversification and political security of a country's energy sources. There is a need for the green energy transition, leading to the challenges in the energy security as it will require investments in infrastructure compatible with the renewable energy and electric mobility era (Kim, 2024). Looking at the energy security, it consists of four dimensions: the geopolitical, the economic, the reliability, and the environmental one; these dimensions were defined by the Global Energy Institute 2020 (GIE) (Tugcu & Menegaki, 2023).

1.5.5 Decarbonization of Hard-to-Abate Sectors

CO_2 removal technologies are needed to compensate the residual emissions in the hard-to-abate sectors, which will directly help in reaching net-zero greenhouse gas targets. Paris Agreement has made the 196 parties to agree onto holding the increasing global average temperature to be below 2°C above the pre-industrial level. This can only happen when CO_2 emissions reach the net-zero around 2050. Heavy industry, international transport, agriculture, and buildings are those four HtA (hard-to-abate) sectors, facing both common and specific challenges in the climate change mitigation. Emissions from the transport sector are projected to initially rise but will fall below 2020 levels after 2035, achieving an 86% reduction by 2060 compared to 2020 levels. In contrast, the industry and buildings sectors are expected to reach 90% and 78% reductions, respectively, by 2060. The agricultural sector shows a slower decline, with a 7% reduction by 2060 (Edelenbosch et al., 2024).

1.5.6 Innovation, Technological Advancement, and Competitiveness

Various drivers of Green Innovations Initiatives (GII), including environmental regulations (ER), market demand (MD), government pressure (GP), competitor pressure (CP), corporate social responsibility (CSR), and employee conduct (EC), rephrase (German et al., 2023). Various advancements have been seen, such as solar energy production using PV, standalone PV systems, and thermal solar energy from the sunlight for electricity or heat. Flowing water kinetics leads to the generation of the hydropower. Biomass energy is an RES derived from organic materials such as wood, agricultural residues, animal waste, and food scraps. These materials can be converted into usable forms of energy, including electricity, heat, and biofuels. The process typically involves burning, fermentation, or other conversion methods to release the stored energy. This approach not only provides an alternative to fossil fuels but also helps in managing waste and reducing greenhouse gas emissions when sourced and utilized sustainably. All of this shows the versatility of renewable energy (Jayabal, 2024).

1.5.7 Public Health and Environmental Benefits

Renewable energy sources have the main advantage of not emitting greenhouse gases, making them very beneficial for public health by reducing the air and water pollution that can lead to various heart diseases, bronchioles, and other similar diseases as well. Use of the electric vehicles EV help in the reduction of carbon emissions (Şenyapar, 2023). In 2019, air pollution—both indoor and outdoor—was responsible for over 5.4 million premature deaths globally, whereas in the year 2018, this pollution resulted in $2.9 trillion additional burdens to healthcare systems, the transition to the clean energy substantially led to the health benefits by less medical cost that increases during the winter months (Li et al., 2023). With the involvement of AI and robots, it has become convenient to replace the people working from the hazardous environments like chemical mining industries, alleviating the exposure to unsafe surroundings. Further, several studies have recently claimed that AI and robotic techniques can diminish the work-related damages to employees. The amalgamation of progressive technologies into public health and professional safety holds noteworthy promise for enhancing outcomes in both domains. By leveraging innovations such as AI, the Internet of Things (IoT), wearable devices, and immersive training tools, organizations can proactively identify risks, improve compliance, and foster safer environments (Yu et al., 2024). For the whole concept of the green industrial transformation, the main idea is the interaction between renewable energy and industrial value-add, which provides synergy to the financial operations to boost environment-friendly industrial output of the country (Mehmood et al., 2023).

1.5.8 Grid Stability and Flexibility

For the adaption of green industrial energy, it is very essential to stock the energy in order to allow the proper existing energy supply chain. The major reasons for adopting renewable energy take account of enhancing grid stability, promoting cleaner energy sources, enhancing energy efficiency, lessening fossil fuel dependency, and mitigating environmental impacts. Renewables like wind and solar power offer a steady and decentralized energy supply, which can help prevent blackouts and improve grid resilience. These energy alternatives also contribute to minimal GHG emissions, in turn, supplying clean air and water. Additionally, advancements in renewable technologies have led to improved efficiency and decreased operational budgets, making them more competitive with traditional energy sources. By transitioning to renewables, we can decrease the dependency on fossil fuels, leading to lower carbon emissions and decreased environmental degradation associated with fossil fuel extraction and consumption. Overall, renewable energy offers multiple benefits, including enhanced grid stability, environmental protection, and improved energy system efficiency (Li & Deusen, 2025). Energy flexibility contributes significantly to the current energy transition, as it leads to the stabilization of the grid electricity coming from the RESs, which causes volatility in the power supply (Pierri et al., 2021).

1.5.9 Support for Circular Economy and Resource Efficiency

As for macro-level sustainability, the most popular narratives include the green economy, circular economy, and bioeconomy. As stated by D'Amato and Korhonen (2021), a study discovered that after 2018, there had been an exponential growth; the growth drivers were the global sustainability initiatives including the United Nations SDGs and the Paris Agreement. The renewable energy technologies, mainly solar and wind, are the leading research output, while the emerging areas such as green hydrogen and the circular bioeconomy are some of the least-studied issues. According to Abdirahman et al. (2025), the circular economy significantly surrounds the global sustainability. The key agreements, including the Kyoto Protocol, Paris Agreement, and the Cancun Agreement, were signed in 1992 under the United Nations Framework Convention on Climate Change or UNFCCC; in this regard, the Ellen MacArthur Foundation designed a model pathway of the future-oriented system, which can eliminate material waste, reduce environmental harm, and mitigate climate change. The idea of resource efficiency underlies the philosophy, implying that the system of materials and production uses them wisely, consuming minimum water, energy, and other resources. Through the re-use, recycling, and reduction of the primary and secondary material use, industries can reduce the individual environmental impacts. For instance, the process of recycling aluminum and paper materials consumes less energy and emits less gas than the production of energy and products from the virgin resources. The re-use of materials on the stages before and after product life cycle may balance or eliminate the need for the respective raw material production. Water-saving methods, such as rain harvesting and irrigation systems, may be instrumental in the reduction of the primary resource.

1.5.10 Regulatory and Brand Advantages

In general, energy policy and regulation are more important to help nations to achieve Sustainable Development Goal 7, attract new investments, and ensure sustainability in energy. Effective green industrial policies and regulations are imperative to value sustainability goals. Over the years, governments have employed tools such as tax credits, subsidies, loan guarantees, and mandates to encourage investments in green technologies and practices. Climate change and global warming goals are becoming central priorities to governments and other policymakers and international organizations. The global environment has suffered from global warming, adversely affecting ecosystems, human health, and the future of humanity as a whole; "climate issues" have threatened biodiversity exponentially and interfered with food and water systems. The increase in climate-induced disasters and extreme weather events has undermined the well-being of the current generation and the generations to come. In recent years, international organizations have deliberated extensively on the most appropriate and effective policy framework to counter climate threats. A major breakthrough on climate-related issues was achieved during the 27th United Nations Climate Change Conference of the Parties in 2022. An agreement to establish a "loss and damage" fund was enacted. The element aimed to aid the most vulnerable nations in combating the impact of climate-instigated disasters. The subsequent

COP meetings, namely COP28 and COP29, intensified the issue. The COP28 meeting in Dubai was characterized by the operationalization of the fund, as participating countries pledged over 700 million USD. COP29 in Baku reiterated the importance of fast-tracking the financial pledge to supporting the affected nations. The debate and discussion details support the idea that governments must deliver funding to the critical nations impacted by climate challenges and facilitate sustainable development and counter global warming.

1.6 Challenges and Market Gaps

1.6.1 Supply–Demand Mismatch

As we head into 2025, there's every reason to believe that the push for clean energy will keep building—especially if new leadership brings fresh ideas and policies to the table. The real engines behind this growth are cleantech manufacturing, AI, and the carbon industry. These sectors are not just talking about change; they're making it happen. On the demand side, companies building cleantech plants, running data centers, or working on direct air capture are all searching for ways to power their operations with more and more renewable energy. Their appetite for green power is only getting bigger as their projects grow. If considering the supply aspect, these industries are actually helping renewable energy corporations tackle some big challenges. For instance, bringing cleantech manufacturing back home is shaking up how we make solar panels and batteries, making supply chains more resilient and local. AI is stepping in to make these supply chains smarter, faster, and more efficient, while also sparking new innovations in renewable tech. And don't forget about carbon markets—selling carbon credits is giving renewable projects a nice extra boost, helping new ideas get off the ground. So, all in all, it's a pretty exciting time for clean energy, with plenty of momentum and new ways to keep things moving forward.

1.6.2 Energy Storage and Grid Reliability

Energy storage is a crucial part of the energy supply chain at present. There are several critical reasons why Energy Storage System (ESS) are important. Primarily, they help maintain electricity demand. The surplus energy produced during low-demand hours can be stored and used during the high-demand hours in an efficient manner that can increase the overall system efficiency and reliability. Second, ESS is a critical factor in reducing carbon emissions by utilizing more renewable energy. Third, ESS helps address the intermittency of renewable energy so that electricity can flow more consistently into homes and businesses for those beyond stable power grids, especially in rural or off-grid regions where electricity is not universally stable. Energy is stored using several methods, including thermomechanical, chemical, and electrochemical storage technologies. ESS faces numerous challenges, foremost among them is the issue of reliability. It is crucial to manage steady sources of energy for widespread acceptability and trust in these clean technologies. This will aid in the effective reduction of carbon footprints in energy systems, enabling a transition to cleaner and less carbon-intensive futures. Reliable ESS are important to manage the variability and intermittency of renewable active power generation, such as wind and solar power. This may have a positive impact on the stabilization and optimization of renewable generation source usage and energy storage and dispatchment.

1.6.3 Work Force and Skill Shortage

A recent article in *Renewable and Sustainable Energy Reviews* stated that in order to allow the proper working of the energy transition, there is a need to predict the required energy sector workforce, including the construction of transition lines, along with expansion in the transmission infrastructure to facilitate an electricity grid powered by renewable energy (Rutovitz et al., 2025). A study published in *Energy Research & Social Science* tends to analyze the shortage of the skill along with social license in regional areas. Several challenges are present that block the opportunities for local job creations such as

1) Boom–Bust Cycles and Industry Volatility: Fluctuations have been observed in the demand for the renewable energy sector that lead to the stages of rapid growth followed by the slowdowns. These stages of ups and downs lead to the formation of job insecurities, making it very difficult for the workers to transition between projects.
2) Training Capacity Limitations: When talking about the working in the various projects, it is very important to provide the workers with proper infrastructure as well as training sessions, but in the regional areas there have been many loopholes, making it difficult for the work force to come into action. This hampers the ability of upskilling the local workers at a faster pace leading to skill shortages.
3) Competition for Labor: Skilled workers are the need for the industry, but due to competition with other sectors, it tends to become difficult to retain workers, directly affecting the cost-effectiveness as well as the pace of renewable energy projects.
4) Regional Labor Market Characteristics: There has been an observation of declining populations in regional areas, leading to the limited presence of the workforce, which in turn restricts the supply of the local workers for the projects.
5) Quality of Social and Economic Infrastructure: The existing infrastructure in many regional areas may not be sufficient to support the influx of workers associated with renewable energy projects. Issues such as housing shortages, inadequate healthcare facilities, and limited transportation options can deter workers from relocating to these areas (Briggs et al., 2022).

1.6.4 Infrastructure and Supply Chain Constrains

A major obstruction to renewable energy implementation is the limited availability of infrastructure, especially in developing countries. Renewable energy plants are generally located in remote areas, necessitating supplementary transmission lines to connect to the main grid. However, most existing grids were not designed for intermittent RES, requiring costly upgrades or modifications. This grid integration challenge is one of the most significant problems affecting the development of renewable energy projects (Olabi et al., 2023; Seetharaman et al., 2019).

1.6.5 Policy and Regulation Uncertainty

In today's time, it is very important to allow economic growth, but traditionally this has been worked by fossil fuels that tend to be non-eco-friendly due to large emissions of CO_2. This leads to severe consequences in affecting natural factors such as biodiversity, threatening the future of humanity (Ivanovski & Marinucci, 2021). There has been a negative long-term relationship that has been observed in between the economic policy uncertainty and renewable energy consumption, indicating that higher levels of country risk may dampen the uptake of renewables (Ivanovski & Marinucci, 2021). New legislation presented in 2017 (Amendment by Renewable Energy Act) was put forward by Japan, with an aim to enhance the use of renewable energy to 22–24% of its electricity mix by 2030. Because of limited domestic resources, Japan continues to depend on the imports for nearly 90% of its total energy needs (as reported by World Nuclear Association in 2019). Due to the Fukushima disaster in 2011, the country's share from nuclear sources plummeted significantly from 30% to a production target of just 20% by 2030 (Lu et al., 2021). Among so many environmental issues present in the 21st century, global warming holds the majority of the problem. Renewable energy has become a major part of the policies, causing its growth; however, its deployment remains low when compared to the other energy sources (Appiah-Otoo, 2021). It has also been stated that the uncertainty in the economic policy tends to improve the renewable energy innovations in countries with high institutional quality, while the effect becomes insignificant for countries with low institutional quality (Feng & Zheng, 2022).

1.6.6 Geographic and Temporal Mismatches

Resource distribution, land use conflict, and seasonal variability are some of the geographical and temporal disparities that create challenges for grid stability, storage, and transmission etc. A study by Laitinen et al. (2021) examined a densely constructed urban area in Helsinki where physical constraints limited the local energy accessibility. While rooftop solar PV panels were easily installed, wall-integrated PV systems proved difficult to implement due to shadowing caused by the area's high density. The research focused on integrating wind and solar power to meet the region's total electricity requirements, including the energy needed for heating and cooling via electricity driven heat pumps. To manage the inherent variability of renewable generation and achieve full energy self-sufficiency, the study utilized both battery and thermal storage systems (Laitinen et al., 2021). Also, Tong et al. (2021) highlighted the critical need for energy storage and grid flexibility to address supply-demand mismatches. Their findings suggest that solar heavy systems require storage to manage daily cycles, whereas wind-heavy systems benefit from continental-scale transmission to mitigate geographic variability.

1.6.7 High Initial Capital Cost

For the working of the green industrial energy, a model set is needed, but these models in energy economics assess the price of alternative power production technologies. They need a well-tuned hypothesis for the capital cost or concession rates; the cost varies extensively across nations and technologies. By and large, the cost of the capital (CoC) in developing countries is notably higher than in industrialized countries, with long heterogeneity within the groups of industrialized or developing countries (Steffen, 2020). The Intergovernmental Panel on Climate Change (IPCC) Sixth Assessment Report highlights the dependence on finance for accelerating the climate action, but the finance is fundamentally unequal across the countries, causing hindrance to the diminution and adaptation investment. There is a great risk in their high investments for the RES, affecting the CoC for the projects. These factors prove to be key challenges in mobilizing funds for the energy transition in the developing world (Calcaterra et al., 2024).

1.7 Integration of Emerging Technologies

Sustainable development is rooted in three fundamental drivers: Economics, the Environment, and Society. With the change in the whole world, it is ushered into the digital age, affecting almost all spheres of human activities and the energy systems along with it. The energy sector is all improved due to IT, electrification, and dematerialization. Emerging technologies like smart grids are crucial for expanding RESs in the future. Tools such as big data analytics, cloud computing, the Internet of Things (IoT), Industrial Internet of Things (IIoT), digital twins, and AI can enhance the efficiency and reliability of smart systems like smart grids (Yamini et al., 2025). Emerging technologies are the need of the hour in order to allow the energy system decarbonization, which will help in the planning of energy transition along with the operations of smart energy systems. Technologies such as AI, blockchain, quantum computing, next-generation communication technologies, and the metaverse are among such emerging trends which are representing a transformative shift in digital infrastructure and industrial applications (Zhao et al., 2023).

1.8 Conclusion

Green fuel market is one of the fastest-growing markets, but due to some gaps, it proves to be uneven in some sectors, all because of the differences in policy, infrastructure, technology, and cost. This chapter highlights a sector undergoing rapid transformation, fueled by technological innovation, evolving policy frameworks, and surging global demand for sustainable solutions. In 2025, advancements in solar PVs, wind energy, smart grids, and green hydrogen are redefining how industries produce, store, and consume energy, accelerating the shift toward decarbonization and climate resilience. Despite challenges such as supply–demand mismatches, workforce gaps, and infrastructure constraints, the momentum for green industrial energy remains strong. Record investments, expanding industrial clusters, and the integration of AI and digital technologies are enabling industries to enhance productivity, reduce emissions, and foster economic growth. Public–private partnerships and targeted policy initiatives are further supporting the transition, ensuring that the advantages of green energy, such as job creation, better air quality, and energy security, are realized across regions and sectors. Looking ahead, the continued evolution of green industrial energy will depend on scaling breakthrough technologies, strengthening supply chains, and fostering collaboration among stakeholders. As the sector matures, it is poised to play a decisive role in accomplishing net-zero goals, enhancing industrial competitiveness, and building a resilient, sustainable energy future for economies worldwide.

References

Aarthy, M., Saravanan, P., Gowthaman, M. K., Rose, C., & Kamini, N. R. (2014). Enzymatic transesterification for production of biodiesel using yeast lipases: An overview. *Chemical Engineering Research and Design*, *92*(8), 1591–1601. https://doi.org/10.1016/j.cherd.2014.04.008.

Abdirahman, A. A., Asif, M., & Mohsen, O. (2025). Circular economy in the renewable energy sector: A review of growth trends, gaps and future directions. *Energy Nexus*, 100395. https://doi.org/10.1016/j.nexus.2025.100395.

Ahan, M. R., Nambi, A., Ganu, T., Nahata, D., & Kalyanaraman, S. (2021). AI-assisted cell-level fault detection and localization in solar PV electroluminescence images. *Proceedings of the 19th ACM conference on embedded networked sensor systems*. Association for Computing Machinery.

Androniceanu, A., & Sabie, O. M. (2022). Overview of green energy as a real strategic option for sustainable development. *Energies*, *15*(22), 8573. https://doi.org/10.3390/en15228573.

Apertus Group. (2025). *Renewable energy trends*. https://apertusgroup.co.uk/renewable-energy-trends-in-2025/.

Appiah-Otoo, I. (2021). Impact of economic policy uncertainty on renewable energy growth. *Energy Research Letters*, *2*(1), 19444. https://doi.org/10.46557/001c.19444.

Azad, S., Sabrina, F., & Wasimi, S. (2019). Transformation of smart grid using machine learning. *2019 29th Australasian Universities Power Engineering Conference (AUPEC), 26–29 Nov. 2019*, pp. 1–6.

Azhar, S. H. M., Abdulla, R., Jambo, S. A., Marbawi, H., Gansau, J. A., Faik, A. A. M., & Rodrigues, K. F. (2017). Yeasts in sustainable bioethanol production: A review. *Biochemistry and Biophysics Reports*, *10*, 52–61. https://doi.org/10.1016/j.bbrep.2017.03.003.

Baruah, D. C., & Enweremadu, C. C. (2019). Prospects of decentralized renewable energy to improve energy access: A resource-inventory-based analysis of South Africa. *Renewable and Sustainable Energy Reviews*, *103*, 328–341. https://doi.org/10.1016/j.rser.2019.01.006.

BDO UK. (2025). *Renewables annual report*. https://www.bdo.co.uk/en-gb/insights/industries/natural-resources-and-energy/renewables-annual-report.

Belu, R. (2013). Artificial intelligence techniques for solar energy and photovoltaic applications. In S. Anwar, H. Efstathiadis, & S. Qazi (Eds.), *Handbook of research on solar energy systems and technologies*. IGI Global.

Briggs, C., Atherton, A., Gill, J., Langdon, R., Rutovitz, J., & Nagrath, K. (2022). Building a 'Fair and Fast' energy transition? Renewable energy employment, skill shortages and social licence in regional areas. *Renewable and Sustainable Energy Transition*, *2*, 100039. https://doi.org/10.1016/j.rset.2022.100039.

Budzianowski, W. M. (2012). Value-added carbon management technologies for low CO_2 intensive carbon-based energy vectors. *Energy*, *41*(1), 280–297. https://doi.org/10.1016/j.energy.2012.03.008.

Calcaterra, M., Reis, L. A., Fragkos, P., Briera, T., De Boer, H. S., Egli, F., Emmerling, J., Iyer, G., Mittal, S., Polzin, F. H. J., Sanders, M. W. J. L., Schmidt, T. S., Serebriakova, A., Steffen, B., Van De Ven, D. J., Van Vuuren, D. P., Waidelich, P., & Tavoni, M. (2024). Reducing the cost of capital to finance the energy transition in developing countries. *Nature Energy*, *9*(10), 1241–1251. https://doi.org/10.1038/s41560-024-01606-7.

Chakraborty, S., Dash, S. K., Elavarasan, R. M., Kaur, A., Elangovan, D., Meraj, S. T., Kasinathan, P., & Said, Z. (2022). Hydrogen energy as future of sustainable mobility. *Frontiers in Energy Research*, *10*, 893475. https://doi.org/10.3389/fenrg.2022.893475.

Cormos, C. C. (2016). Oxy-combustion of coal, lignite and biomass: A techno-economic analysis for a large scale Carbon Capture and Storage (CCS) project in Romania. *Fuel*, *169*, 50–57. https://doi.org/10.1016/j.fuel.2015.12.005.

Curtis, A. J., & McLellan, B. C. (2023). Potential domestic energy system vulnerabilities from major exports of green hydrogen: A case study of Australia. *Energies, 16*(16), 5881. https://doi.org/10.3390/en16165881.

D'Amato, D., & Korhonen, J. (2021). Integrating the green economy, circular economy and bioeconomy in a strategic sustainability framework. *Ecological Economics, 188*, 107143. https://doi.org/10.1016/j.ecolecon.2021.107143.

Deloitte. (2025). *Renewable Energy Industry Outlook. Renewables race to fill resource gap as demand for clean energy is outpacing supply*. https://www2.deloitte.com/us/en/insights/industry/renewable-energy/renewable-energy-industry-outlook.html.

Derakhshandeh, J. F., AlLuqman, R., Mohammad, S., AlHussain, H., AlHendi, G., AlEid, D., & Ahmad, Z. (2021). A comprehensive review of automatic cleaning systems of solar panels. *Sustainable Energy Technologies and Assessments, 47*, 101518. https://doi.org/10.1016/j.seta.2021.101518.

Doucha, J., & Lívanský, K. (2006). Productivity, CO_2/O_2 exchange and hydraulics in outdoor open high density microalgal (*Chlorella* sp.) photobioreactors operated in a Middle and Southern European climate. *Journal of Applied Phycology, 18*, 811–826. https://doi.org/10.1007/s10811-006-9100.

Edelenbosch, O. Y., Hof, A. F., Van Den Berg, M., De Boer, H. S., Chen, H., Daioglou, V., Dekker, M. M., Doelman, J. C., Elzen, M. G. J. D., Harmsen, M., Mikropoulos, S., Van Sluisveld, M. A. E., Stehfest, E., Tagomori, I. S., Van Zeist, W., & Van Vuuren, D. P. (2024). Reducing sectoral hard-to-abate emissions to limit reliance on carbon dioxide removal. *Nature Climate Change, 14*(7), 715–722. https://doi.org/10.1038/s41558-024-02025-y.

Ellabban, O., Abu-Rub, H., & Blaabjerg, F. (2014). Renewable energy resources: Current status, future prospects and their enabling technology. *Renewable and Sustainable Energy Reviews, 39*, 748–764. https://doi.org/10.1016/j.rser.2014.07.113.

Farjana, S. H., Huda, N., Mahmud, M. P., & Saidur, R. (2018a). Solar industrial process heating systems in operation—Current SHIP plants and future prospects in Australia. *Renewable and Sustainable Energy Reviews, 91*, 409–419. https://doi.org/10.1016/j.rser.2018.03.105.

Farjana, S. H., Huda, N., Mahmud, M. P., & Saidur, R. (2018b). Solar process heat in industrial systems—A global review. *Renewable and Sustainable Energy Reviews, 82*, 2270–2286. https://doi.org/10.1016/j.rser.2017.08.065.

Fattah, I. R., Noraini, M. Y., Mofijur, M., Silitonga, A. S., Badruddin, I. A., Khan, T. Y., Ong, H. C., & Mahlia, T. M. I. (2020). Lipid extraction maximization and enzymatic synthesis of biodiesel from microalgae. *Applied Sciences, 10*(17), 6103. https://doi.org/10.3390/app10176103.

Favaro, L., Jansen, T., & van Zyl, W. H. (2019). Exploring industrial and natural *Saccharomyces cerevisiae* strains for the bio-based economy from biomass: The case of bioethanol. *Critical Reviews in Biotechnology, 39*(6), 800–816. https://doi.org/10.1080/07388551.2019.1619157.

Feng, G., & Zheng, M. (2022). Economic policy uncertainty and renewable energy innovation: International evidence. *Innovation and Green Development, 1*(2), 100010. https://doi.org/10.1016/j.igd.2022.100010.

Fernandez, E. S., Goetheer, E. L., Manzolini, G., Macchi, E., Rezvani, S., & Vlugt, T. J. (2014). Thermodynamic assessment of amine based CO_2 capture technologies in power plants based on European Benchmarking Task Force methodology. *Fuel, 129*, 318–329. https://doi.org/10.1016/j.fuel.2014.03.042.

FirstIgnite. (2024). *Green energy technology advances 2024: AI and solar trends.* https://firstignite.com/exploring-the-latest-green-energy-technology-advancements-in-2024/.

German, J. D., Redi, A. A. N. P., Ong, A. K. S., & Liwanag, J. L. (2023). The impact of green innovation initiatives on competitiveness and financial performance of the land transport industry. *Heliyon, 9*(8), e19130. https://doi.org/10.1016/j.heliyon.2023.e19130.

Gökgöz, F., & Güvercin, M. T. (2018). Energy security and renewable energy efficiency in EU. *Renewable and Sustainable Energy Reviews, 96*, 226–239. https://doi.org/10.1016/j.rser.2018.07.046.

Gopinath, D. (2016). Why a clearer 'green industrial policy 'matters for India: Reconciling growth, climate change and inequality. *Local Economy, 31*, 830–835. https://doi.org/10.1177/0269094216675434.

Green City Times. (2025). *Innovations in energy efficiency for industrial applications.* https://www.greencitytimes.com/innovations-in-energy-efficiency-for-industrial-applications/.

Guo, Y., Chen, Y., Ma, J., Zhu, H., Cao, X., Wang, N., & Wang, Z. L. (2019). Harvesting wind energy: A hybridized design of pinwheel by coupling triboelectrification and electromagnetic induction effects. *Nano Energy, 60*, 641–648. https://doi.org/10.1016/j.nanoen.2019.03.094.

Hasanuzzaman, M., Rahman, M. M., Öztop, H. F., Rahim, N. A., & Saidur, R. (2012). Effects of Lewis number on heat and mass transfer in a triangular cavity. *International Communications in Heat and Mass Transfer, 39*(8), 1213–1219. https://doi.org/10.1016/j.icheatmasstransfer.2012.07.002.

Heß, S., & Oliva, A. (2010). *Solar process heat generation: Guide to solar thermal system design for selected industrial process.* http://www.solar-process-heat.eu/publications/.

Ivanovski, K., & Marinucci, N. (2021). Policy uncertainty and renewable energy: Exploring the implications for global energy transitions, energy security, and environmental risk management. *Energy Research & Social Science, 82*, 102415. https://doi.org/10.1016/j.erss.2021.102415.

Jacobus, A. P., Gross, J., Evans, J. H., Ceccato-Antonini, S. R., & Gombert, A. K. (2021). *Saccharomyces cerevisiae* strains used industrially for bioethanol production. *Essays in Biochemistry, 65*(2), 147–161.

Jayabal, R. (2024). Towards a carbon-free society: Innovations in green energy for a sustainable future. *Results in Engineering, 24*, 103121. https://doi.org/10.1016/j.rineng.2024.103121.

Jha, S. K., Bilalovic, J., Jha, A., Patel, N., & Zhang, H. (2017). Renewable energy: Present research and future scope of artificial intelligence. *Renewable and Sustainable Energy Reviews, 77*, 297–317. https://doi.org/10.1016/j.rser.2017.04.018.

Jiang, J., Wang, J., & Cheng, Y. (2024). The impact of industrial transformation on green economic efficiency: New evidence based on energy use. *Petroleum Science, 21*(5), 3644–3655. https://doi.org/10.1016/j.petsci.2024.03.004.

Jiang, Y., Lv, Y., Wu, R., Sui, Y., Chen, C., Xin, F., Zhou, J., Dong, W., & Jiang, M. (2019). Current status and perspectives on biobutanol production using lignocellulosic feedstocks. *Bioresource Technology Reports, 7*, 100245. https://doi.org/10.1016/j.biteb.2019.100245.

Joshi, G., Pandey, J. K., Rana, S., & Rawat, D. S. (2017). Challenges and opportunities for the application of biofuel. *Renewable and Sustainable Energy Reviews, 79*, 850–866. https://doi.org/10.1016/j.rser.2017.05.185.

Karthick, C., & Nanthagopal, K. (2021). A comprehensive review on ecological approaches of waste to wealth strategies for production of sustainable biobutanol and its suitability in automotive applications. *Energy Conversion and Management*, 239. https://doi.org/10.1016/j.enconman.2021.114219.

Kim, J. (2024). *Energy security and the green transition.* IMF Working Paper, 2024(006), 1. https://doi.org/10.5089/9798400263743.001.

Kumar, A., & Sharma, S. (2008). An evaluation of multipurpose oil seed crop for industrial uses (*Jatropha curcas* L.): A review. *Industrial Crops and Products, 28*(1), 1–10. https://doi.org/10.1016/j.indcrop.2008.01.001.

Kumar, L., Hasanuzzaman, M., & Rahim, N. A. (2019). Global advancement of solar thermal energy technologies for industrial process heat and its future prospects: A review. *Energy Conversion and Management, 195*, 885–908. https://doi.org/10.1016/j.enconman.2019.05.081.

Kylili, A., Fokaides, P. A., Ioannides, A., & Kalogirou, S. (2018). Environmental assessment of solar thermal systems for the industrial sector. *Journal of Cleaner Production, 176*, 99–109. https://doi.org/10.1016/j.jclepro.2017.12.150.

Laitinen, A., Lindholm, O., Hasan, A., Reda, F., & Hedman, Å. (2021). A techno-economic analysis of an optimal self-sufficient district. *Energy Conversion and Management, 236*, 114041. https://doi.org/10.1016/j.enconman.2021.114041.

Lee, Y. C., Lee, K., & Oh, Y. K. (2015). Recent nanoparticle engineering advances in microalgal cultivation and harvesting processes of biodiesel production: A review. *Bioresource Technology, 184*, 63–72. Epub 20 141 105. https://doi.org/10.1016/j.biortech.2014.10.145.

Li, K., & Lin, B. (2016). Impact of energy conservation policies on the green productivity in China's manufacturing sector: Evidence from a three-stage DEA model. *Applied Energy, 168*, 351–363.

Li, M., Geng, Y., Zhou, S., & Sarkis, J. (2023). Clean energy transitions and health. *Heliyon, 9*(11), e21250. https://doi.org/10.1016/j.heliyon.2023.e21250.

Li, Y., Horsman, M., Wu, N., Lan, C. Q., & Dubois-Calero, N. (2008). Biofuels from microalgae. *Biotechnology Progress, 24*(4), 815–820. https://doi.org/10.1021/bp070371k.

Li, Z., & Deusen, D. (2025). Role of energy storage technologies in enhancing grid stability and reducing fossil fuel dependency. *International Journal of Hydrogen Energy, 102*, 1055–1074. https://doi.org/10.1016/j.ijhydene.2024.12.489.

Lu, Z., Zhu, L., Lau, C. K. M., Isah, A. B., & Zhu, X. (2021). The role of economic policy uncertainty in renewable energy-growth nexus: Evidence from the Rossi-Wang Causality Test. *Frontiers in Energy Research, 9*. https://doi.org/10.3389/fenrg.2021.750652.

Malode, S. J., Prabhu, K. K., Mascarenhas, R. J., Shetti, N. P., & Aminabhavi, T. M. (2021). Recent advances and viability in biofuel production. *Energy Conversion and Management: X, 10*, 100070. https://doi.org/10.1016/j.ecmx.2020.100070.

Mao, Y., Liu, K., & Zhou, J. (2019). Evolution of green industrial growth between Europe and China based on the energy consumption model. *Sustainability, 11*(24), 7168. https://doi.org/10.3390/su11247168.

Marouani, I., Guesmi, T., Alshammari, B. M., Alqunun, K., Alzamil, A., Alturki, M., & Hadj Abdallah, H. (2023). Integration of renewable-energy-based green hydrogen into the energy future. *Processes, 11*(9), 2685. https://doi.org/10.3390/pr11092685.

Massarweh, O., Al-khuzaei, M., Al-Shafi, M., Bicer, Y., & Abushaikha, A. S. (2023). Blue hydrogen production from natural gas reservoirs: A review of application and feasibility. *Journal of CO2 Utilization*, *70*, 102438. https://doi.org/10.1016/j.jcou.2023.102438.

Mehmood, S., Zaman, K., Khan, S., Ali, Z., & Khan, H. U. R. (2023). The role of green industrial transformation in mitigating carbon emissions: Exploring the channels of technological innovation and environmental regulation. *Energy and Built Environment*, *5*(3), 464–479. https://doi.org/10.1016/j.enbenv.2023.03.001.

Melcher, F., Paper, M., & Brück, T. (2021). 9 Photosynthetic conversion of CO_2 into bioenergy and materials using microalgae. *Photosynthesis*, 227–254. https://doi.org/10.1515/9783110716979-009.

Metz, B., Davidson, O., De Coninck, H., Loos, M., & Meyer, L. (2005). *Carbon dioxide capture and storage*: IPCC Special Report, Cambridge University Press, Cambridge, UK, 2005.

Mischko, W., Hirte, M., Roehrer, S., Engelhardt, H., Mehlmer, N., Minceva, M., et al. (2018). Modular biomanufacturing for a sustainable production of terpenoid-based insect deterrents. *Green Chemistry*, *20*(11), 2637–2650. https://doi.org/10.1039/c8gc00434j.

Mitchell, D., Blanche, J., Harper, S., Lim, T., Gupta, R., Zaki, O., Tang, W., Robu, V., Watson, S., & Flynn, D. (2022). A review: Challenges and opportunities for artificial intelligence and robotics in the offshore wind sector. *Energy and AI*, *8*, 100146. https://doi.org/10.1016/j.egyai.2022.100146.

Moreno-Brieva, F., Guimón, J., & Salazar-Elena, J. C. (2023). From grey to green and from west to east: The geography and innovation trajectories of hydrogen fuel technologies. *Energy Research & Social Science*, *101*, 103146. https://doi.org/10.1016/j.erss.2023.103146.

Naganathan, H., Chong, W. O., & Chen, X. (2016). Building energy modeling (BEM) using clustering algorithms and semi-supervised machine learning approaches. *Automation in Construction*, *72*, 187–194. https://doi.org/10.1016/j.autcon.2016.08.002.

Olabi, A., Elsaid, K., Obaideen, K., Abdelkareem, M. A., Rezk, H., Wilberforce, T., Maghrabie, H. M., & Sayed, E. T. (2023). Renewable energy systems: Comparisons, challenges and barriers, sustainability indicators, and the contribution to UN sustainable development goals. *International Journal of Thermofluids*, *20*, 100498. https://doi.org/10.1016/j.ijft.2023.100498.

Partheeban, P., Shiva, M., Vishnupriyan, J., Ponnusamy, R., Kumar, T. S., & Anuradha, B. (2022). Solar energy optimisation using IoT and deep learning—A review. *2022 International Conference on Data Science, Agents & Artificial Intelligence (ICDSAAI), 8–10 Dec. 2022*, pp. 1–3.

Pierri, E., Hellkamp, D., Thiede, S., & Herrmann, C. (2021). Enhancing energy flexibility through the integration of variable renewable energy in the process industry. *Procedia CIRP*, *98*, 7–12. https://doi.org/10.1016/j.procir.2020.12.001.

Qian, J., & Jing, X. (2018). Wind-driven hybridized triboelectric-electromagnetic nanogenerator and solar cell as a sustainable power unit for self-powered natural disaster monitoring sensor networks. *Nano Energy*, *52*, 78–87.

Rangel-Martinez, D., Nigam, K. D. P., & Ricardez-Sandoval, L. A. (2021). Machine learning on sustainable energy: A review and outlook on renewable energy systems, catalysis, smart grid and energy storage. *Chemical Engineering Research and Design*, *174*, 414–441. https://doi.org/10.1016/j.cherd.2021.08.013.

RSMUK. (2025). *The landscape for UK renewable energy in 2025*. https://www.rsmuk.com/insights/advisory/the-landscape-for-uk-renewable-energy-in-2025.

Rutovitz, J., Langdon, R., Briggs, C., Mey, F., Dominish, E., & Nagrath, K. (2025). Updated employment factors and occupational shares for the energy transition. *Renewable and Sustainable Energy Reviews*, *212*, 115339. https://doi.org/10.1016/j.rser.2025.115339.

Schmitt, B. (2016). Classification of industrial heat consumers for integration of solar heat. *Energy Procedia*, *91*, 650–660. https://doi.org/10.1016/j.egypro.2016.06.

Scovell, M., & Walton, A. (2024). Blue or green? Exploring Australian acceptance and beliefs about hydrogen production methods. *Journal of Cleaner Production*, *444*, 141151. https://doi.org/10.1016/j.jclepro.2024.141151.

Seetharaman, N., Moorthy, K., Patwa, N., Saravanan, N., & Gupta, Y. (2019). Breaking barriers in deployment of renewable energy. *Heliyon*, *5*(1), e01166. https://doi.org/10.1016/j.heliyon.2019.e01166.

Şenyapar, H. N. D. (2023). A bibliometric analysis on renewable energy's public health benefits. *Journal of Energy Systems*, *7*(1), 132–157. https://doi.org/10.30521/jes.1252122.

Sharma, S. K., & Dubey, S. P. (2022). An approach to green energy, advantages and its application. *International Journal for Research in Applied Science and Engineering Technology*, *10*(12), 62–71. https://doi.org/10.22214/ijraset.2022.47803.

StartUs Insights. (2025). *Green energy market report 2025: Key data & innovations*. Retrieved May 6, 2025, from https://www.startus-insights.com/innovators-guide/green-energy-market-report/.

Steffen, B. (2020). Estimating the cost of capital for renewable energy projects. *Energy Economics*, *88*, 104783. https://doi.org/10.1016/j.eneco.2020.104783.

Swain, A.K. India's green industrial policy: Pursuing clean energy for green growth. *Economic & Political Weekly* 2014, *49*(9), 19–21. http://www.jstor.org/stable/24479168.

Tagliapietra, S., & Veugelers, R. (2023). Developing a green industrial policy for the European Green Deal. In F. J. Díaz López (Ed.), *Handbook on innovation, society and the environment* (pp. 36–50). Edward Elgar Publishing.

Theo, W. L., Lim, J. S., Hashim, H., Mustaffa, A. A., & Ho, W. S. (2016). Review of pre-combustion capture and ionic liquid in carbon capture and storage. *Applied Energy*, *183*, 1633–1663. https://doi.org/10.1016/j.apenergy.2016.09.103.

Tong, D., Farnham, D. J., Duan, L., Zhang, Q., Lewis, N. S., Caldeira, K., & Davis, S. J. (2021). Geophysical constraints on the reliability of solar and wind power worldwide. *Nature Communications*, *12*(1). https://doi.org/10.1038/s41467-021-26355-z.

Tugcu, C. T., & Menegaki, A. N. (2023). The impact of renewable energy generation on energy security: Evidence from the G7 countries. *Gondwana Research*, *125*, 253–265. https://doi.org/10.1016/j.gr.2023.08.018.

Uswitch. (2025). *UK renewable energy statistics*. https://www.uswitch.com/gas-electricity/studies/renewable-statistics/.

Wang, P., & Xu, X. (2025). Green finance and energy efficiency improvement: The role of green innovation and industrial upgrading. *Innovation and Green Development*, *4*(1), 100 200. https://doi.org/10.1016/j.igd.2024.100200.

Wu, Y. K., Lee, C. Y., Chen, C. R., Hsu, K. W., & Tseng, H. T. (2014). Optimization of the wind turbine layout and transmission system planning for a large-scale offshore wind farm by AI technology. *IEEE Transactions on Industry Applications*, *50*, 2071–2080.

Yamini, E., Khazaei, H., Soltani, M., Alfraidi, W., & Kyriakopoulos, G. L. (2025). Emerging technologies in renewable energy: Risk analysis and major investment strategies. *Sustainable Futures*, *100*, 736. https://doi.org/10.1016/j.sftr.2025.100736.

Yano, J., Aoki, T., Nakamura, K., Yamada, K., & Sakai, S.-i. (2015). Life cycle assessment of hydrogenated biodiesel production from waste cooking oil using the catalytic cracking and hydrogenation method. *Waste Management*, *38*, 409–423. https://doi.org/10.1016/j.wasman.2015.01.014.

Yu, S., Hu, M., Li, J., & Yan, X. (2024). Green workspace and urban health: Exploring the impacts of industrial robotics in pollution emissions and public health. *Frontiers in Public Health*, *12*, 1445746. https://doi.org/10.3389/fpubh.2024.1445746.

Zhang, F., Yi, J., Pan, W., & Mei, G. (2022). A new process for the production of second-generation biodiesel from waste oils and fats. *Biomass Conversion and Biorefinery*, *14*, 9013–9022. https://doi.org/10.1007/s13399-022-03149-x.

Zhao, C., Wu, Y., Dai, X., Han, J., Dong, B., & Huang, L.-B. (2022). Calliopsis structure-based triboelectric nanogenerator for harvesting wind energy and self-powered wind speed/direction sensor. *Materials & Design*, *221*, 111005. https://doi.org/10.1016/j.matdes.2022.111005.

Zhao, E., Sun, S., & Wang, S. (2022). New developments in wind energy forecasting with artificial intelligence and big data: A scientometric insight. *Data Science and Management*, *5*, 84–95. https://doi.org/10.1016/j.dsm.2022.05.002.

Zhao, N., Zhang, H., Yang, X., Yan, J., & You, F. (2023). Emerging information and communication technologies for smart energy systems and renewable transition. *Advances in Applied Energy*, *9*(100), 125. https://doi.org/10.1016/j.adapen.2023.100125.

2

Fundamentals of Bioprocesses in Green Energy Production

Avanish Kumar, Girdhari Lal Devnani, and Dan Bahadur Pal

Department of Chemical Engineering, Harcourt Butler Technical University, Kanpur, Uttar Pradesh, India

2.1 Introduction

Bioprocessing is defined as the application of biological materials, such as microorganisms, enzymes, or cells, to carry out chemical transformations or to produce useful products (Shuler & Kargi, 2017). It is a significant branch of biotechnology that incorporates upstream processes like microbial growth and fermentation and downstream processes such as separation, purification, and recovery of products. Bioprocesses are intended to operate under normal process conditions, which are more sustainable and eco-friendly than conventional chemical processes (Stanbury et al., 2016). In the context of green energy, bioprocessing permits the conversion of biomass into fuels and energy carriers, providing a substitute for fossil fuels. The role of bioprocesses in sustainable and green energy is becoming increasingly important in the global pursuit of carbon neutrality and environmental preservation (Kumar et al., 2021). Since the fossil sources are limited, even these all are cause of global point of view for becoming carbon neutrality and environmental preservation (Scarlat et al., 2015). Because bioprocesses often depend on bolted in loop systems where waste streams can be recycled, thus minimizing environmental impact. Biomass—the biological material obtained from living organisms—is fundamental to bioprocess-based green energy production. Biomass can come from various sources, including crops (corn, sugarcane), residues (straw, husks, manure), and aquatic plants like algae (Demirbas, 2009a).

The importance of biomass lies in its abundance, renewability, and ability to be biologically converted into an extensive range of energy products (Pandey et al., 2015). Biological systems such as bacteria, yeast, and fungi are applicable in different bioprocesses to reduce multifarious organic matter into simpler compounds, which can then be converted into biofuels or other useful products. The diversity and adaptability of microorganisms enable the development of targeted bioprocesses tailored to various feedstocks and desired energy outputs, resulting in several key bioenergy products. Bioethanol is produced primarily from sugars by yeast and serves as a renewable alcohol fuel (Balat & Balat, 2009). Biodiesel is formed through transesterification of oils and fats—using microbial or chemical catalysts and can replace or be blended with conventional diesel; common feedstocks include vegetable oils, used cooking oil, or algae (Atabani et al., 2012). Biohydrogen, a clean energy carrier, is produced biologically through dark fermentation, photofermentation, or microbial electrolysis and shows great potential in fuel cell technology and zero-emission transport. In summary, bioprocessing underpins modern bioenergy production by harnessing biological systems to convert biomass into renewable fuels.

2.2 Basic Concepts of Bioprocess Engineering

Bioprocesses are different from traditional chemical processes in their basic approach, operating parameters, and environmental influence. While bioprocess reaction require very less amount of temperature, pressure, reactant concentration to operate a reaction bioprocesses use biological agents such as microorganisms or enzymes to facilitate reactions under slow and eco-friendly approach (Koller, 2020). This means bioprocess reaction require very less amount of temperature, pressure, reactant concentration to operate a reaction-efficient and sustainable, as bioprocesses mostly use renewable resources, whereas chemical processes frequently depend on finite fossil-based materials (Lee, Park, Jang, et al., 2008). These differences make bioprocesses particularly attractive for green energy production, where sustainability and low carbon emissions are priorities. A complete bioprocess consists of two main components: upstream and downstream processes. The upstream phase includes all activities leading up to the actual bioconversion, including raw material preparation, inoculum development, microbial cultivation, and fermentation (Singh et al., 2010). The downstream process begins once the desired bio-product is formed and requires purification, separation, and product recovery (Stanbury et al., 2016). Effective integration of both upstream and downstream components is essential to ensure process efficiency, economic feasibility, and environmental friendly (Doran, 2013).

Microbial reactions or processes play a typical role in bioprocess-based energy production (Chisti, 2007). Bacteria or other microorganisms metabolize various organic substrates through biochemical pathways that release energy-rich products (Madigan et al., 2018). In fermentation, yeast convert sugars into ethanol, while certain bacteria can produce hydrogen gas through metabolic fermentation (Kumar et al., 2022). The diversity of microbial metabolic pathways allows for the conversion of a wide range of biomass feedstocks into different types of bioenergy products. Understanding and controlling these metabolic pathways is important for maximizing yield and process efficiency (Kapdan & Kargi, 2006). Biocatalysts—primarily enzymes and microorganisms—are fundamental to the success of bioprocesses. Enzymes catalyze specific biochemical reactions with high selectivity and efficiency, often at ambient temperatures. Table 2.1 shows the various biomass conversion processes, along with the products, key process parameters, and end-use applications.

Neutral pH, reducing the need for harsh conditions. For instance, cellulases decompose cellulose into glucose, which can then be converted into bioethanol. Similarly, lipases are used in the transesterification of oils for biodiesel production. Microorganisms, on the other hand, serve as living biocatalysts, performing complex transformations and adapting to various feedstocks (Stephanopoulos, 2007). Advances in genetic engineering have enhanced the capabilities of these biocatalysts, allowing for strain improvement and pathway optimization. Their importance cannot be overstated, as they are the driving force behind the biological reactions that enable sustainable, scalable, and eco-friendly green energy solutions (Ahmad et al., 2011).

Table 2.1 Different Biomass Conversion Processes, Along with the Products, Key Process Parameters, and End-Use Applications.

Biomass Conversion Process	Primary Products	Key Process Parameters	Product Applications
Combustion	Heat, flue gases (CO_2, H_2O, ash)	Temperature: 800–1000°C Air/fuel ratio: excess air needed Moisture content: <50%	Electricity generation, space heating, industrial steam
Gasification	Syngas (CO, H_2, CH_4), char, tar	Temp: 800–1200°C Gasifying agents: air, O_2, steam ER (equivalence ratio): 0.2–0.4	Power generation, fuel for engines, synthesis of methanol or ammonia
Pyrolysis	Bio-oil, syngas, biochar	Temp: 350–700°C Atmosphere: inert (e.g., N_2) Heating rate: slow/fast/thermal	Biofuel (bio-oil), carbon sequestration (biochar), soil amendment
Anaerobic digestion	Biogas (CH_4, CO_2), digestate	Temp: 35–55°C (mesophilic/thermophilic) pH: 6.5–7.5 HRT: 15–30 days	Cooking gas, electricity production, organic fertilizer (digestate)
Fermentation	Ethanol, butanol, CO_2	Temp: 30–35°C pH: 4.5–6.5 Substrate: sugars (C_6 or C_5)	Biofuels (ethanol for blending), chemical industry, beverage industry
Transester-ification	Biodiesel, glycerol	Temp: 50–60°C Catalyst: NaOH/KOH Alcohol: methanol/ethanol Reaction time: 1–3 h	Biodiesel for transportation, glycerol for pharmaceuticals and cosmetics
Hydrothermal liquefaction	Biocrude oil, aqueous phase, gas	Temp: 250–374°C Pressure: 10–25 MPa Biomass: wet (no drying required)	Biocrude for refining into fuels, chemicals, and synthetic oils
Hydrolysis + fermentation	Bioethanol, organic acids	Acid/enzyme hydrolysis Temp: 40–50°C (fermentation) pH: 4.5–6.0	Ethanol for energy, food, pharma, solvent industries

(Continued)

Table 2.1 (Continued)

Biomass Conversion Process	Primary Products	Key Process Parameters	Product Applications
Torre faction	Terrified biomass (bio-coal)	Temp: 200–300°C Atmosphere: inert/N_2 Residence time: 10–60 min	Coal substitute for combustion, pellet production
Carbonization	Charcoal, gas	Temp: 300–600°C Anaerobic condition Slow heating Anaerobic condition Slow heating	Fuel, metallurgy, water purification (activated carbon production)

2.3 Feedstock for Bioprocesses

Criteria for ideal feedstock in green energy production and feedstock selection play an important role in determining the efficiency, sustainability, and environmental impact of biofuel and bioenergy processes. Biomass feedstocks are broadly categorized into four generations based on their origin, composition, and technological maturity (Keasling, 2010). Each generation of feedstock presents distinct advantages and challenges in terms of availability, conversion efficiency, competition with food resources, and compatibility with current bioprocessing technologies. Figure 2.1 shows the schematic diagram of different bioconversion processes with production of bioproducts.

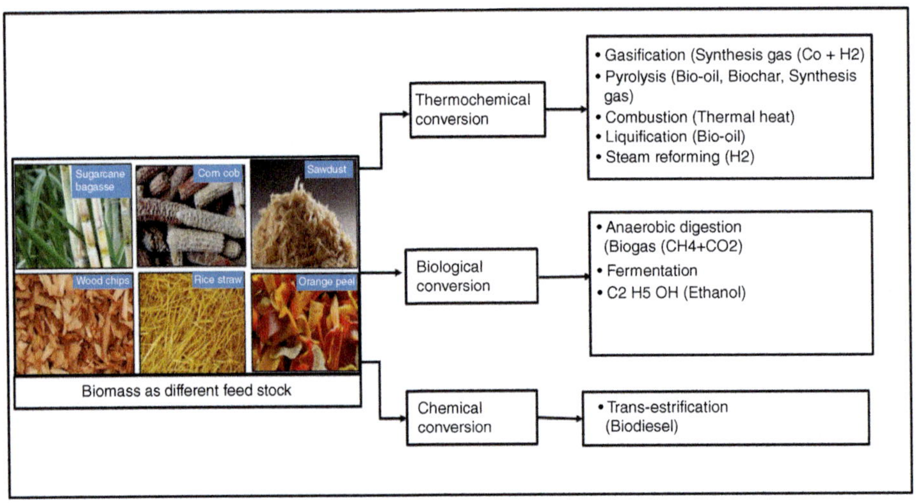

Figure 2.1 Schematic diagram of different bioconversion processes with production of bioproducts (*Source:* (a) MrPreecha/Adobe Stock Photos. (b) Koy Jang/Shutterstock. (d) sprOut7 design/Adobe Stock Photos. (f) Justlight/Adobe Stock Photos).

2.3.1 First-Generation Feedstock

First-generation feedstock is primarily derived from edible food crops such as corn, sugarcane, wheat, and cassava (Hill et al., 2006; Naik et al., 2010). These crops are rich in simple sugars, starches, and vegetable oils, making them easily fermentable or convertible into bioethanol or biodiesel through relatively straightforward biochemical or chemical processes. For instance, sugarcane juice and corn starch can be fermented by yeast to produce bioethanol, while vegetable oils from soybean or palm can be transesterified into biodiesel. However, the major criticism of first-generation feedstocks is the "food vs. fuel" debate. Their use for energy production competes with food supply, potentially increasing food prices and raising ethical and socio-economic concerns (Searchinger et al., 2008).

2.3.2 Second-Generation Feedstock

This type of feedstock is based on non-food lignocellulosic biomass such as wheat straw, rice husks, and dedicated energy crops like switch grass (Kumar et al., 2009). These materials are plentiful and inexpensive and do not participate with the food chain. Switch grass is liginocellulosic grass which is rich in cellulose pre-treatment and enzymatic hydrolysis, which can then be used to produce biofuels like ethanol or butanol (Wyman, 2007). However, second-generation technologies are still evolving due to challenges in efficient biomass pre-treatment and high enzyme costs. Lignin's complex structure makes the process energy-intensive, and commercial-scale operations often face economic and logistical barriers (Humbird et al., 2011).

2.3.3 Third-Generation Feedstock

Third-generation feedstock refers to microalgae, which are considered sources of renewable energy (Brennan & Owende, 2010). Algae can grow rapidly in saline water, thus minimizing freshwater and land usage. They offer high photosynthetic efficiency and yield more biomass per acre than terrestrial plants. Algal biomass can be used to produce biodiesel (from lipids), bioethanol (from carbohydrates), and even biohydrogen under certain conditions. However, technological limitations in large-scale cultivation, harvesting, and processing of algal biomass remain significant. High capital and operating costs currently hinder its commercial viability (Rawat et al., 2011).

2.3.4 Fourth-Generation Feedstock

Fourth-generation feedstock involves genetically modified organisms (GMOs), including engineered crops and microorganisms tailored for improved biofuel yields, resistance to environmental stresses, or enhanced carbon capture. Examples include transgenic algae with higher lipid content or engineered bacteria capable of directly converting biomass into advanced biofuels (Wijffels & Barbosa, 2010). These feedstocks aim to overcome the limitations of previous generations by improving efficiency, reducing processing costs, and enhancing sustainability. However, they also face regulatory, safety, and public acceptance challenges due to concerns over genetic modification and ecological risks.

The ideal feedstock for green energy production should meet several criteria. It should be renewable, non-edible (to avoid food competition), widely available, and low-cost. High energy and biomass yield per unit of land, minimal requirement for water, fertilizers, and pesticides, and compatibility with existing infrastructure are desirable attributes (Sims et al., 2010). Additionally, it should have a low environmental footprint in terms of land use, emissions, and ecological impact. Flexibility in conversion pathways and suitability for various biofuels further enhance its value. Ultimately, a sustainable and economically viable green energy system will likely involve a diversified mix of feedstocks tailored to regional resources and technological capabilities (Cherubini, 2010).

2.4 Microorganisms and Enzymes in Bioprocesses

Microorganisms contribute to various bioprocesses, including biofuel production, wastewater treatment, food processing, and pharmaceutical manufacturing (Nielsen & Keasling, 2016). A wide range of microbial strains are utilized based on their metabolic capabilities and growth characteristics. Among bacteria, *Escherichia coli* is extensively used due to its rapid growth and ease of genetic manipulation, while *Bacillus subtitles* is favored for industrial enzyme production (Lee, Park, Kwon, 2008). In the case of fungi, *Aspergillus niger* is known for citric acid and enzyme production, whereas *Trichoderma reesei* is widely used for producing cellulases essential in biomass hydrolysis. Yeasts such as *Saccharomyces cerevisiae* are central to ethanol fermentation and baking processes, while *Pichia pastoris* is utilized in recombinant protein production. Algae like *Chlorella*, *Spirulina*, and *Scenedesmus* are gaining attention for biofuel generation, carbon capture, and the production of valuable biomolecules (Spolaore et al., 2006).

Enzymes are crucial catalysts in bioprocesses, particularly in hydrolysis and fermentation (Bhat, 2000). During hydrolysis, enzymes such as cellulases, proteases, amylases, and lipases break down complex biomolecules like cellulose, starch, proteins, and fats into simpler, fermentable units such as glucose, amino acids, and fatty acids (Schmid et al., 2001). These enzymes significantly enhance substrate availability and process efficiency. In fermentation, enzymes facilitate the conversion of sugars into desired products like ethanol, organic acids, or antibiotics through metabolic pathways involving alcohol dehydrogenase, lactate dehydrogenase, and other key enzymes (Bailey & Ollis, 2010). These biological catalysts allow reactions to occur under mild conditions, making the processes more sustainable and energy-efficient. Genetic engineering has become a powerful tool to improve microbial strains for industrial bioprocesses (Stephanopoulos et al., 1998). Through gene overexpression, microbes can be engineered to produce higher quantities of desired enzymes, such as cellulases in *Trichoderma* species. Metabolic systems give direction to scientists to redirect biochemical pathways within the cell to increase the yield of target products while reducing by-products. Enhancing microbial tolerance to industrial condition, such as high temperature, pH variations, or toxic compounds—through the use of stress-resistance genes—further improves process healthiness (Keasling, 2008). Advanced tools like CRISPR/Cas9 and synthetic biology approaches now enable precise genome editing and the construction of innovative metabolic pathways, opening new directions in the development of efficient and tailored microbial cell factories (Doudna & Charpentier, 2014). Overall, the integration of microorganisms, enzymes, and genetic engineering are sustainable and high-performance bioprocess technologies.

2.5 Types of Bioprocesses

Anaerobic digestion (AD) is biological reaction which decompose in simple component with out presence of oxygen, primarily carried out by a use of microbe (Appels et al., 2008). This process is mostly used for the treatment of organic waste such as agricultural residues, food waste, animal manure, and sewage sludge. One of its most valuable outcomes is the production of biogas composed of methane (CH_4) and carbon dioxide (CO_2), along with trace amounts of gases like hydrogen sulfide (H_2S) and ammonia (NH_3) (Ward et al., 2008). The AD process occurs in four biological stages—each involving different groups of microorganisms:

2.5.1 Hydrolysis

In this stage, organic polysaccharide is converted into simpler soluble compounds by hydrolytic enzymes (Angelidaki et al., 2018). For example, polysaccharide or organic complex structure can be converted into simpler compound like as water soluble compound. This step is important because only soluble monomers can be further used by microorganisms in the successive stages.

2.5.2 Acidogenisis

In acidogenisis, acidogenic (fermentative) bacteria convert simple sugars, amino acids, and fatty acids into volatile fatty acids (VFAs), alcohols, hydrogen (H_2), and carbon dioxide (CO_2) (Kothari et al., 2014). This stage leads to the formation of acids such as acetic acid, propionic acid, and butyric acid, along with gases like H_2 and CO_2. Figure 2.2 shows the different stages of biogas production.

2.5.3 Acetogenesis

This stage involves the conversion of the volatile fatty acid (VFAs) and alcohols produced in acidogenisis into acetic acid, H_2, and CO_2 by acidogenic bacteria (McInerney et al., 2009). These bacteria are requiring hydrogen producers and work in syntrophic with hydrogen trophic methanogens (Stams & Plugge, 2009). Maintaining a low hydrogen partial pressure is critical at this stage, as it ensures that the reactions remain thermodynamically encouraging.

2.5.4 Methanogens

Methogenogensis is the final stage of anaerobic reaction and is finalized by a specific group of bacteria such as archaea called methano gens. These microorganisms convert acetic acid, or hydrogen and carbon dioxide, into methane (CH_4). This stage is highly sensitive to environmental conditions, which are based on some important parameters such as pH, temperature, and the presence of inhibitors (Angelidaki & Sanders, 2004).

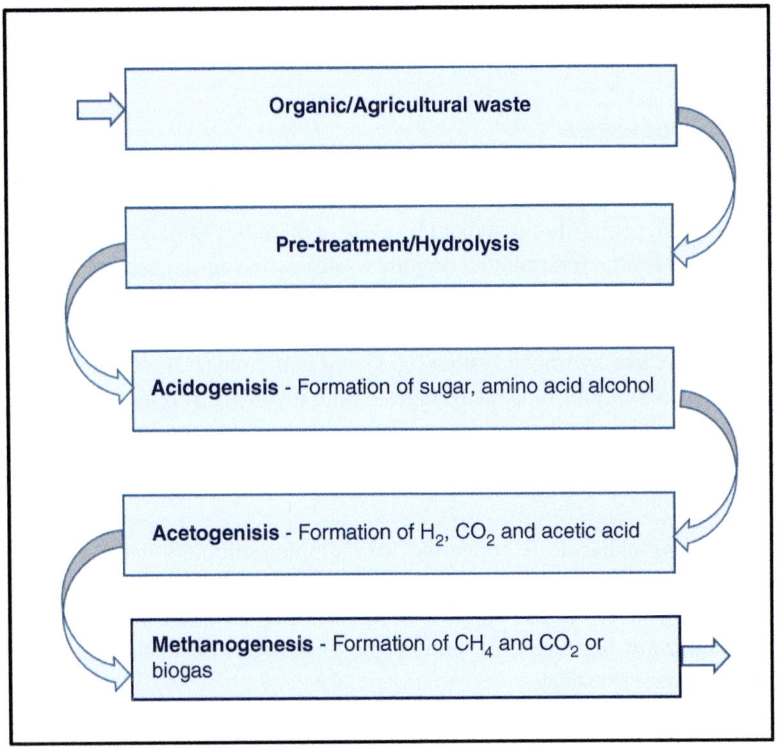

Figure 2.2 Different stages of biogas production.

2.6 Production of Bioprocess Products

2.6.1 Biogas Production

Biogas production is a multi-stage anaerobic process in which organic matter in the absence of oxygen, resulting in a methane-rich gas with the help of microorganism. It begins with hydrolysis, where polysaccharides, proteins, and fats are converted into simple molecules such as amino acids, sugars, and fatty acids (Deublein & Steinhauser, 2011).

Next, in acidogenisis, acidogenic bacteria convert these monomers into VFAs, ammonia, alcohols, hydrogen (H_2), and carbon dioxide (CO_2) (J., 2015). In the acetogenesis stage, fatty acid and alcohols are further converted into acetic acid, which is then converted into CO_2 and H_2 by acidogenic bacteria. Figure 2.3 shows the schematic diagram of biogas production. Finally, in methanogenesis, methanogens convert acetic acid and hydrogen into CO_2 and methane (CH_4). The key reactions include

$$CH_3COOH \rightarrow CH_4 + CO_2$$

and

$$CO_2 + 4\,H_2 \rightarrow CH_4 + 2\,H_2O.$$

The overall simplified reaction is $C_6H_{12}O_6 \rightarrow 3CO_2 + 3CH_4$.

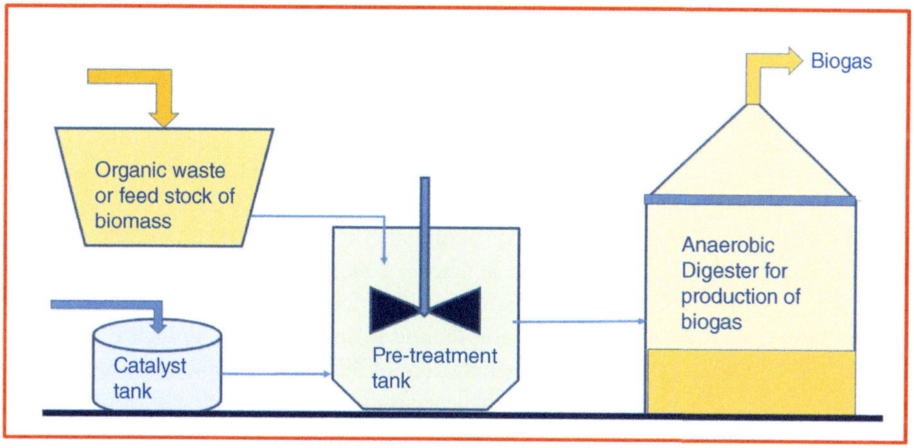

Figure 2.3 Schematic diagram of biogas production.

The resulting biogas typically contains 50–70% CH_4 and 30–50% CO_2, with traces of H_2S, H_2, and moisture, and serves as a non-conventional energy source for heating, electricity, and fuel applications.

2.6.2 Fermentation

This is a metabolic process in which microorganisms converts organic substrates, typically sugars, into simpler compounds—often producing energy-rich products such as ethanol and butanol under anaerobic or semi-anaerobic conditions (Lynd et al., 2002). These bioalcohols are considered important renewable fuels and industrial solvents, and are produced mainly through microbial fermentation using special biocatalysts such as yeast or bacteria.

2.6.2.1 Ethanol Production

Ethanol is most commonly produced by the fermentation of glucose or other fermentable sugars by *Saccharomyces cerevisiae* (baker's yeast). The process involves the glycolysis pathway where glucose is broken down into pyruvate, which is then converted into ethanol and carbon dioxide by the reaction of enzymes like pyruvate decarboxylase and alcohol dehydrogenase (Bhatia et al., 2018). Bioethanol production can be achieved using a variety of feedstocks, broadly categorized into sugar-based, starch-based, and lignocellulosic biomass, each with distinct processing requirements and efficiency levels. These feedstocks are converted into ethanol through fermentation, commonly using yeast such as *S. cerevisiae*. The overall fermentation reaction is:

$$C_6H_{12}O_6 \rightarrow 2C_2H_5OH + 2CO_2,$$

which shows that one molecule of glucose is converted into two moles of ethanol and two moles of carbon dioxide.

2.6.2.1.1 Feedstock for Alcohol Production

2.6.2.1.1.1 Sugar-Based Feedstocks Sugar-based materials like molasses and sugarcane juice are ideal for ethanol production because they contain readily fermentable sugars.

Since the sugars are already in simple forms, no enzymatic hydrolysis is required, making the process faster and more cost-effective (Nigam, 2001). After juice extraction (in the case of sugarcane) or molasses dilution, the sugar solution is directly fermented by yeast into ethanol. This is the dominant process in countries like Brazil, where sugarcane is abundant, and yields are high due to the high sucrose content.

Lignocellulosic Biomass Lignocellulosic materials, like agricultural waste such as wheat straw, wood chips, and energy crops, are abundant and non-edible, making them a highly sustainable feedstock option (Sun & Cheng, 2002). However, these materials are complex in structure and make cellulose accessible. Different pre-treatment methods include physical (grinding), chemical (acid/alkaline hydrolysis), and biological methods (Mosier et al., 2005). Table 2.2 shows the generations of Biomass Feedstock and Bioconversion. Following pre-treatment, enzymatic hydrolysis converts cellulose into fermentable sugars. These sugars are then fermented by yeast or genetically engineered microbes that can also ferment pentoses from hemicellulose. Although promising, this route is currently more expensive due to high enzyme costs and process complexity (Himmel et al., 2007). Figure 2.4 shows the schematic diagram for the production of bioethanol.

Parameters Affecting Fermentation Yield The efficiency of ethanol and butanol production through fermentation is significantly influenced by key parameters such as pH,

Table 2.2 Generations of Biomass Feedstock and Bioconversion.

Generation	Feedstock Type	Examples	Bioconversion Processes	Products	Applications
First generation	Edible food crops (high sugar/oil/starch)	Sugarcane, corn, wheat, soybean, rapeseed	Fermentation (for ethanol), transesterification (for biodiesel)	Bioethanol, biodiesel	Transport fuel (gasoline substitute), diesel engines
second generation	Lignocellulosic biomass (non-edible, waste-based)	Agricultural residues (rice husk, straw), wood, bagasse, water hyacinth	Pre-treatment + enzymatic hydrolysis + fermentation gasification pyrolysis	Ethanol, syngas, bio-oil, biochar	Advanced biofuels, power generation, soil amendment, carbon sequestration
Third generation	Algal biomass	Microalgae, Microalgae	Transesterification (for oil-rich algae), anaerobic digestion, fermentation	Biodiesel, biogas, ethanol	Aviation fuel, biodiesel for transport, electricity, nutraceuticals
Fourth generation	Genetically modified (GM) organisms and synthetic biomass	GM algae, engineered bacteria or yeast	Synthetic biology + photo–bioreactors, fermentation with CO_2 fixation or sunlight, electro fuels	Hydrogen, algae-derived fuels, bioplastics	Carbon-neutral/negative fuels, green chemicals, pharmaceuticals, advanced materials

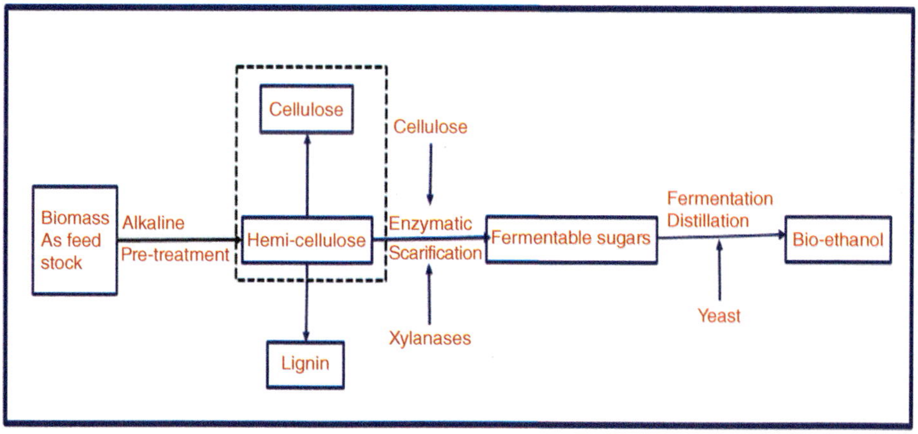

Figure 2.4 Schematic diagram for the production of bioethanol.

temperature, and substrate concentration (van Maris et al., 2006). For ethanol production using *S. cerevisiae*, the optimal pH range is 4.5–5.5, where deviations can hinder microbial growth or encourage contamination. Similarly, for butanol production by *Clostridium* species, an optimal pH range of 5.0–6.0 is critical, as values below 4.5 inhibit solvent genesis, and those above 6.5 disrupt the acid–solvent balance (Jones & Woods, 1986). Temperature also plays a vital role; *S. cerevisiae* functions best at 30–35°C, while *Clostridium* strains prefer 35–37°C. Extremes in temperature can denature enzymes or suppress microbial activity, ultimately reducing yield. Substrate concentration further impacts fermentation performance—low sugar levels limit microbial growth and ethanol/butanol yield, whereas high concentrations can cause osmotic stress, substrate inhibition, and catabolite repression. Ideal sugar concentrations range between 100 and 150 g/L for ethanol and 40–60 g/L for butanol, the latter being more toxic to cells, thus requiring careful control for optimal productivity (Bai et al., 2008).

2.6.2.2 Transesterification (for Biodiesel Production)

Transesterification is the key chemical process where triglycerides (fats and oils) react with an alcohol to form fatty acid methyl esters (FAMEs) or ethyl esters—commonly known as biodiesel—and glycerol as a by-product (Ma & Hanna, 1999). This reaction is catalyzed either chemically or biologically. The main chemical reaction involved in biodiesel production is as follows (Meher et al., 2006).

General Transesterification Reaction:

$$\text{Triglyceride} + 3\,\text{Methanol} \rightarrow 3\,\text{Methyl Esters (Biodiesel)} + \text{Glycerol}$$

Reaction Equation (with methanol):

$$C_{57}H_{104}O_6 + 3\,CH_3OH \rightarrow 3C_{19}H_{36}O_2\,(\text{biodiesel}) + C_3H_8O_3\,(\text{glycerol})$$

where

- $C_{57}H_{104}O_6$ = typical triglyceride molecule
- CH_3OH = methanol

- $C_{19}H_{36}O_2$ = methyl esters (biodiesel)
- $C_3H_8O_3$ = glycerol

This reaction is reversible and typically requires excess alcohol and proper mixing, temperature (~50–60°C), and catalyst concentration to achieve high conversion efficiency. To carry out this reaction, chemical catalysts, such as sodium hydroxide (NaOH), potassium hydroxide (KOH), or sulfuric acid (H_2SO_4), are widely used due to their fast reaction rates and low cost (Lapuerta et al., 2008). Alkaline catalysts are more effective for refined, low free fatty acid (FFA) feedstocks, while acid catalysts are better suited for high-FFA oils. However, chemical methods may lead to soap formation and require extensive purification. Microbial catalysts, specifically lipase enzymes, offer a greener alternative. Enzymatic transesterification operates under milder conditions, tolerates higher FFA levels, and reduces by-product formation (Salis et al., 2010). Lipases from organisms like *Candida antarctica* and *Rhizomucor miehei* are commonly used for this purpose, and their immobilized forms improve reusability and stability. Feedstocks for biodiesel production vary widely, with the most common being vegetable oils such as soybean, canola, palm, sunflower, and jatropha oils. These are preferred for their consistent quality and low FFA content. However, due to economic and sustainability concerns, attention has shifted toward waste cooking oils, animal fats, and non-edible oils. These feedstocks are more abundant and cost-effective, though they may require pre-treatment due to impurities or high FFA levels. Overall, the choice of catalyst and feedstock plays a crucial role in determining the efficiency, cost, and environmental impact of the transesterification process in biodiesel production (Demirbas, 2009b).

2.6.2.3 Biohydrogen Production

This method is an eco-friendly and sustainable method of generating hydrogen gas using biological processes (Das & Veziroglu, 2001). Hydrogen is a clean energy carrier with high energy content and produces only water upon combustion, making it a vital component of future energy systems. Biological hydrogen production employs microorganisms to convert organic matter into hydrogen under anaerobic or semi-anaerobic conditions. The most commonly studied techniques for biohydrogen generation include dark fermentation, photofermentation, and microbial electrolysis cells (MECs)—each offering unique advantages and operating mechanisms (Hallenbeck, 2012).

2.6.2.3.1 Dark Fermentation

This method is a strictly anaerobic process in which microbe convert carbohydrates and organic waste into hydrogen. Microorganisms like *Clostridium* spp. and *Enterobacter* spp. are commonly used due to their hydrogenase enzyme activity [66]. The dark fermentation is its ability to utilize a large range of substrates, including industrial wastewater, agricultural residues, and food waste, making it a cost-effective and scalable option. The process occurs without the need for light and typically at mesophilic (30–37°C) or thermophilic (50–60°C) conditions. However, hydrogen yield in dark fermentation is often limited (usually around 1–2 mol H_2/mole glucose), mainly due to the accumulation of acidic by-products that inhibit microbial activity. Strategies such as pH control, co-culture techniques, and reactor optimization are being explored to enhance hydrogen yield.

2.6.2.3.2 Photofermentation

This is a method carried out by photosynthetic bacteria, particularly purple non-sulfur bacteria like *Rhodobacter* and *Rhodopseudomonas*. These microbes utilize organic acids produced from dark fermentation and convert them into hydrogen using light energy in the presence of nitrogenase enzymes (Logan et al., 2002). The general reaction involves the breakdown of VFAs and water into hydrogen and carbon dioxide under anaerobic and illuminated conditions. While photofermentation has the potential to achieve higher hydrogen yields (up to 9 mol H_2/mole acetate), it requires a continuous light source, which limits its practicality and increases energy input in large-scale systems (Das & Veziroglu, 2008). Additionally, photofermentative systems are sensitive to oxygen, require specific wavelengths of light, and demand high nutrient inputs, making them less economically viable without optimization.

2.6.2.3.3 Microbial Electrolysis Cells (MECs)

MECs represent a hybrid biological–electrochemical method for hydrogen production. In MECs, electroactive bacteria oxidize organic substrates at the anode and release electrons, which are transferred through an external circuit to the cathode, where hydrogen gas is generated by reducing protons with the help of a small external voltage (typically <1.0 V). Microorganisms such as *Geobacter* and *Shewanella* are effective in this process. MECs can operate with a variety of feedstocks, including wastewater, and offer high hydrogen recovery rates (Sharma et al., 2015). The advantages include energy-efficient hydrogen production and wastewater treatment integration. However, challenges remain, such as electrode material cost, system scalability, and biofilm management.

2.6.2.3.4 Algal Bioprocesses: Cultivation Systems and Lipid Extraction for Biodiesel

Algal bioprocesses are considered an important sustainable approach for the production of biodiesel and other biofuels due to the rich content of lipid and fast growth rate of microalgae. Microalgae are photosynthetic microorganisms capable of converting sunlight, carbon dioxide (CO_2), and nutrients into biomass, including valuable lipids, which can be transesterified into biodiesel (Logan & Rabaey, 2012). Algal biofuel production largely depends on two key factors: the cultivation system used for microalgae growth and the lipid accumulation and extraction techniques employed.

2.6.2.3.5 Microalga Cultivation Systems

The primary goal of algal biodiesel production is to maximize lipid accumulation within the algal cells. Lipids, especially triacylglycerols (TAGs), are the key precursors for biodiesel synthesis. Microalgae contain lipid ranging from 20% to over 50% of dry cell weight, depending on the basis of cultivation conditions (Chisti, 2007). Lipid accumulation is typically enhanced by maintaining environmental conditions, such as nitrogen starvation, high salinity, or light intensity. Under stress, microalgae redirect their metabolic pathways to store energy in the form of lipids rather than growing biomass. Species like *Nannochloropsis*, *Chlorella*, and *Botryococcus braunii* are known for high lipid productivity. Once harvested, lipids must be extracted from the algal biomass (Hu et al., 2008). The cell walls of microalgae can be tough, so mechanical, chemical, or enzymatic disruption is often necessary before extraction. Lipid extraction from biomass, particularly microalgae and other oil-rich sources, is a critical step in biodiesel production and can be achieved through various methods, each with its advantages and limitations (Mata et al., 2010). Solvent extraction, commonly using organic solvents like hexane or chloroform–methanol mixtures, is a traditional and widely used technique due to its simplicity and relatively high extraction efficiency; however, it poses environmental and health hazards due to solvent toxicity and flammability (Meng et al., 2009). Supercritical CO_2 extraction offers a cleaner and more sustainable alternative, using carbon dioxide at high pressure and temperature to extract lipids without leaving toxic residues. This method is highly efficient and selective but remains costly due to the need for high-pressure equipment. To enhance lipid recovery while reducing processing time and energy consumption, ultrasound-assisted extraction (UAE) is employed, where ultrasonic waves disrupt cell walls and increase solvent penetration, significantly increasing extraction yield (Shilpi & Mandal, 2018).

2.7 Bioreactors and Process Design

Bioreactors are specific vessels designed to provide controlled environmental conditions for biological reactions involving microorganisms, cells, or enzymes (Shilpi & Mandal, 2018). They are fundamental to bioprocesses, which play a crucial role in industries such as pharmaceuticals, biofuels, food, and waste treatment (Xu et al., 2006). The design and operation of bioreactors directly influence the productivity, efficiency, and scalability of bioprocesses. Understanding the types of bioreactors, key design and operational parameters, and scale-up challenges is essential for successful process development (Shuler & Kargi, 2002).

2.7.1 Types of Bioreactors

Bioreactors are engineered vessels or systems designed to support biologically active environments for the production of microbes, cells, or enzymes to produce valuable bioproducts such as biofuels, pharmaceuticals, enzymes, and chemicals (Lee, Kim, Park, et al., 2019). The choice of bioreactor depends on the process objectives, microorganism behavior, and operational conditions. Below are the major types of bioreactors (Appels et al., 2011)

2.7.1.1 Batch Bioreactor

In batch bioreactors, all nutrients and inoculum are loaded at the beginning of the process, and no additional input is added during cultivation. The reaction proceeds without interference until completion, and the product is produced at the end. These are simple to operate and widely used in small-scale production or lab-scale research (Richmond, 2004). However, product yields can be lower compared to continuous systems due to limited control over microbial growth phases.

2.7.1.2 Fed-Batch Bioreactor

This is a semi-continuous system where nutrients are added intermittently or continuously during the process without removing the culture broth. This approach allows better control of nutrient concentration, preventing substrate inhibition and enhancing product yield (Ugwu et al., 2008). It is particularly useful for producing secondary metabolites and recombinant proteins.

2.7.1.3 Continuous Bioreactor (Chemostat)

In continuous bioreactors, fresh medium is continuously added, and culture broth is simultaneously removed at the same rate, maintaining a constant volume. This provides a steady-state environment and allows prolonged production (Soh & Zimmerman, 2011). Continuous systems are efficient for large-scale operations and ensure consistent product quality but require precise control to prevent contamination and washout.

2.7.1.4 Anaerobic Bioreactor

Anaerobic bioreactors are designed to maintain oxygen-free conditions, essential for processes like AD or butanol fermentation. These bioreactors are sealed to prevent oxygen entry and are often equipped with gas collection systems for biogas (methane and CO_2) recovery. They are crucial in wastewater treatment and renewable energy generation (Shuler, 1999).

2.7.1.5 Photo Bioreactor (PBR)

Photo bioreactors are specialized bioreactors designed for photosynthetic organisms such as microalgae and cyanobacteria. They provide light (natural or artificial), CO_2, and nutrients for biomass growth (Stanbury, 1995). PBRs can be open (e.g., raceway ponds) or closed (e.g., tubular or flat-panel systems). Closed systems allow better control over contamination and environmental factors, improving productivity and biomass quality for applications like biofuel and nutraceutical production (Oh & Zimmerman, 2011).

2.7.1.6 Airlift Bioreactor

Airlift bioreactors utilize air or gas injection to circulate the culture medium and provide aeration without mechanical agitation. They are energy-efficient and suitable for shear-sensitive cells. The design includes a riser and a downcomer section to create natural circulation, enhancing mass and heat transfer (Shuler, 1999).

2.7.1.7 Packed Bed and Fluidized Bed Bioreactors

These reactors are commonly used in immobilized cell or enzyme processes. In packed bed reactors, cells are fixed on a support material, and the substrate flows through the bed. In fluidized bed bioreactors, the support particles are suspended in the upward flow of liquid or gas, enhancing contact and reducing mass transfer limitations (Stanbury, 1995). In conclusion, each bioreactor type serves specific process needs depending on biological requirements, scale, and economic considerations. Advances in bioreactor design and control systems continue to enhance their efficiency, making them central to industrial biotechnology and bioenergy application (Garcia-Ochoa & Gomer, 2009).

2.7.1.8 Key Design and Operational Parameters

Efficient bioprocess operation relies on precise control of key parameters such as mixing, aeration, and temperature (Stanbury et al., 2016). Adequate mixing is essential to prevent concentration gradients and enhance mass transfer, particularly oxygen transfer in aerobic systems. This is typically achieved using mechanical impellers or gas sparging, with careful design to balance mixing efficiency and shear forces to avoid damage to sensitive cells (Shuler & Kargi, 2002). Aeration plays a critical role in supplying oxygen for aerobic processes; maintaining dissolved oxygen levels within optimal ranges is vital for cell metabolism and growth. However, excessive aeration can lead to problems such as foaming and increased shear stress, whereas inadequate oxygen availability can severely limit cell proliferation. Additionally, temperature control is crucial, as most bioprocesses require specific temperature ranges to support enzyme activity and microbial growth. Heating or cooling systems integrated into the bioreactor design help maintain the desired thermal conditions, ensuring consistent and efficient process performance (Sun & Cheng, 2002).

2.8 Process Integration Techniques

Since pre-treatment is an important initial step in the bioconversion of biomass, the process aims to change the complex structural components of the biomass into more accessible sugars that can be further processed in biofuel and biochemical production (Kumar et al., 2009). The main objective of pre-treatment is to disintegrate the tough structure of lignin, cellulose, and hemicellulose in plant biomass, which increases the subsequent enzymatic hydrolysis. There are several types of pre-treatment methods, each tailored to specific biomass characteristics and process requirements. The pre-treatment method increases the surface area available for enzymatic action (Alvira et al., 2010). Chemical pre-treatment involves the use the acids, alkalis, or ionic liquids to reduce the lignocellulosic matrix, making it easier to extract cellulose. Biological pre-treatment employs specific enzymes or microbial agents that degrade lignin or hemicellulose selectively, offering a more environmentally friendly alternative. The importance of pre-treatment lies in its ability to significantly reduce the resistance of biomass, improve sugar yields, and increase the overall efficiency of the bioprocess. From a sustainability perspective, an optimized pre-treatment process minimizes energy consumption, reduces chemical usage, and lowers waste generation, thereby contributing to a more eco-friendly and cost-effective biorefinery system (Singh et al., 2013).

2.8.1 Simultaneous Scarification and Fermentation (SSF)

SSF is an integrated bioprocessing technique in which the enzymatic hydrolysis of biomass and the fermentation of resulting sugars take place concurrently in the same reactor. This approach is particularly effective for converting lignocellulosic biomass into bioethanol. One of the key advantages of SSF is its ability to reduce end-product inhibition, which typically occurs when accumulated sugars inhibit enzyme activity in separate hydrolysis steps. By converting sugars to ethanol during fermentation, SSF maintains a low sugar concentration, which improves enzyme efficiency (Harun et al., 2010). Additionally, this integrated process reduces equipment needs and operational complexity, resulting in lower capital and operating costs. It also increases overall ethanol yield due to the synergistic interaction between hydrolysis and fermentation. However, SSF faces challenges related to the divergent optimal conditions required by enzymes (typically higher temperatures) and fermentative microbes (which prefer lower temperatures), making it necessary to find a compromise operating environment that supports both activities (Lynd et al., 2002). Despite this challenge, SSF remains a widely accepted strategy in industrial bioethanol production due to its process efficiency, cost-effectiveness, and support to sustainable biomass utilization.

2.8.2 Consolidated Bioprocessing (CBP)

Consolidated bioprocessing (CBP) is considered an advanced and highly integrated approach to biomass conversion in which a single microbe and single system are applicable in whole system of bioprocess (Olson et al., 2012). This "one-pot" strategy eliminates the need for separate enzyme production and addition, which are often costly and energy-intensive steps in conventional bioprocessing. By consolidating these stages, CBP significantly reduces operational complexity, equipment needs, and overall processing costs. A notable example of CBP is the use of engineered microbes such as *Clostridium thermocellum*, which possesses both cellulolytic and fermentative capabilities, enabling the direct conversion of cellulose into ethanol (Zhao et al., 2012). CBP supports a more efficient use of resources by simplifying the bioconversion pathway, reducing external inputs, and minimizing energy consumption. This integration not only enhances process economics but also adopts circular economy, where waste is minimized and biological processes are optimized for renewable and eco-friendly production (Peters et al., 2003).

2.8.3 Aspen Plus and Similar Process Simulation Tools

Aspen plays an important role in the design, analysis, and optimization of complex bioprocess systems. These software platforms are equipped with advanced functionalities that allow for the simulation of unit operations, detailed mass and energy balances, and reactor kinetics (Babu & Suresh, 2009). Aspen Plus is widely used for the design and scale-up of critical operations such as fermentation, distillation, and separation, which are essential for bioprocesses. Moreover, it supports the integration of renewable feedstocks and the multiple uses of waste streams, which are key components of sustainable biorefinery design. By modeling material and energy flows within a closed-loop system, Aspen Plus aids in identifying opportunities for resource recovery and reuse, which supports the principles of the circular economy. Through such capabilities, simulation tools not only enhance process efficiency and reduce environmental impact but also contribute to the development of economically and ecologically viability.

2.9 Environmental and Economic Evaluations

2.9.1 Energy Balance

Energy balance is a serious assessment tool in bioprocessing that involves comparing the total energy input required for different processing steps—such as pre-treatment, enzymatic hydrolysis, fermentation, and distillation—with the energy output obtained in the form of biofuels or biochemical (Patel et al., 2016; Zhang et al., 2017). This analysis helps determine the overall energy efficiency of a bioprocess. Which means the energy produced from the final bio-product exceeds the energy consumed during its production (Singh & Sharma, 2018). Achieving a favorable energy balance is essential for validating the sustainability and feasibility of biofuel systems, particularly when competing with fossil fuels. It also guides process optimization by identifying energy-intensive stages that can be improved or integrated with energy recovery strategies, such as using residual biomass for heat or power generation. Achieve a favorable, energy balance analysis supports the development of more.

2.9.2 Life Cycle Assessment (LCA)

LCA is an inclusive evaluation tool used to determine the environmental impacts related with all stages of a product's life, from raw material extraction to production, distribution, use, and final disposal. In the context of bioprocessing, LCA plays a dynamic role in recognizing environmental "hot spots," such as stages with high energy consumption, excessive water use, or the release of harmful emissions (ISO 14040, 2006). It reflects multiple impact categories, including global warming potential, water footprint, and Eco toxicity (harm to aquatic and terrestrial ecosystems). By analyzing these factors, LCA helps stakeholders understand the full environmental cost of bio-based product systems. This information supports informed decision-making aimed at minimizing ecological footprints and selecting process alternatives that are more sustainable (Cucurachi et al., 2018). Finally, the use of LCA in bioprocess design bring into line with circular economy goals by promoting resource efficiency, reducing waste, and guiding the development of cleaner, greener technologies.

2.9.3 Cost–Benefit Analysis (CBA)

CBA is an important economic estimation method that compares the total costs—including capital expenditures (CAPEX) and operational expenditures (OPEX) —with the benefits derived from a bioprocess, such as the market value of bio-based products and potential earnings from carbon credits or environmental enticements. In bioprocessing, CBA is essential for assessing commercial feasibility, guiding whether a proposed technology or process is economically viable in real-world applications (Boardman et al., 2018). This analysis helps to know the return on investment and identifies cost drivers and savings opportunities. It is particularly applicable for emerging technologies like CBP or advanced methods such as magnetic separation, where initial costs may be high, but long-term gains in efficiency and sustainability could compensate them (Li et al., 2020). Beyond industrial decision-making, CBA also has significant policy implications, serving as a critical tool for governments and stakeholders in framing subsidies, employing carbon taxes, and encouraging green investments (Hoffmann, 2011).

2.9.4 Sustainability and the Circular Economy

This approach reduces environmental problems but also adds economic value to agricultural and industrial waste streams (Kirchherr et al., 2017). A key component of the circular bioeconomy is the integration of biorefineries with local agro-industrial systems, which enhances regional sustainability by utilizing locally available biomass and reducing transportation-related emissions. To support this transformation, process integration and simulation play a crucial role in designing efficient and environmentally friendly industrial systems (Koirala et al., 2020). These tools help reduce resource consumption and greenhouse gas emissions by identifying optimal process configurations, enhancing energy and material efficiency, and minimizing the need for landfill disposal These tools help stakeholders assess the sustainability performance of integrated systems and support the transition toward low-carbon, circular, and zero-waste industrial models (Finnveden et al., 2009). Together, these strategies contribute to a resilient and regenerative economy that balances environmental, economic, and social priorities (Geissdoerfer et al., 2017).

2.10 Challenges and Future Directions in Bioenergy Bioprocesses

The transition to bio-based energy systems presents a promising path toward sustainability, but it is accompanied by several challenges that hinder large-scale implementation. These challenges span technical, scientific, economic, and policy dimensions. However, ongoing advancements in biotechnology and system-level innovation are helping to reshape the future of bioenergy production.

2.10.1 Technical Barriers

Since all the enzymes, particularly those used in hydrolysis and fermentation stages of lignocellulosic biomass processing (Balan, 2014). Enzymes like cellulases and hemicellulases are often required in large quantities, and their production, stability, and activity optimization significantly influence process economics. In addition, low product yields due to suboptimal metabolic pathways or inefficient substrate conversion remain a major constraint. Microbial contamination during fermentation, especially in open or large-scale systems, also leads to product loss and reduced consistency (Jang et al., 2018). These barriers necessitate more robust biocatalysts, improved process controls, and integrated bioreactor designs.

2.10.2 Innovations in Synthetic Biology and Metabolic Engineering

Advancements in synthetic biology and metabolic engineering are paving the way to overcome many of these hurdles. By optimizing metabolic pathways, researchers can enhance substrate utilization, increase product yields, and reduce by-product formation (Nielsen & Keasling, 2016). For instance, engineered *Escherichia coli* or *S. cerevisiae* strains are now capable of fermenting both hexose and pentose sugars from lignocellulosic biomass into ethanol or butanol more efficiently. Synthetic biology also enables modular pathway assembly, biosensor design for dynamic regulation, and adaptive control of microbial performance. These tools are essential for developing next-generation biocatalysts tailored to specific feedstocks and process conditions (Lee, Kim, Chae, et al., 2019).

2.10.3 Policy, Funding, and Market Considerations

Despite scientific progress, the policy and economic environment plays an important role in the deployment of bioenergy technologies. Many biofuel projects face financial uncertainty due to fluctuating fossil fuel prices, limited subsidies, and slow regulatory approvals. Stronger policy support, such as feed-in tariffs, renewable energy mandates, and carbon pricing, can make bioenergy more competitive (Carriquiry et al., 2011). Additionally, public and private investment is needed to scale up technologies from the lab to the commercial level, including infrastructure for biomass collection, transportation, and conversion. Market development for bio-based products and by-products, such as biofertilizers, bioplastics, and biochar, also helps improve the overall economic feasibility of bioprocesses (Sadhukhan et al., 2016).

2.10.4 Emerging Trends

Several emerging trends are shaping the future of bioenergy. Bioelectrochemical systems, as MFCs and MECs, have the ability to generate electricity or hydrogen directly from organic waste using electroactive bacteria (Logan & Rabaey, 2012). These systems offer dual benefits of waste treatment and clean energy generation, though their efficiency and scalability are still under development. The application of CRISPR-based gene editing has further revolutionized metabolic engineering by enabling precise, rapid, and multiplexed genetic modifications in microorganisms, greatly enhancing strain development for biofuel production (Jiang et al., 2013). CRISPR can also be used to construct microbial consortia with complementary functionalities, enhancing biomass breakdown and conversion efficiency.

2.11 Conclusion

In conclusion, this chapter underscores the fundamental scientific and engineering principles underpinning bioprocesses for green energy production. By differentiating bioprocesses from conventional chemical methods and exploring the roles of microbial metabolism, enzymes, and diverse biomass feedstocks, it highlights the potential of biological systems in advancing sustainable energy solutions. Key technologies such as AD, fermentation, trans esterification, biohydrogen production, and algal systems are examined in detail, along with their operational parameters and applications. The importance of bioreactor design, process integration strategies like SSF and CBP, and the use of modern simulation tools is emphasized to illustrate the practical implementation of these processes. Furthermore, environmental and economic assessments are discussed from the point of view of sustainability and the circular economy. By addressing current challenges and innovations—including synthetic biology, metabolic engineering, and emerging tools like microbial fuel cells and CRISPR—this chapter provides a comprehensive outlook on the transform.

References

Ahmad, A. L., Yasin, N. H. M., Derek, C. J. C., & Lim, J. K. (2011). Microalgae as a sustainable energy source for biodiesel production: A review. *Renewable and Sustainable Energy Reviews*, *15*(1), 584–593. https://doi.org/10.1016/j.rser.2010.09.018.

Alvira, P., Tomás-Pejó, E., Ballesteros, M., & Negro, M. J. (2010). Pretreatment technologies for an efficient bioethanol production process based on enzymatic hydrolysis: A review. *Bioresource Technology*, *101*(13), 4851–4861. https://doi.org/10.1016/.biortech.2009.11.093.

Angelidaki, I., & Sanders, W. (2004). Assessment of the anaerobic biodegradability of macropollutants. *Reviews in Environmental Science and Biotechnology*, *3*(2), 117–129. https://doi.org/10.1038/rmicro2166.

Angelidaki, I., Treu, L., Tsapekos, P., Luo, G., Campanaro, S., & Kougias, P. G. (2018). Biogas upgrading and utilization: Current status and perspectives. *Biotechnology Advances*, *36*(2), 452–466. https://doi.org/10.1016/j.biortech.2008.02.044.

Appels, L., Baeyens, J., Degrève, J., & Dewil, R. (2008). Principles and potential of the anaerobic digestion of waste-activated sludge. *Progress in Energy and Combustion Science, 34*(6), 755–781. https://doi.org/10.1126/science.1258096.

Appels, L., Lauwers, J., Degrève, J., Helsen, L., Lievens, B., Willems, K., Van Impe, J., & Dewil, R. (2011). Anaerobic digestion in global bio-energy production: Potential and research challenges. *Renewable and Sustainable Energy Reviews, 15*(9), 4295–4301. https://doi.org/10.1016/j.tibtech.2018.11.009.

Atabani, A. E., Silitonga, A. S., Badruddin, I. A., Mahlia, T. M. I., Masjuki, H. H., & Mekhilef, S. (2012). A comprehensive review on biodiesel as an alternative energy resource and its characteristics. *Renewable and Sustainable Energy Reviews, 16*(4), 2070–2093. DOI: 10.1016/j.rser.2012.01.003.

Babu, B. V., & Suresh, K. (2009). Process simulation in biofuels production using Aspen Plus. *International Journal of Chemical Reactor Engineering, 7*(1), 1–16. https://doi.org/10.2202/1542-6580.2074.

Bai, F. W., Anderson, W. A., & Moo-Young, M. (2008). Ethanol fermentation technologies from sugar and starch feedstocks. *Biotechnology Advances, 26*(1), 89–105. https://doi.org/10.1016/j.biotechadv.2007.09.002.

Bailey, J. E., & Ollis, D. F. (2010). *Biochemical engineering fundamentals* (2nd ed.). McGraw-Hill.

Balan, V. (2014). Current challenges in commercial-scale lignocellulosic biofuel production. *Biofuels, Bioproducts and Biorefining, 8*(5), 495–501. https://doi.org/10.1002/bbb.1503.

Balat, M., & Balat, H. (2009). Recent trends in global production and utilization of bio-ethanol fuel. *Applied Energy, 86*(11), 2273–2282. https://doi.org/10.1016/j.apenergy.2009.03.015.

Bhat, M. K. (2000). Cellulases and related enzymes in biotechnology. *Biotechnology Advances, 18*(5), 355–383. https://doi.org/10.1016/S0734-9750(00)00041-0.

Bhatia, S. K., Kim, J., Yang, Y. H., & Lee, J. K. (2018). *Biotechnological production of ethanol: Methods and applications* (1st ed.). Springer.

Boardman, A. E., Greenberg, D. H., Vining, A. R., & Weimer, D. L. (2018). *Cost-benefit analysis: Concepts and practice* (5th ed., pp. 20–45). Cambridge University Press.

Brennan, L., & Owende, P. (2010). Biofuels from microalgae—A review of technologies for production, processing, and extractions. *Renewable and Sustainable Energy Reviews, 14*(2), 557–577. https://doi.org/10.2172/1013269.

Carriquiry, M. A., Du, X., & Timilsina, G. R. (2011). Second generation biofuels: Economics and policies. *Energy Policy, 39*(7), 4222–4234. https://doi.org/10.1016/j.enpol.2011.04.036.

Cherubini, F. (2010). The biorefinery concept: Using biomass instead of oil for producing energy and chemicals. *Energy Conversion and Management, 51*(7), 1412–1421. https://doi.org/10.1126/science.1189003.

Chisti, Y. (2007). Biodiesel from microalgae. *Biotechnology Advances, 25*(3), 294–306. https://doi.org/10.1016/j.biotechadv.2007.02.001.

Choi, J., Park, S. Y., & Kim, J. H. (2014). Ultrasound-assisted extraction of lipids from microalgae. *Ultrasonics Sonochemistry, 21*(3), 1144–1148. https://doi.org/10.1016/j.ultsonch.2013.10.020.

Cucurachi, S., van der Giesen, C., & Guinée, J. B. (2018). Life cycle assessment for sustainable bio-based product development: Challenges and perspectives. *Sustainability, 10*(11), 4143. https://doi.org/10.1017/9781108235599.

Das, D., & Veziroglu, T. N. (2001). Hydrogen production by biological processes: A survey of literature. *International Journal of Hydrogen Energy*, *26*(1), 13–28. https://doi.org/10.1002/bit.22502.

Das, D., & Veziroglu, T. N. (2008). Photofermentation of organic acids for hydrogen production: A review. *International Journal of Hydrogen Energy*, *33*(21), 6330–6340. https://doi.org/10.1126/science.1217412.

Demirbas, A. (2009a). *Biofuels: Securing the planet's future energy needs*. Springer. https://doi.org/10.1016/j.enconman.2009.05.010.

Demirbas, A. (2009b). Biofuels securing the planet's future energy needs. *Energy Conversion and Management*, *50*(9), 2239–2249. https://doi.org/10.1002/bit.22502.

Deublein, D., & Steinhauser, A. (2011). *Biogas from waste and renewable resources: An introduction* (1st ed.). Wiley-VCH.

Doran, P. M. (2013). *Bioprocess engineering principles* (2nd ed.). Academic Press.

Doudna, J. A., & Charpentier, E. (2014). The new frontier of genome engineering with CRISPR-Cas9. *Science*, *346*(6213), 1258096. https://doi.org/10.1021/cb7002438.

Finnveden, G., Hauschild, M. Z., Ekvall, T., Guinée, J., Heijungs, R., Hellweg, S., Koehler, A., Pennington, D., & Suh, S. (2009). Recent developments in life cycle assessment. *Journal of Environmental Management*, *91*(1), 1–21. https://doi.org/10.1016/j.jenvman.2009.06.018.

Garcia-Ochoa, F., & Gomer, E. (2009). Bioreactor scale-up and oxygen transfer rate in microbial processes: An overview. *Biotechnology Advances*, *27*(2), 153–176. https://doi.org/10.1016/j.biotechadv.2008.10.006.

Geissdoerfer, M., Savaget, P., Bocken, N. M. P., & Hultink, E. J. (2017). The circular economy – A new sustainability paradigm? *Journal of Cleaner Production*, *143*, 757–768. https://doi.org/10.1016/j.jenvman.2009.06.018.

Hallenbeck, P. C. (2012). *Modern biohydrogen production* (1st ed.). Springer.

Harun, R., Danquah, M. K., & Forde, G. M. (2010). Microalgal biomass as a fermentation feedstock for bioethanol production. *Biotechnology Advances*, *28*(1), 1–10. doi.org/10.1016/j.biotechadv.2009.09.003.

Hill, J., Nelson, E., Tilman, D., Polasky, S., & Tiffany, D. (2006). Environmental, economic, and energetic costs and benefits of biodiesel and ethanol biofuels. *Proceedings of the National Academy of Sciences*, *103*(30), 11206–11210. https://doi.org/10.1126/science.1193990.

Himmel, M. E., Ding, S. Y., Johnson, D. K., Adney, W. S., Nimlos, M. R., Brady, J. W., & Foust, T. D. (2007). Biomass recalcitrance: Engineering plants and enzymes for biofuels production. *Science*, *315*(5813), 804–807. https://doi.org/10.1016/j.biortech.2004.06.025.

Hoffmann, U. (2011). Policy instruments for sustainable bioeconomy development. *Energy Policy*, *39*(10), 5777–5785. https://doi.org/10.1016/j.copbio.2011.11.021.

Hu, Q., Sommerfeld, M., Jarvis, E., Ghirardi, M., Posewitz, M., Seibert, M., & Darzins, A. (2008). Microalgal triacylglycerols as feedstocks for biofuel production: Perspectives and advances. *The Plant Journal*, *54*(4), 621–639. https://doi.org/10.1111/j.1365-313X.2008.03492.x.

Humbird, D., Davis, R., Tao, L., Kinchin, C., Hsu, D., Aden, A., Schoen, P., Lukas, J., Olthof, B., Worley, M., & Sexton, D. (2011). *Process design and economics for biochemical conversion of lignocellulosic biomass to ethanol*. National Renewable Energy Laboratory (NREL).

ISO 14040. (2006). *Environmental management – Life cycle assessment – Principles and framework*. International Organization for Standardization.

J. (2015). Acidogenesis in anaerobic digestion. *Annual Review of Microbiology*, *69*(1), 65–84. https://doi.org/10.1146/annurev-micro-091014-104222. Retrieved June 5, 2025. https://doi.org/10.1016/j.rser.2014.08.043.

Jang, Y.-S., Malaviya, A., Lee, J., Jung, G. Y., & Kim, B.-G. (2018). Recent advances in consolidated bioprocessing for biofuel production. *Trends in Biotechnology*, *36*(7), 642–654. https://doi.org/10.1002/bbb.1503.

Jiang, W., Bikard, D., Cox, D., Zhang, F., & Marraffini, L. A. (2013). RNA-guided editing of bacterial genomes using CRISPR-Cas systems. *Nature Biotechnology*, *31*(3), 233–239. https://doi.org/10.1038/nbt.2508.

Jones, D. T., & Woods, D. R. (1986). Acetone-butanol fermentation revisited. *Microbiological Reviews*, *50*(4), 484–524. https://doi.org/10.1016/j.biotechadv.2007.09.002.

Kapdan, I. K., & Kargi, F. (2006). Bio-hydrogen production from waste materials. *Enzyme and Microbial Technology*, *38*(5), 569–582. doi.org/10.1016/j.rser.2021.111682.

Keasling, J. D. (2008). Synthetic biology for synthetic chemistry. *ACS Chemical Biology*, *3*(1), 64–76. https://doi.org/10.1016/B978-012666260-3/50002-1.

Keasling, J. D. (2010). Manufacturing molecules through metabolic engineering. *Science*, *330*(6009), 1355–1358. https://doi.org/10.1126/science.1193990.

Kirchherr, J., Reike, D., & Hekkert, M. (2017). Conceptualizing the circular economy: An analysis of 114 definitions. *Resources, Conservation and Recycling*, *127*, 221–232. https://doi.org/10.1016/j.biortech.2020.123548.

Koirala, S., Kumar, A., Bandyopadhyay, S., Zainudin, N., & Samali, B. (2020). Process integration and simulation for sustainable bio-refinery design. *Renewable Energy*, *145*, 2343–2357. https://doi.org/10.1016/j.renene.2019.07.124.

Koller, M. (2020). Bioprocessing and bioenergy: Present and future potential. *Bioengineering*, *7*(4), 133. DOI:10.1007/s13762-023-05211-1.

Kothari, R., Pandey, A., Kumar, S., & Tyagi, V. V. (2014). Different aspects of dry anaerobic digestion for bio-energy: An overview. *Renewable and Sustainable Energy Reviews*, *39*, 174–195. https://doi.org/10.1016/j.biotechadv.2018.01.011.

Kumar, P., Barrett, D. M., Delwiche, M. J., & Stroeve, P. (2009). Methods for pretreatment of lignocellulosic biomass for efficient hydrolysis and biofuel production. *Industrial & Engineering Chemistry Research*, *48*(8), 3713–3729. https://doi.org/10.1126/science.1151861.

Kumar, A., Sharma, S., & Yadav, B. (2021). Role of bioprocessing in sustainable green energy systems. *Renewable and Sustainable Energy Reviews*, *144*, 111012. https://doi.org/10.1021/acssuschemeng.1c03114.

Kumar, S., Shrestha, P., & Ogunlana, T. (2022). Microbial metabolism in bioenergy production. *Renewable and Sustainable Energy Reviews*, *152*, 111682. https://doi.org/10.1016/j.rser.2021.111682.

Lapuerta, M., Armas, O., & Rodriguez-Fernandez, J. (2008). Effect of biodiesel fuels on diesel engine emissions. *Progress in Energy and Combustion Science*, *34*(2), 198–223. https://doi.org/10.1016/j.rser.2004.09.002.

Lee, S. Y., Kim, H. U., Chae, T. U., Cho, J. S., Kim, J. W., Shin, J. H., Kim, D. I., Ko, Y.-S., Jang, W. D., Jang, Y. S., Chung, J. H., Park, S. J., & Kim, S. W. (2019). A comprehensive metabolic map for production of bio-based chemicals. *Nature Catalysis*, *2*, 18–33. https://doi.org/10.1016/j.tibtech.2018.11.009.

Lee, S. Y., Kim, H. U., Park, J. H., Park, J. M., & Kim, T. Y. (2019). Microbial cell factories for industrial biotechnology. *Trends in Biotechnology*, *37*(4), 388–391. https://doi.org/10.1016/j.tibtech.2018.11.009.

Lee, S. Y., Park, J. H., Jang, S. H., Nielsen, L. K., Kim, J., & Jung, K. S. (2008). Fermentative butanol production by clostridia. *Biotechnology and Bioengineering*, *101*(2), 209–228. https://doi.org/10.1016/j.tibtech.2018.11.009.

Lee, J. W., Park, J. Y., & Kwon, H. (2008). Development of *Clostridium* strains for butanol production. *Biotechnology Advances*, *26*(6), 591–605. https://doi.org/10.1016/j.tibtech.2018.11.009.

Li, X., Li, Y., Zeng, S., & Liu, Z. (2020). Economic feasibility analysis of advanced biofuel production technologies. *Bioresource Technology*, *312*, 123548. https://doi.org/10.1016/j.tibtech.2018.11.009.

Logan, B. E., Oh, S. E., Kim, I. S., & Van Ginkel, S. (2002). Biological hydrogen production measured in batch anaerobic respirometers. *Environmental Science & Technology*, *36*(11), 2530–2535. https://doi.org/10.1126/science.1217412.

Logan, B. E., & Rabaey, K. (2012). Conversion of wastes into bioelectricity and chemicals by using microbial electrochemical technologies. *Science*, *337*(6095), 686–690. https://doi.org/10.1126/science.1217412.

Lynd, L. R., Weimer, P. J., van Zyl, W. H., & Pretorius, I. S. (2002). Microbial cellulose utilization: Fundamentals and biotechnology. *Microbiology and Molecular Biology Reviews*, *66*(3), 506–577. https://doi.org/10.1146/annurev-micro-091014-104222.

Ma, F., & Hanna, M. A. (1999). Biodiesel production: A review. *Bioresource Technology*, *70*(1), 1–15. https://doi.org/10.1016/j.biortech.2004.06.025.

Madigan, M. T., Bender, K. S., Buckley, D. H., Sattley, W. M., & Stahl, D. A. (2018). *Brock biology of microorganisms* (15th ed.). Pearson.

Mata, T. M., Martins, A. A., & Caetano, N. S. (2010). Microalgae for biodiesel production and other applications: A review. *Renewable and Sustainable Energy Reviews*, *14*(1), 217–232. https://doi.org/10.1016/j.rser.2009.07.020.

McInerney, M. J., Sieber, J. R., & Gunsalus, R. P. (2009). Syntrophy in anaerobic global carbon cycles. *Current Opinion in Biotechnology*, *20*(6), 623–632. https://doi.org/10.1016/j.rser.2014.07.033.

Meher, L. C., Sagar, D. V., & Naik, S. N. (2006). Technical aspects of biodiesel production by transesterification—A review. *Renewable and Sustainable Energy Reviews*, *10*(3), 248–268. https://doi.org/10.1016/S0960-8524(99)00025-5.

Meng, X., Yang, J., Xu, X., Zhang, L., Nie, Q., & Xian, M. (2009). Biodiesel production from oleaginous microorganisms. *Renewable Energy*, *34*(1), 1–5. https://doi.org/10.1016/j.renene.2008.04.014.

Mosier, N., Wyman, C., Dale, B., Elander, R., Lee, Y. Y., Holtzapple, M., & Ladisch, M. (2005). Features of promising technologies for pretreatment of lignocellulosic biomass. *Bioresource Technology*, *96*(6), 673–686. https://doi.org/10.1016/j.renene.2008.04.014.

Naik, S. N., Goud, V. V., Rout, P. K., & Dalai, A. K. (2010). Production of first and second generation biofuels: A comprehensive review. *Renewable and Sustainable Energy Reviews*, *14*(2), 578–597. https://doi.org/10.1126/science.1193990.

Nielsen, J., & Keasling, J. D. (2016). Engineering cellular metabolism. *Cell*, *164*(6), 1185–1197. https://doi.org/10.1126/science.1193990.

Nigam, J. N. (2001). Ethanol production from molasses using immobilized yeast cells. *Bioresource Technology*, *76*(1), 17–22. https://doi.org/10.1016/S0960-8524(00)00085-2.

Oh, K. C., & Zimmerman, W. B. (2011). Multiphase bioreactors for microalgae production: A review. *Chemical Engineering Science*, *66*(7), 1573–1584. https://doi.org/10.1016/S0960-8524(00)00085-2.

Olson, D. G., McBride, J. E., Joe Shaw, A., & Lynd, L. R. (2012). Recent progress in consolidated bioprocessing. *Current Opinion in Biotechnology*, *23*(3), 396–405. https://doi.org/10.1016/j.copbio.2011.11.021.

Pandey, A., Bhaskar, T., Stöcker, M., & Sukumaran, R. K. (2015). *Recent advances in thermo-chemical conversion of biomass*. Elsevier.

Patel, A. K., Arora, N., & Arora, S. (2016). Application of Aspen Plus in biomass conversion processes: A review. *Renewable and Sustainable Energy Reviews*, *65*, 445–460. https://doi.org/10.1016/j.jclepro.2018.01.114.

Peters, M. S., Timmerhaus, K. D., & West, R. E. (2003). *Plant design and economics for chemical engineers* (5th ed., pp. 320–350). McGraw-Hill. https://doi.org/10.2202/1542-6580.2074.

Rawat, I., Kumar, R. R., Mutanda, T., & Bux, F. (2011). Dual role of microalgae: Phycoremediation of domestic wastewater and biomass production for sustainable biofuels production. *Applied Energy*, *88*(10), 3411–3424. https://doi.org/10.1016/j.rser.2009.10.009.

Richmond, A. (2004). Principles for attaining maximal microalgal productivity in photobioreactors: An overview. *Hydrobiologia*, *512*, 33–37. https://doi.org/10.1016/j.rser.2011.07.121.

Sadhukhan, J., Ng, K. C., & Martinez-Hernandez, E. (2016). Waste biorefinery models for sustainable chemical and energy production from biomass. *Current Opinion in Chemical Engineering*, *13*, 14–22. https://doi.org/10.1126/science.1217412.

Salis, A., Pisu, M., Monduzzi, M., Cossu, M., Ricca, E., & Vianello, R. (2010). Lipase-catalyzed biodiesel production from waste cooking oil. *Biotechnology and Bioengineering*, *105*(2), 209–217. https://doi.org/10.1016/j.pecs.2007.07.001.

Scarlat, N., Dallemand, J. F., Monforti-Ferrario, F., & Nita, V. (2015). The role of biomass and bioenergy in a carbon-neutral future. *Energy Policy*, *87*, 1–10. https://doi.org/10.1021/acs.energyfuels.1c04046.

Schmid, A., Dordick, J. S., Hauer, B., Kiener, A., Wubbolts, M., & Witholt, B. (2001). Industrial biocatalysis today and tomorrow. *Nature*, *409*(6817), 258–268. https://doi.org/10.1016/S0734-9750(00)00041-0.

Searchinger, T., Heimlich, R., Houghton, R. A., Dong, F., Elobeid, A., Fabiosa, J., Tokgoz, S., Hayes, D., & Yu, T. H. (2008). Use of U.S. croplands for biofuels increases greenhouse gases through emissions from land-use change. *Science*, *319*(5867), 1238–1240. https://doi.org/10.1073/pnas.0604600103.

Sharma, S., Sharma, S., & Kumar, A. (2015). Photofermentation: A potential process for biological hydrogen production. *Renewable and Sustainable Energy Reviews*, *41*, 206–216. https://doi.org/10.1016/j.rser.2014.08.043.

Shilpi, A., & Mandal, B. K. (2018). Solvent extraction methods for oil recovery from microalgae biomass: An overview. *Energy Sources, Part A: Recovery, Utilization, and Environmental Effects*, *40*(12), 1454–1464. https://doi.org/10.1080/15567036.2018.1488012.

Shuler, M. L. (1999). Immobilized cell bioreactors. *Biotechnology and Bioengineering*, *1*, 9–25. https://doi.org/10.1016/j.biortech.2007.01.046.

Shuler, M. L., & Kargi, F. (2002). *Bioprocess engineering: Basic concepts* (2nd ed., pp. 65–85). Prentice Hall.

Shuler, M. L., & Kargi, F. (2017). *Bioprocess engineering: Basic concepts* (3rd ed.). Prentice Hall.

Sims, R. E. H., Mabee, W., Saddler, J. N., & Taylor, M. (2010). An overview of second generation biofuel technologies. *Bioresource Technology*, *101*(6), 1570–1580. https://doi.org/10.1126/science.1189003.

Singh, A., Chen, C., & Ravindra, P. (2013). Simultaneous saccharification and fermentation: Advances and applications. *Biofuels, Bioproducts and Biorefining*, *7*(5), 507–521. https://doi.org/10.1002/bbb.1428.

Singh, A., Pant, D., & Korres, N. E. (2010). Sustainable resource recovery from waste biomass: A review of current research trends. *Bioresource Technology*, *101*(14), 5350–5364. https://doi.org/10.3403/30143651.

Singh, A., & Sharma, M. (2018). Optimizing energy efficiency in bioethanol production: An energy balance approach. *Journal of Cleaner Production*, *180*, 57–64. https://doi.org/10.3403/30143651.

Soh, K. C., & Zimmerman, W. B. (2011). Multiphase bioreactors for microalgae production: A review. *Chemical Engineering Science*, *66*(7), 1573–1584. https://doi.org/10.1016/j.biortech.2007.01.046.

Spolaore, P., Joannis-Cassan, C., Duran, E., & Isambert, A. (2006). Commercial applications of microalgae. *Journal of Bioscience and Bioengineering*, *101*(2), 87–96. https://doi.org/10.1016/j.biotechadv.2008.07.001.

Stams, A. J. M., & Plugge, C. M. (2009). Electron transfer in syntrophic communities of anaerobic bacteria and archaea. *Nature Reviews Microbiology*, *7*(8), 568–577. https://doi.org/10.1016/j.copbio.2009.10.001.

Stanbury, P. F. (1995). Bioreactor design: New directions. *Trends in Biotechnology*, *13*(3), 109–115. https://doi.org/10.1002/(SICI)1097-0290(1999).

Stanbury, P. F., Whitaker, A., & Hall, S. J. (2016). *Principles of fermentation technology* (3rd ed., pp. 115–140). Elsevier.

Stephanopoulos, G. (2007). Challenges in engineering microbes for biofuels production. *Science*, *315*(5813), 801–804. https://doi.org/10.1016/j.rser.2021.111682.

Stephanopoulos, G., Aristidou, A. A., & Nielsen, J. (1998). *Metabolic engineering: Principles and methodologies*. Academic Press.

Sun, Y., & Cheng, J. (2002). Hydrolysis of lignocellulosic materials for ethanol production: A review. *Bioresource Technology*, *83*(1), 1–11. https://doi.org/10.1016/S0960-8524(00)00085-2.

Ugwu, C. U., Aoyagi, H., & Uchiyama, H. (2008). Photobioreactors for mass cultivation of algae. *Bioresource Technology*, *99*(10), 4021–4028. https://doi.org/10.1023/B:HYDR.0000020360.92782.52.

van Maris, A. J. A., Winkler, A. A., Kuyper, M., de Laat, W. T. A. M., van Dijken, J. P., & Pronk, J. T. (2006). Development of efficient xylose fermentation in *Saccharomyces cerevisiae*: Characterization of evolved strains. *Applied and Environmental Microbiology*, *72*(3), 2051–2056. https://doi.org/10.1016/S0167-7799(00)88979-187.

Ward, A. J., Hobbs, P. J., Holliman, P. J., & Jones, D. L. (2008). Optimisation of the anaerobic digestion of agricultural resources. *Bioresource Technology*, *99*(17), 7928–7940. https://doi.org/10.1002/bit.22003.

Wijffels, R. H., & Barbosa, M. J. (2010). An outlook on microalgal biofuels. *Science*, *329*(5993), 796–799. https://doi.org/10.1016/j.apenergy.2010.11.025.

Wyman, C. E. (2007). What is (and is not) vital to advancing cellulosic ethanol. *Trends in Biotechnology*, *25*(4), 153–157. https://doi.org/10.1016/S0960-8524(01)00212-7.

Xu, H., Miao, X., & Wu, Q. (2006). High quality biodiesel production from a microalga Chlorella protothecoides by microwave-assisted extraction and transesterification. *Bioresource Technology*, *97*(6), 1222–1228. https://doi.org/10.1016/j.biortech.2005.05.008.

Zhang, Y., Cui, Z., Cheng, J., & Wu, X. (2017). Energy balance analysis in biofuel production from lignocellulosic biomass. *Renewable and Sustainable Energy Reviews*, *79*, 950–960. https://doi.org/10.1016/j.jclepro.2018.01.114.

Zhao, X., Liu, D., & Zhang, Y.-H. P. (2012). Consolidated bioprocessing of lignocellulosic biomass to ethanol with thermophilic bacteria. *Current Opinion in Biotechnology*, *23*(3), 364–371. https://doi.org/10.1016/j.copbio.2011.11.021.

3

Biochemical Processes to Enhance Bioenergy Production

Madhavi Tiwari

Department of Applied Sciences, Shri Rawatpura Sarkar University, Raipur, Chhattisgarh, India

3.1 Introduction

The urgent need to cut back on fossil fuel use and confront climate change has turned global attention toward renewable energy alternatives. Among these, bioenergy stands out for its renewable nature, carbon neutrality, and ability to integrate with existing energy systems (International Energy Agency [IEA], 2022). It includes a broad range of energy forms derived from biological sources like crop residues, forestry by-products, and organic waste. A variety of fuels, including bioethanol, biodiesel, biogas, and biohydrogen, can be produced from these resources. However, increasing the production of bioenergy is not without challenges, despite its potential (Figure 3.1).

When it comes to energy efficiency, traditional methods frequently fall short. High pretreatment expenses, ineffective fermentation processes, and feedstock quality unpredictability are still major obstacles (Chandel et al., 2018; Demirbas, 2011).

Scientists are increasingly using sophisticated biochemical and biotechnological techniques to address these issues. This chapter examines some of the most exciting advancements in the industry, such as omics-based optimization, co-culture tactics, metabolic and enzyme engineering, and process intensification (PI). By overcoming the drawbacks of outdated bioenergy systems, these strategies hope to open the door to more effective, scalable, and environmentally and financially sustainable solutions.

3.1.1 Advanced Microbial Strain Development

The creation of modified microorganisms that can more efficiently convert biomass into fuel has been one of the most significant advances in bioenergy research. Many naturally occurring microorganisms lack the metabolic range and durability required for intricate conversion processes, making them unsuitable for the manufacture of biofuel at an industrial scale. In order to overcome this, scientists have resorted to metabolic engineering, which involves modifying particular genes to get around obstacles or redirect carbon

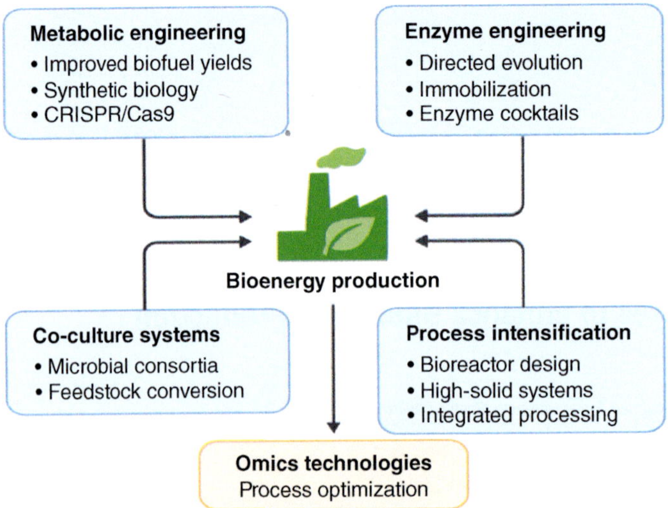

Figure 3.1 Biochemical strategies for enhancing bioenergy production. The flowchart highlights key processes–metabolic engineering, enzyme engineering, co-culture systems, and process intensification–that drive efficient bioenergy production. Omics technologies support these strategies through process optimization.

Table 3.1 A Comparative Analysis of Biochemical Techniques to Improve the Generation of Bioenergy.

Biochemical Strategy	Key Benefits	Tools/Techniques Used
Metabolic engineering	Improved substrate utilization, higher yields	CRISPR, pathway engineering
Enzyme engineering	Increased enzyme stability and activity	Directed evolution, rational design
Co-culture systems	Synergistic substrate conversion, enhanced stability	Synthetic consortia, mixed cultures
Process intensification	Higher product output, reduced time and cost	High-solid fermentation, ISPR
Omics integration	System-level process optimization	Genomics, proteomics, metabolomics

flow in the direction of desired products like hydrogen, ethanol, or butanol (Hasunuma & Kondo, 2012). Key tactics including metabolic engineering, enzyme engineering, co-culture systems, and the related resources and advantages are compiled in Table 3.1.

These efforts have been greatly enhanced by the introduction of CRISPR/Cas9. Multiple genes can be changed with relative simplicity and precision thanks to this genome editing tool (Liu et al., 2020). Consider the popular laboratory strains *Saccharomyces cerevisiae* and *Escherichia coli*, which have been genetically modified to break down pentose sugars, which are plentiful in lignocellulosic feedstocks but usually ineffective for wild-type bacteria (Kumar et al., 2019).

Developments in synthetic biology are allowing researchers to create completely new pathways in addition to modifying already-existing ones. Even in conditions with elevated amounts of toxins or inhibitors, engineered strains can sustain high fuel productivity by using synthetic DNA constructions or innovative gene circuits.

Adapting thermophilic and extremophilic microorganisms to operate at higher temperatures and tolerate higher feedstock and product concentrations is another exciting avenue. Overall, these modifications may result in bioenergy processes that are more resilient and effective (Singh et al., 2021a, 2021b).

3.1.2 Enzyme Engineering and Catalytic Enhancement

Complex biomass must be broken down by enzymes in order for bacteria to ferment the sugars. However, many enzymes in their native state are inadequate for industrial settings; they may lose activity, degrade rapidly, or react poorly to inhibitors. To get around this, scientists have improved the efficiency, stability, and selectivity of enzymes by employing strategies including directed evolution and rational design (Bayer et al., 2007).

Consider *xylanases* and *cellulases*, for example—two key enzymes in biomass processing. Through **protein structure-programmed mutagenesis**, scientists have been able to tweak specific amino acids, improving these enzymes' heat tolerance and catalytic function (Singhania et al., 2013). Another promising approach involves anchoring enzymes to **nanocarriers or polymer supports**, which not only prolongs their functional lifespan but also makes it easier to separate and reuse them (Sheldon & van Pelt, 2013).

More recently, specialized enzyme cocktails—well-balanced mixtures of ligninases, hemicellulases, and cellulases—have become a viable way to more thoroughly break down plant material. Together, these combinations accelerate the process and lower the overall enzyme dose needed, which lowers the cost of converting biomass.

3.1.3 Co-Culture and Consortium-Based Strategies

Depending on a specific microorganism to process biomass often leads to limitations. Complex feedstocks and multi-step conversions are better handled by **microbial communities**, where different organisms work together—just like they do in nature. For instance, one species might break down cellulose into glucose, while another ferments the glucose into ethanol.

Researchers have started to design **synthetic consortia** with defined roles for each member, optimizing them for stress tolerance and metabolic cooperation. These systems not only make the overall process more resilient but also help prevent bottlenecks caused by toxic by-products and expand the types of materials that can be used (Zuroff & Curtis, 2012).

One real-world example is **anaerobic digestion**, which relies on a series of microbial groups—hydrolytic, acidogenic, acetogenic, and methanogenic species—working in sequence. Engineering and stabilizing these consortia has led to improved **biogas production** and more consistent performance, even when feedstock quality or environmental conditions fluctuate (Angelidaki et al., 2018).

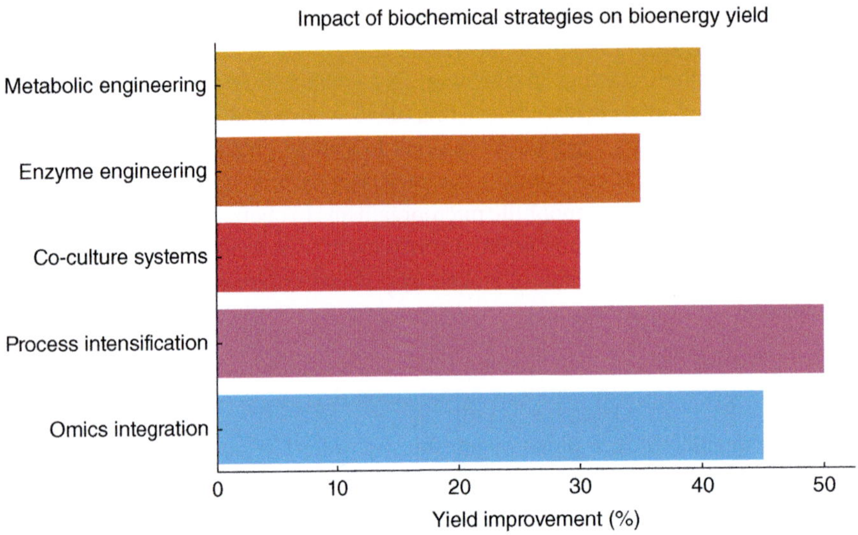

Figure 3.2 Comparative impact of key biochemical strategies on bioenergy yield. Process intensification and omics integration show the highest improvements based on current research trends.

3.1.4 Process Intensification Techniques

Bridging the gap between lab success and large-scale bioenergy production demands systems that are not only efficient but also economically viable. That's where **process intensification** (PI) comes in. The goal is to rethink how different steps in the production pipeline are designed and integrated—ultimately boosting output and cutting down on energy and time. Figure 3.2 displays the estimated yield improvement (%) using various biochemical strategies in bioenergy production. This visual shows how PI and metabolic engineering have the highest impact.

Common strategies include **high-solid fermentation (HSF)**, **fed-batch and continuous bioreactor setups**, and **simultaneous saccharification and fermentation (SSF)**. These methods help speed up the overall process and reduce the number of separate stages involved (Oskarsson et al., 2018).

Emerging reactor technologies, such as **membrane bioreactors** and systems using **immobilized microbial cells**, allow for higher microbial concentrations and longer fermentation times. The result? Better yields and more stable production.

In addition, techniques like **in situ product recovery (ISPR)**—which involve removing biofuels as they're being produced using methods such as **gas stripping**, **pervaporation**, or **adsorption**—help prevent product buildup that can inhibit fermentation. These integrated systems increase overall efficiency and maintain smooth operations (Gòdia & Pirt, 2021).

3.1.5 Omics-Driven Process Optimization

Advances in **omics technologies**—including genomics, transcriptomics, proteomics, and metabolomics—have dramatically improved our ability to understand and fine-tune bioenergy systems at the molecular level. These tools allow researchers to track gene expression, monitor enzyme behavior, and identify key metabolic roadblocks. By integrating these insights with **systems biology models**, scientists can simulate metabolic pathways, predict flux distributions, and make more informed decisions in strain engineering and process optimization (Nielsen, 2017).

3.1.6 Commercialization and Scale-Up Challenges

Despite tremendous advancements in the laboratory, there are still several obstacles in the way of bringing bioenergy technologies to the industrial scale. Variability in feedstock composition, which can impact process consistency, is one enduring problem. Furthermore, commercial adoption is frequently slowed by the expense and scalability of enzyme production as well as the requirement for reliable performance in industrial settings (Lynd et al., 2008). In addition to technical improvements, economic and logistical innovation are also needed to close this gap.

3.1.7 Future Prospects and Emerging Technologies

Both cutting-edge technologies and interdisciplinary innovation are likely to influence the direction of bioenergy in the future. With the ability to directly transform biological energy into electrical power, bioelectrochemical systems—such as microbial fuel cells—represent an exciting new frontier. Cell-free synthetic biology platforms, on the other hand, are simplifying manufacturing while preserving high specificity by creating new opportunities for finely regulated reactions without the need for living microbial systems (Silverman et al., 2020).

With the use of potent instruments like metabolic and enzyme engineering, synthetic microbial communities, and omics-guided approaches, bioenergy research has entered a new era. These developments are contributing to the scalability and sustainability of bioenergy in addition to increasing yields and efficiency. These technologies have the potential to produce reliable, economically feasible green energy systems that can handle the demands of a changing global environment as they develop further and as new ones appear.

3.2 Two Microbial Biofactories of the Future for the Production of Energy

It has long been known that microorganisms possess extraordinary biochemical capabilities. However, they must be developed beyond their inherent potential in order to satisfy the needs of producing bioenergy on an industrial scale. Often called next-generation biofactories, these sophisticated microbial systems are meticulously designed to process a broad variety of raw materials, produce high product yields, and withstand a variety of environmental challenges. Modern methods including metabolic engineering, CRISPR/Cas9 gene editing, and synthetic biology are combined to achieve this level of performance. With the use of these resources, scientists can create and refine microbial strains that effectively transform a variety of feedstocks into biofuels and other substances that are high in energy. Figure 3.3 lists the main tactics used to build microbial biofactories.

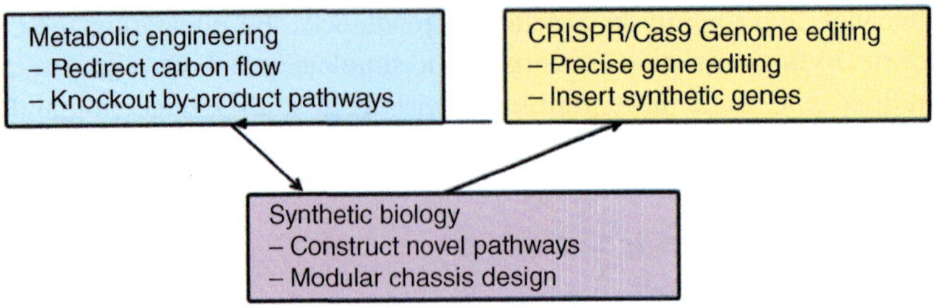

Figure 3.3 Metabolic engineering, CRISPR/Cas9 genome editing, and synthetic biology are the three main approaches for designing next-generation microbial biofactories in order to increase the production of bioenergy.

3.2.1 Metabolic Engineering: Redirecting Cellular Machinery

Metabolic engineering focuses on reshaping the internal biochemical pathways of microorganisms to improve how they convert carbon and energy into valuable biofuels like ethanol, butanol, biodiesel, or hydrogen.

This process typically involves steps such as boosting the activity of key enzymes, shutting down pathways that compete with biofuel production, or fine-tuning the balance of cellular cofactors to make the process more efficient.

A good example is *S. cerevisiae* (brewer's yeast), which doesn't naturally ferment certain plant sugars like xylose and arabinose. However, by introducing specific genes—such as those encoding xylose isomerase from bacteria or fungi—scientists can enable the yeast to use these sugars effectively for ethanol production (Liu et al., 2020). Table 3.2 highlights microbial strains modified through metabolic engineering and the resulting improvements in biofuel production.

3.2.2 Genome Editing: Precision with CRISPR/Cas9

By providing scientists with a potent tool to precisely and specifically alter microbial DNA, CRISPR/Cas9 has transformed genome editing. This technique gives researchers

Table 3.2 Engineered Microbial Strains and Their Enhanced Bioenergy Outcomes.

Microorganism	Modification Strategy	Target Product	Outcome Achieved
Saccharomyces cerevisiae	Xylose pathway introduction	Ethanol	Xylose utilization from lignocellulose
Clostridium acetobutylicum	ABE pathway overexpression	Butanol	Higher butanol:acetone ratio
E. coli	ldhA knockout + alcohol dehydrogenase overexpression	Ethanol	Increased ethanol production by 60%
Zymomonas mobilis	Pyruvate decarboxylase enhancement	Ethanol	Elevated ethanol titers

exquisite control over microbial metabolism by enabling them to alter numerous genes simultaneously.

Eliminating undesirable metabolic pathways, such as those that lead to the production of by-products, is one of its main uses. For instance, removing the ldhA gene from *E. coli* stops lactic acid accumulation and directs more carbon toward the biosynthesis of ethanol. Additionally, new pathways for the production of biofuels that microorganisms do not naturally produce, such as hydrogen and isobutanol, are introduced via CRISPR.

Furthermore, sophisticated variants of the system, such as CRISPRi (interference) and CRISPRa (activation), enable adjusting the levels of gene expression without irreversibly changing the DNA. CRISPR is especially helpful for industrial strain development, where genetic stability and performance are crucial, because it allows for "scarless" editing.

3.2.3 Synthetic Biology: Designing Novel Biofuel Pathways

By creating completely new biological systems, synthetic biology goes beyond genetic engineering. This method entails putting together unique DNA sequences, including operons, promoters, and regulatory components, to produce whole novel activities in bacteria rather than merely altering preexisting pathways.

By using this method, scientists can give microbes non-native metabolic pathways to produce sophisticated biofuels. For instance, modified strains of *Clostridium ljungdahlii* and *E. coli* have been created to generate isobutanol and hydrogen from non-traditional feedstocks such as syngas and carbon dioxide (Nielsen & Keasling, 2016).

Additionally, modular "chassis" organisms—microbes designed to readily accept additional genetic circuits—are introduced by synthetic biology. This adaptability makes it simpler to customize microorganisms for particular industrial processes and expedites the strain design process. Furthermore, it is possible to incorporate programmable control systems and feedback-resistant enzymes into these artificial networks to maintain stable and effective production even in the face of fluctuating circumstances.

The creation of effective biofactories from microbes represents a significant advancement in the creation of sustainable and clean energy sources. Scientists can now create unique strains that efficiently transform a range of feedstocks into profitable biofuels by combining metabolic engineering, CRISPR/Cas9 genome editing, and synthetic biology.

In addition to being more productive, these modified microorganisms are also more versatile, able to flourish in a variety of industrial settings. As a result, they are positioned to play a vital role in the future of green energy, making bioenergy production more scalable, economical, and ecologically friendly.

3.3 Tailoring Enzymes for Efficient Bioenergy Conversion

By facilitating the breakdown of complex biomass into fermentable sugars or directly converting intermediates into biofuels, enzymes are essential to the creation of bioenergy. However, because of their limits in stability, specificity, or activity, native enzymes frequently perform poorly in industrial settings. Scientists have created sophisticated methods to modify enzymes with improved characteristics appropriate for bioenergy operations in order to overcome these obstacles.

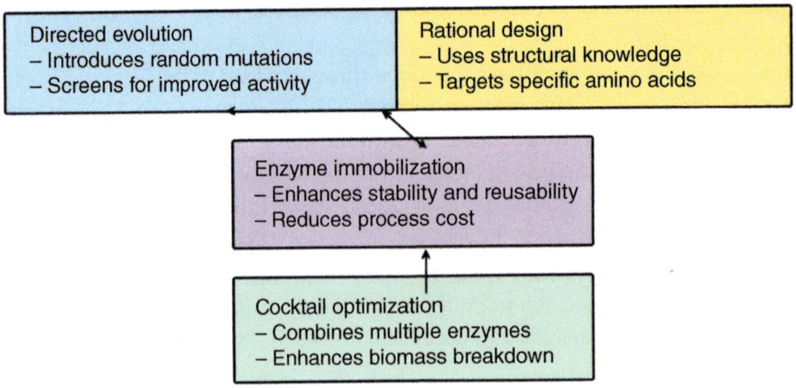

Figure 3.4 Strategies in enzyme engineering for bioenergy.

Four main strategies are highlighted in this section: cocktail optimization, directed evolution, rational design, and enzyme immobilization. The efficiency, resilience, and financial feasibility of enzyme-driven bioenergy systems are all enhanced by these techniques. Figure 3.4 shows the important approaches to enzyme engineering for bioenergy applications, such as cocktail optimization, directed evolution, rational design, and enzyme immobilization.

3.3.1 Directed Evolution: Mimicking Nature's Selection

Directed evolution is the process of randomly altering an enzyme's gene and looking for variations with better characteristics. In a controlled laboratory setting, this technique speeds up the process while simulating natural selection. It has proved very effective in increasing the catalytic efficiency, solvent tolerance, and thermal stability of enzymes.

Cellulases developed using this method, for example, have demonstrated noticeably greater hydrolytic activity in industrial settings (Arnold, 1998). Because it enables the identification of superior variants from vast mutant libraries, high-throughput screening is crucial to this approach.

3.3.2 Rational Design: Engineering with Structure in Mind

Rational design depends on in-depth understanding of the three-dimensional structure and active site chemistry of an enzyme, in contrast to directed evolution, which is random. To forecast how particular amino acid modifications would improve activity or stability, researchers employ molecular dynamics and computational models.

One prominent example is the modification of surface residues to enable glucose oxidase to withstand acidic pH, hence enhancing its efficiency in the production of bioethanol (Bornscheuer & Pohl, 2001). Although this approach necessitates fewer rounds, precise structural data is crucial.

3.3.3 Enzyme Immobilization: Stability and Reusability

In commercial bioprocesses, immobilizing enzymes on solid supports—like silica, polymers, or nanoparticles—offers a number of advantages. According to Sheldon and van Pelt (2013), these include enhanced operational and thermal stability, simpler recovery and reuse, and greater adaptability for continuous bioreactor operations.

Cellulases and lipases are two examples of enzymes that have been effectively immobilized for use in the hydrolysis of lignocellulosic biomass and the generation of biodiesel, respectively. Large-scale biorefineries find this technology appealing since it drastically cuts down on enzyme usage, lowers production costs, and prolongs the shelf life of enzymes (Datta et al., 2013).

3.3.4 Cocktail Optimization: Synergistic Enzyme Blends

The complex structure of lignocellulosic biomass, which is made up of cellulose, hemicellulose, and lignin, cannot be efficiently broken down by a single enzyme. In order to enhance overall biomass breakdown, cocktail optimization entails combining cellulases, hemicellulases, and lignin-degrading enzymes in certain ratios (Singhania et al., 2013).

In addition to improving the release of sugar from biomass, these improved enzyme blends also lower the amount of enzyme required, which is a significant cost element in the manufacture of industrial biofuel (Bischof et al., 2016). For the deconstruction and saccharification processes to be effective, these enzymes must work in concert.

3.4 Co-Culture and Microbial Consortium Strategies for Bioenergy

Microorganisms do not exist in isolation in natural settings; rather, they are part of complex communities. Through metabolic connections, they are dependent on one another for survival and growth. This natural synergy has sparked interest in co-culture and microbial consortium techniques for the production of bioenergy (Angenent et al., 2004). Multiple species of microbes work together to transform biomass into energy-rich products like bioethanol, biogas, or biohydrogen rather than relying on a single germ.

Metabolic collaboration, a process in which each microorganism assumes a specific task, such as cellulose degradation, sugar fermentation, or methanogenesis, is the foundation of this approach. These systems are more durable, efficient, and able to respond to complicated substrates such as wastewater or agricultural waste since they divide the effort (Kumar et al., 2008).

3.4.1 Synergistic Microbial Interactions

The combined activity of well-chosen and paired microorganisms can surpass that of individual species. For instance, complex polysaccharides can be broken down into sugars by

Table 3.3 Examples of Microbial Consortia in Bioenergy Production.

Co-culture/ Consortium	Microbial Members	Target Biofuel	Feedstock Used	Advantage
Anaerobic digestion consortium	Hydrolytic, acidogenic, methanogenic bacteria	Biogas	Organic waste, sludge	Complete biomass breakdown
Cellulolytic fungi + yeast	*Trichoderma reesei* + *S. cerevisiae*	Ethanol	Lignocellulosic biomass	Efficient saccharification and fermentation
Algal-bacterial consortium	*Chlorella vulgaris* + *Bacillus subtilis*	Biohydrogen	Wastewater	Nutrient recycling and enhanced hydrogen yield
Co-culture of *Clostridium* spp.	*C. butyricum* + *C. acetobutylicum*	Butanol	Glucose, xylose	High product selectivity

cellulolytic fungi like *Trichoderma reesei*, which are subsequently fermented by organisms that produce ethanol, including *S. cerevisiae* (Lynd et al., 2002).

This kind of collaboration leads to shorter processing times, increased product yields, and quicker substrate conversion. In industrial ethanol production systems that use simultaneous SSF, the synergy is very helpful. Various examples of microbial consortia used in bioenergy production, along with their specific roles and advantages are summarized in Table 3.3.

3.4.2 Division of Metabolic Labor

In a well-functioning microbial consortia, each species makes a unique contribution to the process of turning biomass into energy. For example, certain bacteria or fungi may break down lignin and cellulose, and other microbes may ferment the resultant sugars to form alcohols or acids. Finally, solventogenic bacteria or methanogens can convert these acids into ethanol, hydrogen, or methane (Brenner et al., 2008).

This work division reduces the metabolic load on each individual, resulting in a more efficient and effective system overall. This type of configuration mimics natural ecosystems and is very helpful for handling complex, mixed-feedstock substrates.

3.4.3 Community Stability and Resilience

The capacity of microbial consortia to adjust to changes in their surroundings is one of their main advantages. These microbial communities may frequently self-regulate and sustain performance regardless of changes in the feedstock composition, pH, temperature, or nutrient levels (Konopka, 2009).

In industrial bioenergy settings, where conditions are rarely stable, this kind of resilience is essential. Even under less-than-ideal operational conditions, cooperative interactions and feedback loops within the consortium support long-term stability and sustained biofuel production (Zuroff & Curtis, 2012).

Table 3.4 Summary of Process Intensification Techniques in Bioenergy.

Technique	Key Feature	Benefits	Common Application
High-solid fermentation	Low water use; high substrate load	Lower energy cost; higher productivity	Lignocellulosic biomass
Simultaneous saccharification and fermentation (SSF)	Combines hydrolysis and fermentation	Shorter process time; fewer inhibitors	Bioethanol from cellulose
Fed-batch and continuous systems	Substrate fed gradually or continuously	Stable productivity; better control	Ethanol, biogas
In situ product recovery (ISPR)	Extracts product during fermentation	Reduces inhibition; enables continuous process	Ethanol, butanol, hydrogen

3.5 Advanced Process Integration for Enhanced Biofuel Production

Industrial-scale bioenergy production frequently faces difficulties, such as low conversion rates, protracted processing periods, and expensive operating expenses. PI, a collection of techniques intended to increase the efficiency, compactness, and affordability of bioenergy systems, has been used by researchers to address these problems (Gòdia & Pirt, 2021). By merging several phases into a single, seamless process and optimizing conditions, these techniques seek to enhance the conversion of biomass into energy.

HSF, fed-batch and continuous fermentation, simultaneous SSF, and ISPR are some of the most successful PI methods currently employed in bioenergy. These methods enable continuous biofuel extraction, reduce enzyme and water consumption, and increase yields. Table 3.4 displays the summary of various process intensification techniques in bioenergy.

3.5.1 High-Solid Fermentation (HSF)

HSF works with more concentrated biomass and requires less water. According to, this configuration lowers the energy expenses associated with drying or recovering water. Additionally, it increases the amount of product that can be produced in a specific area, which is perfect for large-scale enterprises. It does, however, present mixing and oxygen flow issues that need to be resolved for efficient operation.

3.5.2 Simultaneous Saccharification and Fermentation (SSF)

SSF is a clever method in which biomass is broken down into sugars and then those sugars are transformed into biofuels in the same tank. This improves the process faster and more effective overall and helps prevent sugar accumulation, which can limit enzyme function (Oskarsson et al., 2018). It reduces the need for additional equipment and energy between phases and is commonly utilized in the production of ethanol from cellulose.

3.5.3 Fed-Batch and Continuous Fermentation

By adding substrates gradually, a fed-batch system helps keep microorganisms in optimal conditions without overburdening them. Continuous fermentation is an ongoing process in which new material enters and products exit simultaneously. This method is perfect for producing bioethanol and biogas over the long term since it boosts output and reduces downtime.

3.5.4 In Situ Product Recovery (ISPR)

While fermentation is still taking place, ISPR extracts the product, such as butanol or ethanol, from the mixture. By doing this, chemicals that could slow down the microorganisms are kept from building up. Methods such as gas stripping, adsorption, or membrane filtration are employed to collect the fuel as it forms and maintain the smooth operation of fermentation.

3.6 Integrative Omics Approaches for Optimizing Bioenergy Yields

One of the main goals of modern biotechnology research is increasing the productivity and efficiency of microorganisms employed in the generation of bioenergy. Utilizing omics technologies—specifically, proteomics, metabolomics, transcriptomics, and genomics—is one of the most promising methods to achieve this goal. These methods are essential for maximizing the potential of biological systems for the production of clean energy since they provide a thorough, system-wide understanding of microbial behavior (Nielsen, 2017).

Scientists can monitor how microorganisms react to various biomass types, environmental stressors, or fermentation conditions by using omics. With the help of these insights, metabolic pathways can be changed, the best microbial strains can be selected, and bioenergy production processes can be adjusted for increased yields and more stable operations. Figure 3.5 clearly depicts the overview of the four main omics tools—genomics, transcriptomics, proteomics, and metabolomics—and their role in optimizing microbial bioenergy systems.

3.6.1 Genomics

Genomics includes decoding the complete DNA sequence of bacteria. This aids in the identification of genes involved in biomass breakdown, biofuel production, and stress tolerance. For instance, several genes involved in the manufacture of butanol have been discovered through genome analysis of *Clostridium* species (Choi et al., 2010). When choosing or creating high-performing strains, these insights are quite helpful.

3.6.2 Transcriptomics

By identifying the genes that are actively expressed, transcriptomics reveals what is happening, whereas genomics reveals what might occur. This aids researchers in understanding

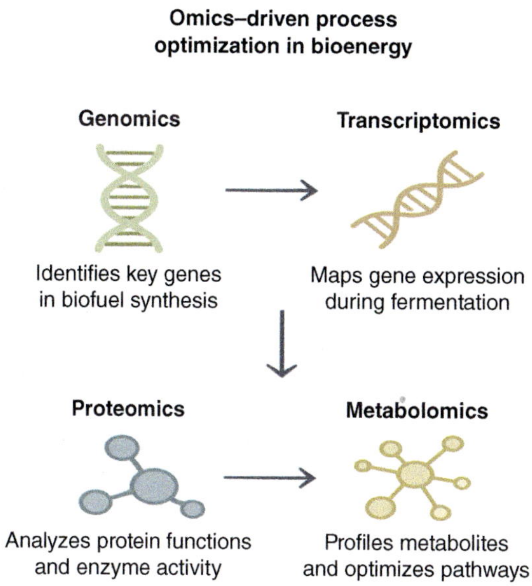

Figure 3.5 Omics approaches in bioenergy.

which genes are essential for the production of chemicals rich in energy and how bacteria respond in various environments (Singh et al., 2021a, 2021b). With this information, gene activity can be adjusted to increase productivity.

3.6.3 Proteomics

The study of proteomics focuses on the proteins—including enzymes that perform vital reactions—that microorganisms really make. This method shows how genes are translated into action in addition to their expression. It's especially helpful for locating weak spots or bottlenecks in metabolic pathways, such as enzymes that might be present in excess or not working correctly (Wilkins et al., 2002).

3.6.4 Metabolomics

Metabolomics examines the different small molecules and intermediates that are the final products of microbial metabolism. Monitoring these substances gives an overview of what is (or is not) happening during a fermentation process. For instance, an accumulation of a particular intermediate may indicate a route blockage. According to Scalcinati et al. (2012), metabolomics aids in identifying these problems and directing advancements. The function of omics in bioenergy research is shown in Table 3.5.

Omics-based technologies provide a powerful toolkit for improving microbial strains and fermentation processes in bioenergy production. By integrating information across the

Table 3.5 How Omics Supports Bioenergy Research.

Omics Layer	Focus Area	Key Applications	Example Outcome
Genomics	DNA sequencing, gene identification	Strain selection, genome mining	Found genes for butanol synthesis
Transcriptomics	Gene expression profiling	Stress response, pathway control	Enhanced expression of biofuel-related genes
Proteomics	Protein abundance and activity	Pathway analysis, enzyme identification	Pinpointed rate-limiting enzymes
Metabolomics	Metabolite profiling and fluxes	Bottleneck detection, flux optimization	Solved product inhibition issues

genomic, transcriptomic, proteomic, and metabolomic levels, researchers can better understand how microbes function and how to make them more productive. This systems-level insight is critical for advancing sustainable, large-scale bioenergy solutions.

3.7 Commercialization and Scale-Up Challenges in Bioenergy Production

While laboratory-scale bioenergy innovations have shown immense potential, moving these breakthroughs from the bench to industrial-scale production remains a significant hurdle. The **commercialization and scale-up** of bioenergy processes involve a complex interplay of technical, economic, regulatory, and social factors that can limit the adoption of even the most promising technologies.

3.7.1 Technical Barriers

One of the key challenges in scaling up bioenergy production lies in maintaining **process efficiency and consistency** at a larger volume. Reactions that perform optimally in controlled lab settings often behave unpredictably when transferred to pilot- or commercial-scale bioreactors. Parameters such as **oxygen transfer, heat dissipation, shear stress, and substrate mixing** become more difficult to manage. For instance, high-solid lignocellulosic feedstocks tend to clog reactors or inhibit microbial growth due to mass transfer limitations (Lynd et al., 2008).

Furthermore, the **robustness of microbial strains** at an industrial scale is critical. Strains engineered for high-yield biofuel production must tolerate environmental fluctuations, contamination risks, and inhibitory compounds generated during biomass breakdown (Alper & Stephanopoulos, 2009).

3.7.2 Economic Constraints

From a commercial perspective, bioenergy must compete with fossil fuels and other renewable sources on **cost-effectiveness**. The high costs of enzyme production, feedstock preprocessing, fermentation infrastructure, and downstream purification can make large-scale biofuel production financially unviable. Additionally, the lack of **economies of scale** in early-stage biorefineries makes capital investment risky (Gnansounou & Dauriat, 2010).

Another issue is the **low market price of biofuels** like ethanol or biogas, which makes it difficult to recover production costs without government subsidies or carbon pricing mechanisms. Cost-effective pretreatment and integrated bioprocessing systems are urgently needed to reduce expenses and improve the energy return on investment.

3.7.3 Regulatory and Policy Challenges

Navigating the **regulatory landscape** is another major obstacle. Bioenergy products and technologies must comply with safety, environmental, and quality standards that vary across regions. For genetically modified organisms (GMOs) used in biofuel production, additional layers of approval and monitoring may apply, potentially delaying commercialization.

Moreover, **policy uncertainty**—such as fluctuating biofuel mandates, tax incentives, or import tariffs—can discourage investment. Long-term policy support is essential for encouraging public–private partnerships and innovation in the bioenergy sector (IEA, 2020).

3.7.4 Infrastructure and Logistics

Scaling up also requires **logistical infrastructure** for feedstock supply, transportation, and biofuel distribution. Most bioenergy systems depend on bulky, low-density agricultural residues or organic waste, which are expensive to transport and store. Decentralized biorefineries near feedstock sources could be beneficial, but this calls for infrastructure development and regional planning.

3.7.5 Social Acceptance and Public Perception

A minor but important factor in the marketing of bioenergy is public acceptance. Environmental or civil society organizations may oppose the use of GMOs due to concerns about land use, food security, water consumption, and other issues. To increase public confidence in bioenergy technologies, open communication, lifecycle assessments, and public involvement are crucial.

Even though microbial engineering and bioenergy technologies have advanced significantly, there are still many obstacles to large-scale commercialization that go beyond science. It will take interdisciplinary cooperation, infrastructure investment, supportive legislative environments, and proactive stakeholder engagement to overcome these obstacles. Bioenergy systems can only be a sustainable and scalable part of the world's energy mix with these combined efforts.

3.8 Conclusion

Bioenergy has emerged as a viable substitute for fossil fuels due to the increasing demand for clean and renewable energy. This chapter examined a number of cutting-edge biochemical techniques that are assisting in increasing the scalability, cost-effectiveness, and efficiency of bioenergy production.We discussed about how scientists are creating potent microorganisms that can convert a range of raw materials into biofuels using microbial strain engineering, which is made possible by technologies like synthetic biology and CRISPR. Similarly, difficult biomass materials are becoming easier to break down for fuel production because to developments in enzyme engineering, including enzyme immobilization and customized enzyme mixes.

Additionally, we examined the ways in which microbial consortia—carefully selected microbial communities—can cooperate to enhance overall process effectiveness. In addition, industrial-scale cost reduction and yield improvement are facilitated by PI methods such as simultaneous SSF and HSF. The significance of omics technologies—genomics, transcriptomics, proteomics, and metabolomics—was also emphasized, demonstrating how they aid in the optimization of each stage of the bioenergy production process and offer profound insights into microbial functions. There are still difficulties in spite of these developments. Critical areas that require attention include navigating regulatory obstacles, ensuring economic viability, building public trust, and scaling up from lab to industry.

In conclusion, the methods and technologies discussed in this chapter provide a solid basis for developing the upcoming generation of sustainable bioenergy systems. These biological processes have the potential to help satisfy the world's increasing energy demands in a cleaner and more sustainable manner with sustained research, innovation, and industry and policymaker support.

References

Alper, H., & Stephanopoulos, G. (2009). Engineering for biofuels: Exploiting innate microbial capacity or importing biosynthetic potential? *Nature Reviews Microbiology*, *7*(10), 715–723. https://doi.org/10.1038/nrmicro2186.

Angelidaki, I., Treu, L., Tsapekos, P., Luo, G., Campanaro, S., Wenzel, H., & Kougias, P. G. (2018). Biogas upgrading and utilization: Current status and perspectives. *Biotechnology Advances*, *36*(2), 452–466. https://doi.org/10.1016/j.biotechadv.2018.01.011.

Angenent, L. T., Karim, K., Al-Dahhan, M. H., Wrenn, B. A., & Domíguez-Espinosa, R. (2004). Production of bioenergy and biochemicals from industrial and agricultural wastewater. *Trends in Biotechnology*, *22*(9), 477–485. https://doi.org/10.1016/j.tibtech.2004.07.001.

Arnold, F. H. (1998). Design by directed evolution. *Accounts of Chemical Research*, *31*(3), 125–131. https://doi.org/10.1021/ar960017f.

Bayer, E. A., Shoham, Y., & Lamed, R. (2007). Cellulose-decomposing bacteria and their enzyme systems. *The Prokaryotes*, *1*, 578–617. https://doi.org/10.1007/978-0-387-69061-7_35.

Bischof, R. H., Ramoni, J., & Seiboth, B. (2016). Cellulases and beyond: The first 70 years of the enzyme producer *Trichoderma reesei. Microbial Cell Factories, 15*. Article 106. https://doi.org/10.1186/s12934-016-0507-6.

Bornscheuer, U. T., & Pohl, M. (2001). Improved biocatalysts by directed evolution and rational protein design. *Current Opinion in Chemical Biology, 5*(2), 137–143. https://doi.org/10.1016/S1367-5931(00)00186-6.

Brenner, K., You, L., & Arnold, F. H. (2008). Engineering microbial consortia: A new frontier in synthetic biology. *Trends in Biotechnology, 26*(9), 483–489. https://doi.org/10.1016/j.tibtech.2008.05.004.

Chandel, A. K., Garlapati, V. K., Singh, A. K., Antunes, F. A. F., & da Silva, S. S. (2018). The path forward for lignocellulose biorefineries: Bottlenecks, solutions, and perspective on commercialization. *Bioresource Technology, 264*, 370–380. https://doi.org/10.1016/j.biortech.2018.06.004.

Choi, S., Song, C. W., Shin, J. H., & Lee, S. Y. (2010). Biorefineries for the production of top building block chemicals and their derivatives. *Metabolic Engineering, 12*(6), 487–505. https://doi.org/10.1016/j.ymben.2010.05.003.

Datta, S., Christena, L. R., & Rajaram, Y. R. S. (2013). Enzyme immobilization: An overview on techniques and support materials. *3 Biotech, 3*, 1–9. https://doi.org/10.1007/s13205-012-0071-7.

Demirbas, A. (2011). Competitive liquid biofuels from biomass. *Applied Energy, 88*(1), 17–28. https://doi.org/10.1016/j.apenergy.2010.07.016.

Gnansounou, E., & Dauriat, A. (2010). Techno-economic analysis of lignocellulosic ethanol: A review. *Bioresource Technology, 101*(13), 4980–4991. https://doi.org/10.1016/j.biortech.2009.11.096.

Gòdia, F., & Pirt, S. J. (2021). Process intensification for advanced biofuels: Status and future directions. *Renewable and Sustainable Energy Reviews, 143*, 110847. https://doi.org/10.1016/j.rser.2021.110847.

Hasunuma, T., & Kondo, A. (2012). Consolidated bioprocessing and simultaneous saccharification and fermentation of lignocellulose to ethanol with thermotolerant yeast strains. *Process Biochemistry, 47*(9), 1287–1294. https://doi.org/10.1016/j.procbio.2011.11.002.

International Energy Agency (IEA). (2020). *Renewables 2020: Analysis and forecast to 2025*. IEA Publications. https://www.iea.org/reports/renewables-2020.

International Energy Agency (IEA). (2022). *Renewables 2022: Analysis and forecast to 2027*. IEA Publications. https://www.iea.org/reports/renewables-2022.

Konopka, A. (2009). What is microbial community ecology? *The ISME Journal, 3*, 1223–1230. https://doi.org/10.1038/ismej.2009.88.

Kumar, V., Ashok, S., & Park, S. (2019). Recent advances in biological production of pentanol: A promising fuel. *Biotechnology Advances, 37*(1), 107409. https://doi.org/10.1016/j.biotechadv.2018.12.003.

Kumar, P., Barrett, D. M., Delwiche, M. J., & Stroeve, P. (2008). Methods for pretreatment of lignocellulosic biomass for efficient hydrolysis and biofuel production. *Industrial & Engineering Chemistry Research, 48*(8), 3713–3729. https://doi.org/10.1021/ie801542g.

Liu, R., Liang, L., Lu, M., Zhang, L., Wang, M., & Zhou, Z. (2020). CRISPR–Cas9: A powerful tool for gene function study and genome editing in industrial microorganisms. *Applied*

Microbiology and Biotechnology, *104*, 6381–6394. https://doi.org/10.1007/s00253-020-10733-0.

Lynd, L. R., van Zyl, W. H., McBride, J. E., & Laser, M. (2008). Consolidated bioprocessing of cellulosic biomass: An update. *Current Opinion in Biotechnology*, *19*(3), 230–238. https://doi.org/10.1016/j.copbio.2008.05.002.

Lynd, L. R., Weimer, P. J., van Zyl, W. H., & Pretorius, I. S. (2002). Microbial cellulose utilization: Fundamentals and biotechnology. *Microbiology and Molecular Biology Reviews*, *66*(3), 506–577. https://doi.org/10.1128/MMBR.66.3.506-577.2002.

Nielsen, J. (2017). Systems biology of metabolism: A driver for developing personalized and precision medicine. *Cell Metabolism*, *25*(3), 572–579. https://doi.org/10.1016/j.cmet.2017.02.002.

Nielsen, J., & Keasling, J. D. (2016). Engineering cellular metabolism. *Cell*, *164*(6), 1185–1197. https://doi.org/10.1016/j.cell.2016.02.004.

Oskarsson, E., Rova, U., & Christakopoulos, P. (2018). Evaluation of different process configurations for bioethanol production from spruce. *Biotechnology for Biofuels*, *11*, 42. https://doi.org/10.1186/s13068-018-1034-y.

Scalcinati, G., Knuf, C., Partow, S., Chen, Y., Maury, J., Schalk, M., Daviet, L., Nielsen, J., & Siewers, V. (2012). Dynamic control of gene expression in *Saccharomyces cerevisiae* engineered for the production of plant sesquiterpene α-santalene in a fed-batch mode. *Metabolic Engineering*, *14*(2), 91–103. https://doi.org/10.1016/j.ymben.2012.01.002.

Sheldon, R. A., & van Pelt, S. (2013). Enzyme immobilisation in biocatalysis: Why, what and how. *Chemical Society Reviews*, *42*(15), 6223–6235. https://doi.org/10.1039/c3cs60075k.

Silverman, A. D., Karim, A. S., & Jewett, M. C. (2020). Cell-free gene expression: An expanded repertoire of applications. *Nature Reviews Genetics*, *21*, 151–170. https://doi.org/10.1038/s41576-019-0186-3.

Singh, R., Kumar, M., Mittal, A., & Mehta, P. K. (2021a). Microbial thermotolerance and its biotechnological applications. *World Journal of Microbiology and Biotechnology*, *37*. Article 60. https://doi.org/10.1007/s11274-021-03000-3.

Singh, R., Kumar, M., Mittal, A., & Mehta, P. K. (2021b). Microbial enzymes: Industrial progress in 21st century. *3 Biotech*, *11*, 272. https://doi.org/10.1007/s13205-021-02808-z.

Singhania, R. R., Patel, A. K., Sukumaran, R. K., Larroche, C., & Pandey, A. (2013). Role and significance of beta-glucosidases in the hydrolysis of cellulose for bioethanol production. *Bioresource Technology*, *127*, 500–507. https://doi.org/10.1016/j.biortech.2012.09.012.

Wilkins, M. R., Sanchez, J. C., Gooley, A. A., Appel, R. D., Humphery-Smith, I., Hochstrasser, D. F., & Williams, K. L. (2002). Progress with proteome projects: Why all proteins expressed by a genome should be identified and how to do it. *Biotechnology and Genetic Engineering Reviews*, *19*(1), 19–50. https://doi.org/10.1080/02648725.2002.10648033.

Zuroff, T. R., & Curtis, W. R. (2012). Developing symbiotic consortia for lignocellulosic biofuel production. *Applied Microbiology and Biotechnology*, *93*, 1423–1435. https://doi.org/10.1007/s00253-011-3754-1.

4

Bioprocess Technology: Commercial Challenges for Application

Muhammad Islam[1], Muhammad Zubair Ali[1], Yuliya Kulikova[2], Olga Babich[2], Afsana Huseynova Anvar[3], Hafiz Abdullah Shakir[4], Muhammad Khan[4], Titin Haryati[5], Marcelo Franco[6], and Muhammad Irfan[1]

[1]*Department of Biotechnology, University of Sargodha, Sargodha, Pakistan*
[2]*Research and Education Center "Industrial Biotechnologies", Immanuel Kant Baltic Federal University, Kaliningrad, Russia*
[3]*Department of Molecular Biology and Biotechnology, Baku State University, Baku, Azerbaijan*
[4]*Institute of Zoology, University of the Punjab, Lahore, Pakistan*
[5]*Research Center for Applied Microbiology, National Research and Innovation Agency, Bogor, West Java, Indonesia*
[6]*Department of Exact Science, State University of Santa Cruz, Ilheus, Brazil*

4.1 Introduction

Rapid industrial growth, advancements in transportation, and overconsumption of traditional energy resources (non-renewable energy resources) cause depletion of fossil-based fuels (Johnsson et al., 2019). Moreover, the burning of fossil-based fuels causes severe environmental problems such as climate change, rising global temperatures, and rising sea levels. This situation has alarmed the scientific community to explore alternative energy resources, which are commonly referred to as renewable or biomass based energy (biofuels) (Deora et al., 2022). Bioprocessing technology is a fermentation-based industrial biotechnology technique that utilizes biomass from numerous sources and converts it into value-added biofuels including bio-ethanol, bio-butanol, and bio-diesel (Raganati & Procentese, 2024). Biofuels production from biomass using bioprocessing strategies is a highly sustainable approach to bring this world toward a greener future. This chapter provides a comprehensive review of green biofuels production from bioprocessing technology and its commercial challenges that create hindrances in its applications (Dwivedi et al., 2019). Moreover, it sheds light on types of biofuels, their production pathways, and their significance in reducing the increasing global temperature. In addition, it describes certain economic, environmental, and commercial challenges that need to be addressed and resolved to attain maximum benefit from this approach. Similarly, it sheds light on existing research gaps, technology advancements, and hybrid techniques which are highly valuable to achieve energy security in future (El-Araby, 2024). Bioprocessing technology includes applying advancements in technology to living systems for the production of value-added goods and biofuels (Raganati & Procentese, 2024). It is a hybrid strategy that involves modern biotechnology processes along with traditional practices.

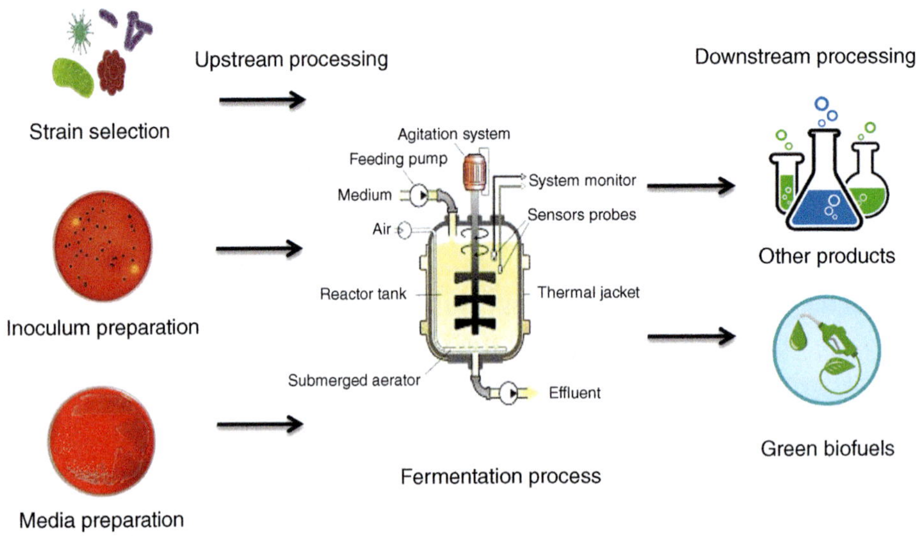

Figure 4.1 Steps in bioprocess technology for green biofuels productions (*Source:* Lindskog (2018) / with permission of Elsevier).

Recent advancements in genetic engineering and biocatalysis have had a great impact on green biofuels production (Nandy, 2016). The process of bioprocessing technology consists of three major stages including upstream processing, fermentation process, and downstream processing as shown in Figure 4.1. The first step involves the choice of a specific microbial strain, inoculum preparation, media preparation, and sterilization. Moreover, equipment cleaning and sterilization also take place in this step (Lindskog, 2018). The second step is the fermentation process through which raw substrate contents are converted into value-added products and biofuels in the presence of microorganisms (Raganati & Procentese, 2022). Similarly, the last step involves product harvesting and other procedures for further purification and packaging of specified products. All three steps require highly optimized conditions for achieving maximum growth rate and final yield (Kaligotla & Gottschalk, 2023). In this way, bioprocessing technology can pave the way to achieve UN SDG 7 (sustainable development goal) by 2030. SDG 7 is based on the transition of world economies from fossil-based fuels to renewable biofuels. This is a highly sustainable approach to bring sustainable development across the globe (Li, Jia, et al., 2023). There are different types of biofuels which are produced through bioprocessing technology, including liquid and gaseous forms. Similarly, numerous pathways exist based on the category of substrate used for biofuels production, such as edible crops, lignocellulosic waste material, and non-food waste material. Primary biofuels including animal and plant waste material, are directly used for different applications (Shahid et al., 2021). However, secondary biofuels are a category of biomass-based biofuels which are produced through bioprocessing of biomass into value-added biofuels such as bio-ethanol, bio-diesel, bio-butanol, and biogas. Different types of feedstocks are utilized for the production of such types of biofuels (Deora et al., 2022). In this way, we can further classify them into numerous types which are commonly called first-generation, second-generation, third–generation, and fourth-generation biofuels. First-generation biofuels are produced from edible crops, while second-generation biofuels are produced from non-food feedstock, usually lignocellulosic waste material, which is highly rich in cellulose. Similarly, third-generation biofuels are produced from numerous algae strains, while fourth-generation biofuels are produced via genetically modified microorganisms in order to maximize their production (El-Araby, 2024). Similarly, they are classified into three major categories based on the forms in which they exist, such as gaseous, liquid, and solid biofuels. All these types of biofuels have their own specific applications. Solid biofuels are utilized directly for bioenergy applications. There is no need for further bioprocessing of these types of biofuels (Murthy et al., 2021). Liquid biofuels include bio-ethanol, bio-diesel, and bio-butanol. These are produced via different bioprocessing techniques in the presence of specific microbial strains (Hafeez et al., 2020). Similarly, gaseous biofuels include biogas, which is mainly produced from animal and plant waste material through the process of methanogenesis. Numerous kinds of microbial strains play a significant role in the production of biogas from biomass or waste material (Pandey et al., 2019). In addition, different microorganisms possess different metabolic pathways through which they convert raw substrate into valuable biofuels. Major bioprocessing procedures include biochemical processes, thermochemical methods, and other chemical approaches. All these approaches are used based on the category of substrate material used in the bioprocessing technology (El-Araby, 2024). Biofuels have tremendous applications in the energy sector, and they represent a highly sustainable approach for renewable energy production. As bio-ethanol is used in blended form with petrol to fulfill increasing demand and reduces green-house gases emission to a significant level. In the European Union, a 5% blend of petrol and bio-ethanol is used, while in the United States it is used up to 10%. Similarly, in South America (Brazil), a 27% blend of petrol and bio-ethanol is used (Khoo et al., 2023). Recent research investigations have found that bio-butanol can be utilized without blending, via certain modifications in the existing engines (El-Araby, 2024). Similarly, bio-diesel produced from algal cells possesses tremendous energetic potential and can be used to replace fossil-based bio-diesel (Brahma et al., 2022). Likewise, advancements in the genetic engineering tools such as CRISPR-Cas9 system have redefined industrial biotechnology landscapes. It is highly specific approach to genetically engineer microbial strains to enhance biofuels production and reduce bioprocessing costs to a significant level (Wei & Li, 2023). Similarly, advancements in the artificial intelligence tools and automated bioprocessing technologies can eliminate chances of microbial cell contaminations and maximize biofuels production in order to fulfill increasing demand. This hybrid approach overcomes all the issues faced by existing fermentation-based bioprocessing techniques (Guan, 2024). Other problems faced by bioprocessing techniques include feedstock availability and the quality of substrate material. Most of feedstocks create food and land competition that limits their potential bioprocessing applications (Mat Aron et al., 2020). Similarly, conversion of feedstock into suitable substrates is a difficult and costly process. Lignocellulosic waste material consists of complex lignin that needs to be removed before further bioprocessing applications (Jayachandran et al., 2022). Chemicals used for the removal of lignin from lignocellulosic waste material and hydrolysis cause severe damage to environment. However, using enzymes for these applications is a highly ecofriendly approach but still there exists certain cost challenges that need to be addressed and resolved to enhance biofuels production (Li,

Table 4.1 Comparative Analysis of Final Biofuels Yields from Different Substrates.

Sr. No.	Green Biofuel	Feedstock/ Substrate	Approach	Final Yield	References
1	Bio-ethanol	Rice straw	Biochemical approach	116.5 L/kg dried biomass	El-Araby (2024)
2	Bio-ethanol	Palm oil residue	Biochemical strategy	110,600–172,100 kg dried biomass	El-Araby (2024)
3	Biogas	Fruits and vegetable wastes	Methanogenesis	180–732 mL/g volatile solids	Nair et al. (2022)
4	Biogas	Orange peel	Methanogenesis	217 mL/g volatile solids	Nair et al. (2022)
5	Biogas	Cattle manure	Methanogenesis	298 mL/g volatile solids	Nair et al. (2022)
6	Bio-hydrogen	Sugarcane bagasse	Biochemical approach	6980 mL H_2/L	Eloffy et al. (2022)
7	Bio-hydrogen	Corn stoker	Biochemical approach	49–68 mL H_2/g	Eloffy et al. (2022)
8	Biodiesel	Algal biomass	Biochemical strategy	—	El-Araby (2024)
9	Bio-ethanol	Corn meal	Biochemical approach	9.77 ± 0.12 kg dried biomass	El-Araby (2024)
10	Bio-hydrogen	Potato waste	Biochemical approach	298.11 mL H_2/g kg dried biomass	Eloffy et al. (2022)

Wang, et al., 2023). Other commercial challenges include the lack of policy framework and regulations that limit their availability at the international level (El-Araby, 2024). The absence of standardization creates issues in the international trade market (Marzocchella et al., 2022). Moreover, the high cost compared to fossil-based fuels limits its applications to a greater level. There is a dire need to optimize bioprocessing techniques via integration of synthetic biology and artificial intelligence to overcome all these challenges. Green biofuels production using bioprocessing technology is a highly sustainable approach that can bring sustainability across the globe (Rame et al., 2023; Table 4.1).

4.2 Biomass Conversion Strategies

4.2.1 Biochemical Approach

In this strategy, microorganisms are used to convert raw biomass into value-added products and biofuels (Leonard et al., 2020). Anaerobic digestion and fermentation are two important techniques that are used for the conversion of biomass into green biofuels. During the fermentation process, sugars are broken down into valuable products such as biofuels in the

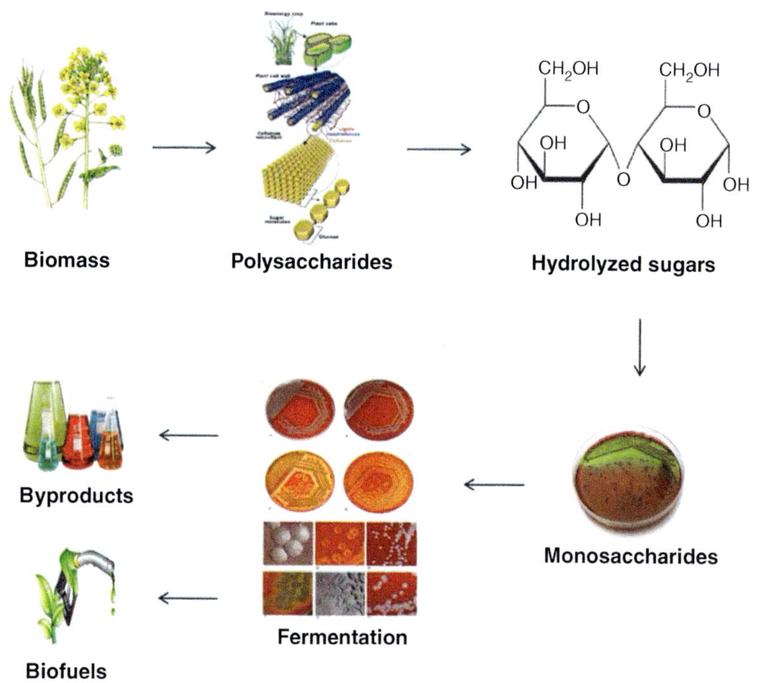

Figure 4.2 Biomass conversion into green biofuels through biochemical approach (*Source:* Adapted from Deng et al. (2020) / Elsevier / CC BY 4.0).

presence of specific microbial strains as shown in Figure 4.2. It is widely used method for the production of bio-ethanol from sugar-based substrates such as sugarcane and maize. Microbial strains convert only simple sugar molecules instead of complex ones such as lignocellulosic waste material, which requires certain pretreatment. Both chemical and enzymatic treatments are used depending upon the reaction conditions. Similarly, polysaccharides are hydrolyzed first into simple sugars before moving toward the main fermentation process. Likewise, microorganisms consist of complex metabolic pathways that are utilized for the production of specific kinds of products from biomass (Muritala et al., 2020; Siddiki et al., 2022). In the second approach, bacterial strains are used to break down raw organic material into methane-rich biogas in the absence of oxygen. This process is commonly called as anaerobic digestion. Biogas possesses numerous applications in the electricity and biofuels sector. The process remains are widely used as fertilizers and bio-solids for soil conditioning (Deng et al., 2020).

4.2.2 Thermochemical Approach

There is a wide range of thermochemical strategies used for the production of green biofuels such as pyrolysis, gasification, and supercritical fluid conversion. In pyrolysis, biomass is heated in the absence of oxygen in order to produce bio-oil, bio-char, and syngas as shown in Figure 4.3. Catalysts can be used to stimulate process and enhance its efficiencies (Rame

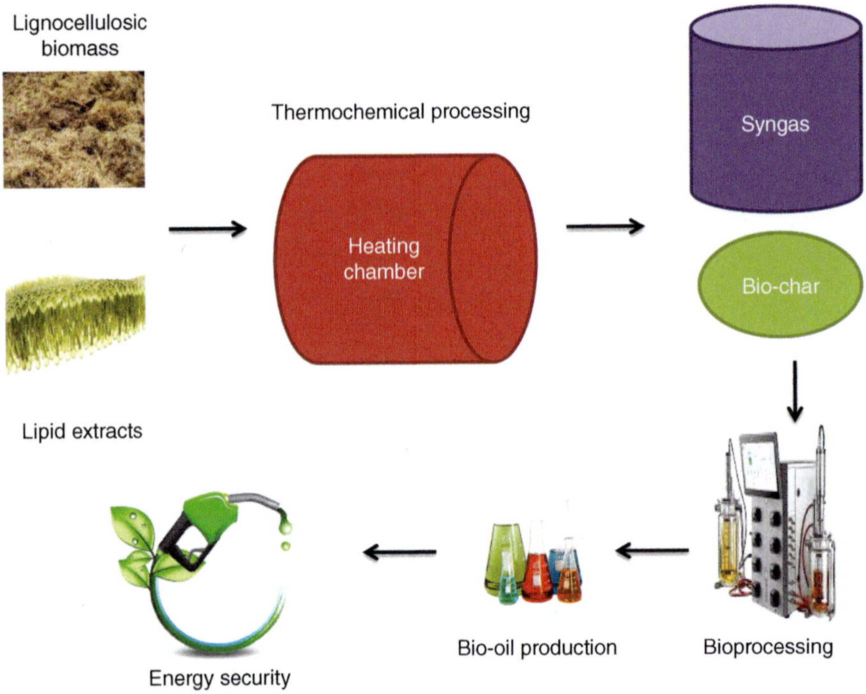

Figure 4.3 Biomass conversion into green biofuels through the thermochemical approach (*Source:* Rame et al. (2023) / with permission of Elsevier).

et al., 2023). Similarly, in the gasification process, biomass is heated to produce syngas, which mainly consists of hydrogen and carbon monoxide. Likewise, syngas can be further processed into green biofuels (Muritala et al., 2020). Similarly, in the hydrothermal liquefaction process, biomass is processed into bio-oil under extremely high pressure and temperature conditions. Co-solvents and catalysts can be used to enhance production and optimize process efficiency (Friedemann, 2021). In supercritical fluid conversion, supercritical ethanol and water are processed into green biofuels (Nema et al., 2023).

4.3 Green Biofuels Produced via Bioprocess Technology

4.3.1 First-Generation Biofuels

On the basis of substrates, biofuels are classified into four different generations as shown in Figure 4.4. These are kinds of biofuels that are produced from edible crops. Major substrates include palm oil, sugarcane, sugar, corn, vegetable oil, wastes, and animal fats. Biodiesel and bioethanol are important categories of green biofuels that are produced by utilizing the aforementioned substrates in bioprocessing technology (Khan et al., 2021). Biodiesel is produced through the process of transesterification. Similarly, simple sugars undergo fermentation and are converted into bioethanol (Mahapatra et al., 2021). However, first-generation biofuels cannot fulfill increasing energy demands due to rapid population

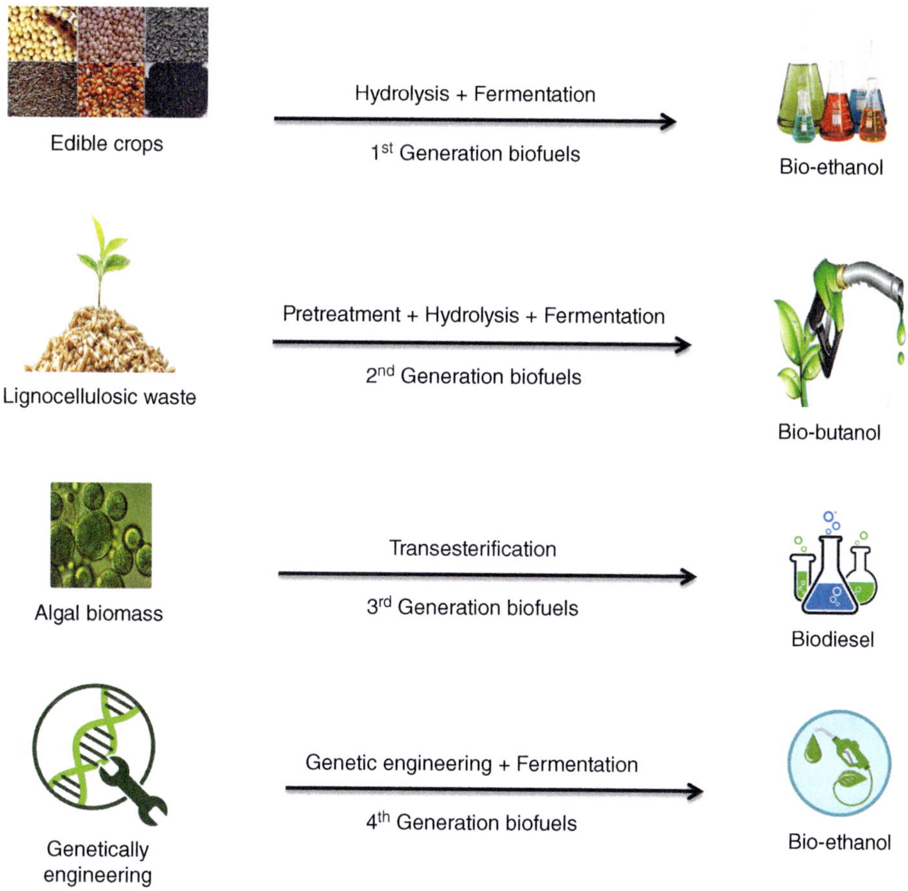

Figure 4.4 Different generations of green biofuels based on the substrates used for their production (*Source:* Adapted from El-Araby (2024) / Springer Nature / CC BY 4.0).

growth and uncontrolled industrialization. Moreover, they create food shortage issues that limit their widespread potential applications. Careful measures need to be taken in order to maintain balance among food security, environmental aspects, and energy security (Kumar et al., 2019). Recent research studies have found that instead of using whole food crops, their residues and waste extracts can be utilized for the production of green biofuels. In this way, it has redefined large-scale biofuels production. Current researchers must focus on the exploration of nutrient-rich substrates to optimize green biofuels production (Nisar et al., 2024).

4.3.2 Second-Generation Biofuels

As with first-generation biofuels, production creates numerous food security problems, whereas second-generation biofuels can be produced from agricultural waste material (lignocellulosic waste material). In this way, they can overcome challenges faced by first-generation bioprocessing technologies. Lignocellulosic waste material consists of cellulose, hemicellulose, and lignin, which require certain kinds of pretreatment to remove unrequired materials such as lignin. Both chemical and enzymatic pretreatment strategies exist; however, enzymatic pretreatment is the most eco-friendly approach. Similarly, complex polysaccharides require a hydrolysis process before being processed into biofuels. There exists a few cost and environmental issues that create hindrances in large-scale green biofuels production. But still it is highly sustainable approach through which plant waste extracts are converted into numerous kinds of biofuels (Aslam & Mahmood, 2024).

4.3.3 Third-Generation Biofuels

First- and second-generation biofuels production strategies face certain challenges, such as food and land competition. However, third-generation biofuels are produced from various algae strains, which overcome all the aforementioned challenges. Algae possess highly nutrient-rich composition and do not require large areas for their cultivation. Moreover, algae constitute of diverse genera that can grow everywhere. Moreover, blue sequestration has reduced almost 40% of the global atmospheric carbon level. The development of closed pond system provides highly optimized conditions for algal cell growth and enhances green biofuels production (Aslam & Mahmood, 2024). Advancement in synthetic biology tools, such as CRISPR CAS9, has paved the way toward metabolic engineering of algae strains to reduce process costs and maximize biofuels productions. Similarly, metabolomics techniques can be used for the exploration of new algae strains (Wei & Li, 2023).

4.3.4 Fourth-Generation Biofuels

Green biofuels produced from genetically modified organisms are referred as fourth-generation biofuels. The major aim is to lower greenhouse gas emissions and enhance atmospheric carbon sequestration to maintain global temperature. Numerous synthetic biology approaches have been discovered and are widely used for the metabolic engineering of microbial strains. Advanced genetic engineering techniques include the CRISPR CAS9 system, RNA interference, and various adaptive laboratory evolution strategies (Mat Aron et al., 2020).

4.4 Green Biofuels Examples

4.4.1 Gaseous Biofuels

Different kinds of gaseous biofuels, such as bio-methane, are produced from biomass, which reduces greenhouses gases emission to a significant level as shown in Figure 4.5. Bio-hydrogen is a category of green biofuels produced from waste extracts. It is widely used as an alternative to traditional fossil-based gaseous fuels (Zhao et al., 2024). Numerous microorganisms possess significant metabolic pathways, such as fermentation or photolysis, through which these microbial strains transform hydrogen into valuable chemical energy (Pachaiappan et al., 2024). Moreover, it is very easy to produce bio-hydrogen from

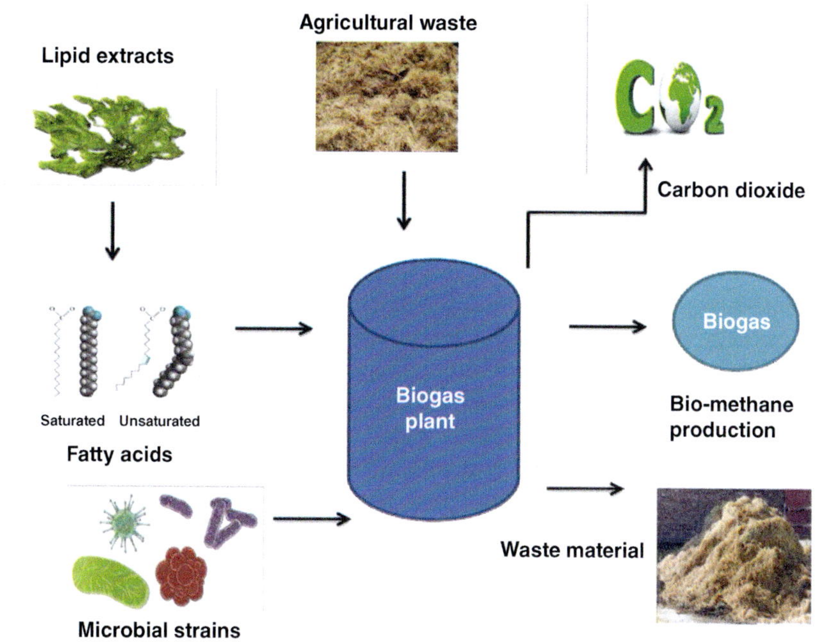

Figure 4.5 Examples of green biofuels produced from biomass via bioprocess technology.

biomass. In addition, bio-hydrogen is source of clean energy that can prevent environmental emission problems (Mogili et al., 2022). Along with biochemical approach other techniques including electrolytic, chemical and thermochemical strategies are utilized for its large scale productions. Similarly, through the process of methanogenesis, biogas is produced that is widely used as a source of energy for both small- and large-scale applications (El-Araby, 2024).

4.4.2 Liquid Biofuels

Liquid biofuels include bio-methanol, bio-ethanol, bio-butanol, and biodiesel. Different substrates and various biochemical and thermochemical processes are used for the large-scale production of the aforementioned liquid biofuels. Bioethanol is mainly produced through the fermentation process, in which the *Saccharomyces cerevisiae* strain converts sugar-based substrates (maize, molasses, rice, potatoes, and sugarcane) into the desired product (Mat Aron et al., 2020). Different bacterial strains, including *Escherichia coli*, *Clostridium*, and *Bacillus* strains, are also used for this purpose (Liew et al., 2016). Similarly, bio-methanol is the simplest hydrocarbon containing biofuel produced via a biochemical approach. Traditionally, it was used for power autos in Germany during World War II. It is also called as "wood al-cool," which is produced through the partial oxidation of natural gas (Liu et al., 2021). Biodiesel is another category of green biofuels, which is used as alternative to fossil-based diesel. It is mainly produced through the transesterification process. Major substrates in biodiesel production include animal fat, vegetable oil, and microalgae-derived triglycerides (Brahma et al., 2022). There are numerous catalysts that exist and are used to catalyze the transesterification process in order to maximize the final product yield. Alkali-catalyzed transesterification completes in 60 minutes instead of 90 minutes. It can maximize the final product yield to approximately 98%, which is a highly sustainable approach to enhance energy security across the world (Sankar et al., 2017).

4.4.2.1 Recent Advancements in Bioprocessing Technology

Genetic engineering and genome editing tools have redefined industrial landscapes. Conventional synthetic biology tools lack precision, and numerous challenges are faced by these techniques, including off-target site effects. However, advanced genome editing tools have overcome all these issues. The CRISPR-CAS9 system is a newly discovered synthetic biology tool which can be widely used for metabolic engineering of all major life domains, including prokaryotes, eukaryotes, and archaea (Zetsche et al., 2017). CRISPR CAS9 consists of vibrant genetic engineering mechanism, as shown in Figure 4.6 (Wei & Li, 2023). It possesses advanced genome editing capabilities, including CRISPR interference, CRISPR activation, and regulation of gene transcription (Bikard et al., 2013). Numerous industrially significant microbial strains, including *E. coli* strains, have been manipulated genetically to enhance the production of green biofuels. Research investigations have found that CRISPR-CAS9 system possesses capabilities to introduce approximately 65%

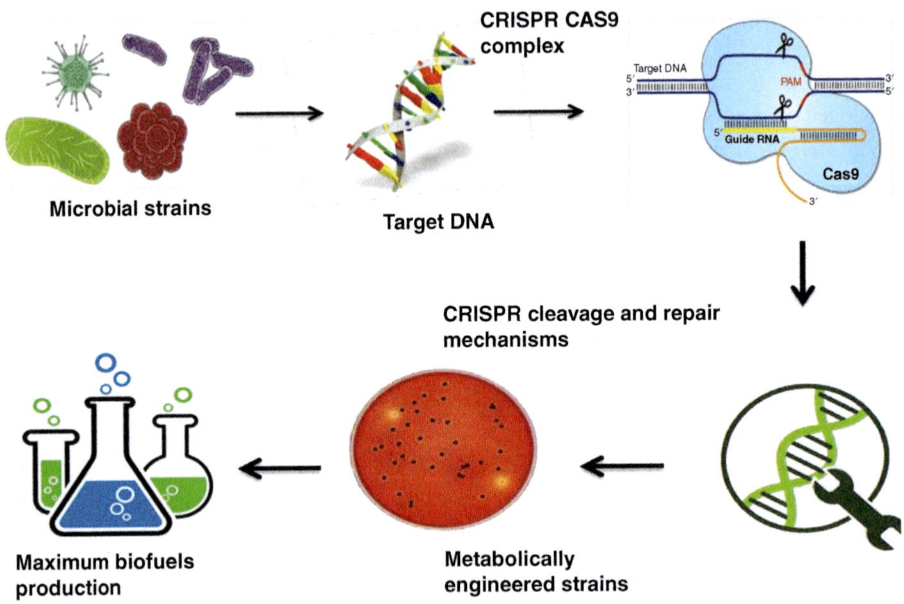

Figure 4.6 Mechanism of genetic engineering of microbial strains by the CRISPR-CAS9 system (*Source:* Adapted from Wei and Li (2023)).

mutation rate in *E. coli* strains (Li et al., 2015). Moreover, CRISPR CAS9 system genome modification approaches possess almost 100% target efficiencies (Lan et al., 2021). In addition, *Corynebacterium glutamicum* strains have also been metabolically engineered to enhance the production of green biofuels and amino acids. Similarly, *Bacillus subtilis* strains have tremendous metabolic pathways that can be targeted using CRISPR-CAS9 system. In this way, advanced genome editing tools can be utilized to optimize bioprocess technology to maximize the production of green biofuels. This hybrid approach can prevent all major issues faced by bioprocessing technology, including lower product yield and high production cost (Wei & Li, 2023).

4.5 Challenges

4.5.1 Sustainability and Availability of Feedstock

The production of biofuel is significantly hampered by the lack of abundant and sustainable feedstocks. Concerns regarding land-use change, deforestation, and biodiversity loss may arise from conflicting demands for land, water, and resources between biofuel crops, conservation, and food production. Growing biofuel feedstocks in an environmentally conscious and sustainable manner is essential (Mat Aron et al., 2020; Rutz & Janssen, 2007). Many developed nations have not changed their minds despite the arguments for investing a sizable amount of money in algae. Both public and private investors are drawn to it because of its growth and oil-rich composition. However, it will cost a lot of money to stop using petroleum and reduce CO_2 emissions from algae biofuels.

4.5.2 Diversity of Feedstock

Biofuel production's dependence on a limited range of feedstocks, including soybeans, maize, and sugarcane, could lead to competition with food production and ensuing price rises. Reducing reliance on food-based feedstocks and increasing the sustainability of biofuel production need the development of alternative feedstocks, such as non-food crops, agricultural waste, and leftovers (Omer, 2008; Rame et al., 2023).

4.5.3 Efficiency of Conversion

Problems with efficiency are common when turning feedstocks into biofuels. Thermochemical conversion, fermentation, and enzymatic hydrolysis are all intricate procedures that require optimization. In order to increase the economic viability of producing biofuel, ongoing research aims to increase conversion efficiency, reduce energy inputs, and restrict by-product creation (Jayachandran et al., 2022; Li, Jia, et al., 2023; Li, Wu, et al., 2023; Li et al., 2023a). Algal fuels still seem to be processed at an unprofitable rate. Although there are a number of algae businesses, none can produce enough for commercial use at prices that are competitive. Production costs must be lowered by at least two orders of magnitude before algae fuels might potentially be commercialized. Algae can be cultivated in a bioreactor, which is an open or closed system. Compared to closed systems, open systems offer a greater advantage.

4.5.4 Developments in Technology

To advance conversion technologies and make them more commercially viable, further research and development are needed. This entails improving the stability and performance of enzymes, refining catalysts, creating novel reactor designs, and expanding procedures from lab to commercial production (Shahid et al., 2021). It will be necessary to remove several obstacles preventing the large-scale manufacturing of algal biofuel. Large-scale algae cultivation, increasing its oil-rich content, and fuel processing are areas that the industrialized sectors will need to concentrate on more. If we want to completely embrace the usage of algae biofuels, we must look at all of this (Hannon et al., 2010).

4.5.5 High Production Costs and Initial Investment

Due to the high cost of production, closed systems—such as cultivating algae in bioreactors—have not shown themselves to be the best alternative means of manufacturing biofuels (Foley et al., 2011; Scott et al., 2010; Vasudevan & Briggs, 2008). In a closed system, the cost of manufacturing algal biofuels is $32.16 per kilogram. According to numerous studies, closed-system production of algal biofuels can only rival crude oil if the price of crude oil drops to $5 per liter. Numerous sectors have noted that growing algae for biofuels alone is not completely feasible for its widespread commercial application (Table 4.2).

Table 4.2 Comparative Analysis of Conventional Fuels and Green Biofuels Produced Through Bioprocess Technology.

Sr. No.	Type of Fuels	Energy Density (MJ/L)	Production Cost (USD/L)	Cost Per Energy Unit (USD/GJ)	GHG Emissions (g CO_2e/MJ)
1	Conventional gasoline	34.2	0.5–0.65	14.6–19	—
2	Conventional diesel	38.6	0.55–0.70	—	95.1
3	Cellulosic ethanol	21.2	0.8–1.20	36.7–57.7	—
4	Renewable diesel (HVO)	34.4	0.70.1.00	—	29.2
5	Advanced biofuels (e.g., algal)	35	1.2–2.00	34.3–57.1	—
6	Sugarcane ethanol	21.2	0.45–0.65	21.2–30.7	—
7	Biodiesel (soybean)	33.3	0.70–0.90	21–27	—
8	Ethanol (sugarcane)	21.20	0.45–0.65	—	24.3
9	Biodiesel (waste cooking oil)	33.3	0.60–0.80	18–24	—
10	Bio-jet (HEFA)	34.3	0.80–1.20	—	25

Source: Adapted from Renewables (2021).

4.6 Frameworks of Regulations and Policies for the Development of Biofuels

Frameworks for regulations and policies are essential for promoting the creation and use of biofuels. To support the industry and ensure that biofuels are produced and utilized sustainably, they set standards, incentives, and regulations. The policy and regulatory frameworks for the development of biofuels include the following crucial elements.

4.6.1 Blending Goals and Renewable Fuel Standards (RFS)

Numerous countries have implemented blending objectives or renewable fuel standards that mandate a specific percentage of biofuels to be used as transportation fuels. These policies give producers long-term stability while also raising the market demand for biofuels. Blending goals may be based on a percentage (e.g., a specific percentage of biofuels in the overall fuel mix) or volume (e.g., a specific amount of biofuel to be mixed) (Grangeia et al., 2022; Heo et al., 2024).

4.6.2 Regulations for Feedstock

Feedstock-related issues like certification, sustainability standards, and land use are addressed by policies and laws. These systems ensure the sustainable supply of biofuel feedstocks, preventing deforestation, biodiversity loss, and adverse social effects. Requirements for sustainable feedstock production are provided by certification programs like the International Sustainability and Carbon Certification (ISCC) and the Roundtable on Sustainable Biomaterials (RSB) (Ma et al., 2024; Yu et al., 2020).

4.6.3 Monetary Rewards

To promote the development and use of biofuels, governments usually provide financial incentives. The higher costs of generating biofuels compared to fossil fuels can be mitigated by tax credits, grants, loan guarantees, and subsidies. Financial incentives are used to increase the use of biofuels, draw in investment, and remove barriers to the market (Osman et al., 2023; Yu et al., 2020).

4.6.4 Support for Research and Development

Research and development initiatives pertaining to biofuels may be funded by governments. This support promotes the creation of new feedstocks, improvements in efficiency, and technical innovation. Prominent strategies for promoting innovation in biofuel technology include public–private partnerships, research funding, and collaboration with academic institutions (Balali & Stegen, 2021).

4.6.5 Policies for Imports and Exports

Regulations governing international trade may have an impact on the biofuel sector. Trade agreements, quotas, and tariffs all affect the import and export of feedstocks and biofuels. Governments can create preferential trade agreements to promote the export of biofuels or impose trade restrictions to protect domestic companies (Ifaei et al., 2022; Petrović-Ranđelović et al., 2020).

4.6.6 Rules Pertaining to the Environment

The use of biofuel may be impacted by environmental regulations, such as those pertaining to emissions and air quality standards. Due to their lower greenhouse gas emissions, biofuels are typically chosen over fossil fuels. To promote the use of biofuels, governments may establish carbon pricing plans or emission-reduction targets (Hakeem et al., 2022; Nageswara-Rao & Soneji, 2018).

4.6.7 Support for Infrastructure

The development of biofuel infrastructure, such as distribution networks, storage facilities, and filling stations, may be partially funded by governments. Overcoming market barriers and boosting the use of biofuels are made possible by funding infrastructure expansion and modernization (Grangeia et al., 2022; Peixoto & Temmes, 2019; Sheikh et al., 2023; Shouse et al., 2021).

4.6.8 Public Education and Awareness

The development of biofuels is governed by different legal and policy frameworks in each nation and region, which reflect regional objectives, available resources, and market dynamics. Governments usually collaborate with businesses, academic institutions, and environmental organizations to create effective policies that strike a balance between the goals of sustainability, energy security, and economic development. To adapt to shifting market conditions, technological advancements, and environmental concerns, policy frameworks must be regularly reviewed and changed (Heo et al., 2024; Hossain, 2023).

4.7 Conclusion

As the world's energy demands are increasing at a quick pace and fossil-based energy reserves are depleting day by day, there is a dire need to find alternative energy sources. Likewise, bioprocess technology proves to be the best way for the production of green biofuels. It is a highly eco-friendly approach that can bring energy security across the globe. It consists of a three-step process: upstream processing (media preparation, microbial strains selection, media sterilization, inoculum preparation), main fermentation process, and downstream processing (filtration, centrifugation, product harvesting, and packaging). There are four different generations of green biofuels, classified based on the substrate used for their production. First-generation biofuels are produced from edible food crops. Second-generation biofuels are produced from lignocellulosic feedstock. Third-generation biofuels are produced from algal biomass, while fourth-generation biofuels are produced from genetically modified organisms. Different types of biofuels, such as bio-ethanol, bio-methanol, bio-butanol, and biodiesel, are produced using bioprocessing technology. However, certain commercial and environmental challenges still exist, which needs current research focus to overcome these issues. Moreover, an efficient policy framework can be formulated to provide uniform access to green biofuels across the globe.

References

Aslam, L., & Mahmood, S. (2024). Biofuels for a green future: A pathway to environmental and economic sustainability: Biofuels production and sustainability: A mini review. *Nature's Symphony*, *2*(2), 53–62. https://doi.org/10.69547/ns.22.05.

Balali, Y., & Stegen, S. (2021). Review of energy storage systems for vehicles based on technology, environmental impacts, and costs. *Renewable and Sustainable Energy Reviews*, *135*, 110185. https://doi.org/10.1016/j.rser.2020.110185.

Bikard, D., Jiang, W., Samai, P., Hochschild, A., Zhang, F., & Marraffini, L. A. (2013). Programmable repression and activation of bacterial gene expression using an engineered CRISPR-Cas system. *Nucleic Acids Research*, *41*(15), 7429–7437. https://doi.org/10.1093/nar/gkt520.

Brahma, S., Nath, B., Basumatary, B., Das, B., Saikia, P., Patir, K., & Basumatary, S. (2022). Biodiesel production from mixed oils: A sustainable approach towards industrial biofuel production. *Chemical Engineering Journal Advances*, *10*, 100284. https://doi.org/10.1016/j.ceja.2022.100284.

Deng, C., Lin, R., Kang, X., Wu, B., O'Shea, R., & Murphy, J. D. (2020). Improving gaseous biofuel yield from seaweed through a cascading circular bioenergy system integrating anaerobic digestion and pyrolysis. *Renewable and Sustainable Energy Reviews*, *128*, 109895. https://doi.org/10.1016/j.rser.2020.109895.

Deora, P. S., Verma, Y., Muhal, R. A., Goswami, C., & Singh, T. (2022). Biofuels: An alternative to conventional fuel and energy source. *Materials Today: Proceedings*, *48*, 1178–1184. https://doi.org/10.1016/j.matpr.2021.08.227.

Dwivedi, G., Pillai, S., & Shukla, A. K. (2019). Study of performance and emissions of engines fueled by biofuels and its blends. In A. Agarwal, A. Gautam, N. Sharma, & A. Singh (Eds.), *Methanol and the alternate fuel economy* (pp. 77–106). Springer. https://doi.org/10.1007/978-981-13-3287-6_5.

El-Araby, R. (2024). Biofuel production: Exploring renewable energy solutions for a greener future. *Biotechnology for Biofuels and Bioproducts*, *17*(1), 129. https://doi.org/10.1186/s13068-024-02571-9.

Eloffy, M., Elgarahy, A. M., Saber, A. N., Hammad, A., El-Sherif, D. M., Shehata, M., Mohsen, A., & Elwakeel, K. Z. (2022). Biomass-to-sustainable biohydrogen: Insights into the production routes, and technical challenges. *Chemical Engineering Journal Advances*, *12*, 100410. https://doi.org/10.1016/j.ceja.2022.100410.

Foley, P. M., Beach, E. S., & Zimmerman, J. B. (2011). Algae as a source of renewable chemicals: Opportunities and challenges. *Green Chemistry*, *13*(6), 1399–1405. https://doi.org/10.1039/c1gc00015b.

Friedemann, A. (2021). *Life after fossil fuels: A reality check on alternative energy*. Springer-Lecture Notes in Energy. https://doi.org/10.1007/978-3-030-70335-6.

Grangeia, C., Santos, L., & Lazaro, L. L. B. (2022). The Brazilian biofuel policy (RenovaBio) and its uncertainties: An assessment of technical, socioeconomic and institutional aspects. *Energy Conversion and Management: X*, *13*, 100156. https://doi.org/10.1016/j.ecmx.2021.100156.

Guan, W. (2024). Bioprocess engineering and intelligent biomanufacturing. *Bioprocess Engineering*, *1*, 20. https://doi.org/10.1007/s00449-024-03039-8.

Hafeez, S., Al-Salem, S., Manos, G., & Constantinou, A. (2020). Fuel production using membrane reactors: A review. *Environmental Chemistry Letters*, *18*(5), 1477–1490. https://doi.org/10.1007/s10311-020-01024-7.

Hakeem, K. R., Bandh, S. A., Malla, F. A., & Mehmood, M. A. (2022). *Environmental sustainability of biofuels: Prospects and challenges*. Elsevier. https://doi.org/10.1016/b978-0-323-91159-7.00010-2.

Hannon, M., Gimpel, J., Tran, M., Rasala, B., & Mayfield, S. (2010). Biofuels from algae: Challenges and potential. *Biofuels*, *1*(5), 763–784. https://doi.org/10.4155/bfs.10.44.

Heo, S., Byun, J., Ifaei, P., Ko, J., Ha, B., Hwangbo, S., & Yoo, C. (2024). Towards mega-scale decarbonized industrial park (Mega-DIP): Generative AI-driven techno-economic and environmental assessment of renewable and sustainable energy utilization in petrochemical industry. *Renewable and Sustainable Energy Reviews*, *189*, 113933. https://doi.org/10.1016/j.rser.2023.113933.

Hossain, E. (2023). *The sun, energy, and climate change*. Springer. https://doi.org/10.1007/978-3-031-22196-5.

Ifaei, P., Charmchi, A. S. T., Loy-Benitez, J., Yang, R. J., & Yoo, C. (2022). A data-driven analytical roadmap to a sustainable 2030 in South Korea based on optimal renewable microgrids. *Renewable and Sustainable Energy Reviews*, *167*, 112752. https://doi.org/10.1016/j.rser.2022.112752.

Jayachandran, M., Gatla, R. K., Rao, K. P., Rao, G. S., Mohammed, S., Milyani, A. H., Azhari, A. A., Kalaiarasy, C., & Geetha, S. (2022). Challenges in achieving sustainable development goal 7: Affordable and clean energy in light of nascent technologies. *Sustainable Energy Technologies and Assessments*, *53*, 102692. https://doi.org/10.1016/j.seta.2022.102692.

Johnsson, F., Kjärstad, J., & Rootzén, J. (2019). The threat to climate change mitigation posed by the abundance of fossil fuels. *Climate Policy*, *19*(2), 258–274. https://doi.org/10.1080/14693062.2018.1483885.

Kaligotla, H., & Gottschalk, U. (2023). Innovations in downstream processing. *BioPharm International*, *36*(9), 30–34. https://doi.org/10.1201/9781315164953-15.

Khan, N., Sudhakar, K., & Mamat, R. (2021). Role of biofuels in energy transition, green economy and carbon neutrality. *Sustainability*, *13*(22), 12374. https://doi.org/10.3390/su132212374.

Khoo, K. S., Ahmad, I., Chew, K. W., Iwamoto, K., Bhatnagar, A., & Show, P. L. (2023). Enhanced microalgal lipid production for biofuel using different strategies including genetic modification of microalgae: A review. *Progress in Energy and Combustion Science*, *96*, 101071. https://doi.org/10.1016/j.pecs.2023.101071.

Kumar, N., Sonthalia, A., Pali, H. S., & Sidharth. (2019). Next-generation biofuels—Opportunities and challenges. In A. Gupta, A. De, S. Aggarwal, A. Kushari, & A. Runchal (Eds.), *Innovations in sustainable energy and cleaner environment* (pp. 171–191). Springer. https://doi.org/10.1007/978-981-13-9012-8_8.

Lan, Y.-J., Tan, S.-I., Cheng, S.-Y., Ting, W.-W., Xue, C., Lin, T.-H., Cai, M.-Z., Chen, P.-T., & Ng, I.-S. (2021). Development of *Escherichia coli* Nissle 1917 derivative by CRISPR/Cas9 and application for gamma-aminobutyric acid (GABA) production in antibiotic-free system. *Biochemical Engineering Journal*, *168*, 107952. https://doi.org/10.1016/j.bej.2021.107952.

Leonard, M. D., Michaelides, E. E., & Michaelides, D. N. (2020). Energy storage needs for the substitution of fossil fuel power plants with renewables. *Renewable Energy*, *145*, 951–962. https://doi.org/10.1016/j.renene.2019.06.066.

Li, C., Jia, T., Wang, H., Wang, X., Negnevitsky, M., Hu, Y.-J., Zhao, G., & Wang, L. (2023). Assessing the prospect of deploying green methanol vehicles in China from energy, environmental and economic perspectives. *Energy*, *263*, 125967. https://doi.org/10.1016/j.energy.2022.125967.

Li, S., Wu, Y., Dao, M. U., Dragoi, E.-N., & Xia, C. (2023). Spotlighting of the role of catalysis for biomass conversion to green fuels towards a sustainable environment: Latest innovation avenues, insights, challenges, and future perspectives. *Chemosphere*, *318*, 137954. https://doi.org/10.1016/j.chemosphere.2023.137954.

Li, Y., Lin, Z., Huang, C., Zhang, Y., Wang, Z., Tang, Y.-J., Chen, T., & Zhao, X. (2015). Metabolic engineering of *Escherichia coli* using CRISPR–Cas9 meditated genome editing. *Metabolic Engineering*, *31*, 13–21. https://doi.org/10.1016/j.ymben.2015.06.006.

Li, Y., Wang, X., Imran, A., Aslam, M. U., & Mehmood, U. (2023). Analyzing the contribution of renewable energy and natural resources for sustainability in G-20 countries: How gross capital formation impacts ecological footprints. *Heliyon*, *9*(8), e18882. https://doi.org/10.1016/j.heliyon.2023.e18882.

Liew, F., Martin, M. E., Tappel, R. C., Heijstra, B. D., Mihalcea, C., & Köpke, M. (2016). Gas fermentation—A flexible platform for commercial scale production of low-carbon-fuels and chemicals from waste and renewable feedstocks. *Frontiers in Microbiology*, *7*, 694. https://doi.org/10.3389/fmicb.2016.00694.

Lindskog, E. K. (2018). The upstream process: Principal modes of operation. In G. Jagschies, E. Lindskog, K. Łącki, & P. Galliher (Eds.), *Biopharmaceutical processing: Development, design, and implementation of manufacturing processes* (pp. 625–635). Elsevier. https://doi.org/10.1016/b978-0-08-100623-8.00031-1.

Liu, Y., Cruz-Morales, P., Zargar, A., Belcher, M. S., Pang, B., Englund, E., Dan, Q., Yin, K., & Keasling, J. D. (2021). Biofuels for a sustainable future. *Cell*, *184*(6), 1636–1647. https://doi.org/10.1016/j.cell.2021.01.052.

Ma, X., Pan, Y., Zhang, M., Ma, J., & Yang, W. (2024). Impact of carbon emission trading and renewable energy development policy on the sustainability of electricity market: A Stackelberg game analysis. *Energy Economics*, *129*, 107199. https://doi.org/10.1016/j.eneco.2023.107199.

Mahapatra, S., Kumar, D., Singh, B., & Sachan, P. K. (2021). Biofuels and their sources of production: A review on cleaner sustainable alternative against conventional fuel, in the framework of the food and energy nexus. *Energy Nexus*, *4*, 100036. https://doi.org/10.1016/j.nexus.2021.100036.

Marzocchella, A., Chirone, R., Paulillo, A., Lettieri, P., & Salatino, P. (2022). Sustainability assessment of biotechnological processes: LCA and LCC of second-generation biobutanol production. In C. Teodosiu, S. Fiore, & A. Hospido (Eds.), *Assessing progress towards sustainability* (pp. 365–382). Elsevier. https://doi.org/10.1016/b978-0-323-85851-9.00015-8.

Mat Aron, N. S., Khoo, K. S., Chew, K. W., Show, P. L., Chen, W. H., & Nguyen, T. H. P. (2020). Sustainability of the four generations of biofuels—A review. *International Journal of Energy Research, 44*(12), 9266–9282. https://doi.org/10.1002/er.5557.

Mogili, N. V., Murugesan, N., Ayothiraman, S., Gautam, R., Deshavath, N. N., & Erva, R. R. (2022). Biohydrogen production from wastewater and organic solid wastes. In C. M. Hussain, S. Singh, & L. Goswami (Eds.), *Waste-to-energy approaches towards zero waste* (pp. 165–195). Elsevier. https://doi.org/10.1016/b978-0-323-85387-3.00009-4.

Muritala, I. K., Guban, D., Roeb, M., & Sattler, C. (2020). High temperature production of hydrogen: Assessment of non-renewable resources technologies and emerging trends. *International Journal of Hydrogen Energy, 45*(49), 26022–26035. https://doi.org/10.1016/j.ijhydene.2019.08.154.

Murthy, G. S., Gnansounou, E., Khanal, S. K., & Pandey, A. (2021). *Biomass, biofuels, biochemicals: Green-economy: Systems analysis for sustainability*. Elsevier. https://doi.org/10.1016/b978-0-12-819242-9.00020-8.

Nageswara-Rao, M., & Soneji, J. (2018). *Advances in biofuels and bioenergy*. BoD–Books on Demand. https://doi.org/10.5772/intechopen.70022.

Nair, L. G., Agrawal, K., & Verma, P. (2022). An overview of sustainable approaches for bioenergy production from agro-industrial wastes. *Energy Nexus, 6*, 100086. https://doi.org/10.1016/j.nexus.2022.100086.

Nandy, S. (2016). Bioprocess technology governs enzyme use and production in industrial biotechnology: An overview. *Enzyme Engineering, 144*(5), 1–5. https://doi.org/10.4172/2329-6674.1000144.

Nema, A., Prasad, R., Kumar, R., Sharma, D., Sonaviya, D. K. R., Maurya, N. S., & Kumari, A. (2023). Conventional and the recent advances in technologies for the production of different generations of biofuels. *Biofuels*, 39–64. https://doi.org/10.1201/9781003197737-5.

Nisar, A., Hashum, K., Bashir, M., Mubeen, N., Younus, S., Mehmood, S., Haq, F., & Haroon, M. (2024). Advancing sustainable biofuel production from agricultural residues: A comprehensive mini-review. *Sustainable Chemical Engineering*, 115–128. https://doi.org/10.37256/sce.5120243763.

Omer, A. M. (2008). Energy, environment and sustainable development. *Renewable and Sustainable Energy Reviews, 12*(9), 2265–2300. https://doi.org/10.1016/j.rser.2007.05.001.

Osman, A. I., Chen, L., Yang, M., Msigwa, G., Farghali, M., Fawzy, S., Rooney, D. W., & Yap, P.-S. (2023). Cost, environmental impact, and resilience of renewable energy under a changing climate: A review. *Environmental Chemistry Letters, 21*(2), 741–764. https://doi.org/10.1007/s10311-022-01532-8.

Pachaiappan, R., Cornejo-Ponce, L., Sagade, A. A., Mani, M., Aroulmoji, V., Rajan, V. F., & Manavalan, K. (2024). A concise review of recent biohydrogen production technologies. *Sustainable Energy Technologies and Assessments, 62*, 103606. https://doi.org/10.1016/j.seta.2024.103606.

Pandey, A., Larroche, C., Gnansounou, E., Khanal, S. K., Dussap, C.-G., & Ricke, S. (2019). *Biomass, biofuels, biochemicals: Biofuels: Alternative feedstocks and conversion processes for the production of liquid and gaseous biofuels*. Academic Press. https://doi.org/10.1016/b978-0-12-816856-1.09989-9.

Peixoto, I., & Temmes, A. (2019). Market organizing in the European Union's biofuels market: Organizing for favouring, acceptability, and future preferences. *Journal of Cleaner Production, 236*, 117476. https://doi.org/10.1016/j.jclepro.2019.06.307.

Petrović-Ranđelović, M., Kocić, N., & Stojanović-Ranđelović, B. (2020). The importance of renewable energy sources for sustainable development. *Economics of Sustainable Development, 4*(2), 15–24. https://doi.org/10.5937/esd2002016p.

Raganati, F., & Procentese, A. (2022). Special issue on "Bioreactor system: Design, modeling and continuous production process". *Processes, 10*, 1936. https://doi.org/10.3390/pr10101936.

Raganati, F., & Procentese, A. (2024). Special issue on "Advances in bioprocess technology". *Processes, 12*, 1154. https://doi.org/10.3390/pr12061154.

Rame, R., Purwanto, P., & Sudarno, S. (2023). Biotechnological approaches in utilizing agro-waste for biofuel production: An extensive review on techniques and challenges. *Bioresource Technology Reports, 24*, 101662. https://doi.org/10.1016/j.biteb.2023.101662.

Renewables, I. (2021). Renewables 2021: Analysis and forecast to 2026. International Energy Agency. https://doi.org/10.1787/6dcd2e15-en.

Rutz, D., & Janssen, R. (2007). *Biofuel technology handbook* (Vol. *95*). WIP Renewable Energies. https://doi.org/10.1081/e-eee2-120053752.

Sankar, S., Manivarma, V. K., & Thampi, A. (2017). Performance and emission analysis of a CI engine fuelled with JOME-JOEE-diesel blends. *Applied Mechanics and Materials, 867*, 261–271. https://doi.org/10.4028/www.scientific.net/amm.867.261.

Scott, S. A., Davey, M. P., Dennis, J. S., Horst, I., Howe, C. J., Lea-Smith, D. J., & Smith, A. G. (2010). Biodiesel from algae: Challenges and prospects. *Current Opinion in Biotechnology, 21*(3), 277–286. https://doi.org/10.1016/j.copbio.2010.03.005.

Shahid, M. K., Batool, A., Kashif, A., Nawaz, M. H., Aslam, M., Iqbal, N., & Choi, Y. (2021). Biofuels and biorefineries: Development, application and future perspectives emphasizing the environmental and economic aspects. *Journal of Environmental Management, 297*, 113268. https://doi.org/10.1016/j.jenvman.2021.113268.

Sheikh, Z. U. D., Bajar, S., Devi, A., Rose, P. K., Suhag, M., Yadav, A., Yadav, D. K., Deswal, T., Kaur, J., Kothari, R., Pathania, D., Rani, N., & Singh, A. (2023). Nanotechnology based technological development in biofuel production: Current status and future prospects. *Enzyme and Microbial Technology, 171*, 110304. https://doi.org/10.1016/j.enzmictec.2023.110304.

Shouse, K. C., Lattanzio, R. K., Leggett, J. A., Procita, K., Ramseur, J. L., Clark, C. E., Genevieve K. Croft, Rena S. Miller, Westphal, M. I. (2021). *US Climate Change Policy. Congressional Research Service (CRS) reports and issue briefs*. https://doi.org/10.1080/14693062.2013.777604.

Siddiki, S. Y. A., Mofijur, M., Kumar, P. S., Ahmed, S. F., Inayat, A., Kusumo, F., Badruddin, I. A., Yunus Khan, T. M., Nghiem, L. D., Ong, H. C., & Mahlia, T. M. I. (2022). Microalgae biomass as a sustainable source for biofuel, biochemical and biobased value-added products: An integrated biorefinery concept. *Fuel, 307*, 121782. https://doi.org/10.1016/j.fuel.2021.121782.

Vasudevan, P. T., & Briggs, M. (2008). Biodiesel production—Current state of the art and challenges. *Journal of Industrial Microbiology and Biotechnology, 35*(5), 421. https://doi.org/10.1007/s10295-008-0312-2.

Wei, J., & Li, Y. (2023). CRISPR-based gene editing technology and its application in microbial engineering. *Engineering Microbiology*, *3*(4), 100101. https://doi.org/10.1016/j.engmic.2023.100101.

Yu, X., Wu, Z., Wang, Q., Sang, X., & Zhou, D. (2020). Exploring the investment strategy of power enterprises under the nationwide carbon emissions trading mechanism: A scenario-based system dynamics approach. *Energy Policy*, *140*, 111409. https://doi.org/10.1016/j.enpol.2020.111409.

Zetsche, B., Heidenreich, M., Mohanraju, P., Fedorova, I., Kneppers, J., DeGennaro, E. M., DeGennaro, E. M., Winblad, N., Choudhury, S. R., Abudayyeh, O. O., Gootenberg, J. S., Wu, W. Y., Scott, D. A., Severinov, K., van der Oost, J., & Zhang, F. (2017). Multiplex gene editing by CRISPR–Cpf1 using a single crRNA array. *Nature Biotechnology*, *35*(1), 31–34. https://doi.org/10.1038/nbt.3737.

Zhao, Z.-T., Ding, J., Wang, B.-Y., Bao, M.-Y., Liu, B.-F., Pang, J.-W., Ren, N.-Q., & Yang, S.-S. (2024). Advances in the biomass valorization in dark fermentation systems: A sustainable approach for biohydrogen production. *Chemical Engineering Journal*, *481*, 148444. https://doi.org/10.1016/j.cej.2023.148444.

5

Consolidated Bioprocess Technology: Principles and Applications

Gursharan Singh[1] *and Shailendra Kumar Arya*[2]

[1] *School of Allied Medical Sciences, Lovely Professional University, Phagwara, India*
[2] *UIET, Biotechnology Branch, Panjab University, Chandigarh, India*

5.1 Introduction

Gradual depletion of energy sources like crude oil and natural gases leads to the discovery of bio-based fuels, especially those that may be synthesized from agro-waste, non-food biomass generally known as lignocellulosic biomass (LCB), such as corn straw, wheat straw, rice straw, peel of vegetables, and fruits. The key components present in LCB are cellulose, hemicelluloses, and lignin. Cellulose is made of glucose, which is the major carbon source that can be consumed by yeasts to produce alcohols, while xylose and other sugars in LCB like mannose are not preferable on a priority basis (Figure 5.1). Lignin is the key polymer that creates an enormous resistance before the extraction of cellulose; lignin provides recalcitrance as well as strength to LCB. In literature, numerous chemical and physical methods are available to remove lignin from LCB, but the best and cost-effective method of lignin removal is yet to be discovered (Singh & Arya, 2019; Singh et al., 2024).

First-generation biofuels are easy to produce through direct exploitation of sugars derived from starch and sugar cane. First-generation biofuels, mainly ethanol, are produced incredibly in the United States, Brazil, and China, but ethically, their competition with food resources that are primarily for human consumption is a matter of debate. On the other hand, second-generation biofuels can be synthesized from agro-based waste, non-food LCB, forest debris, or municipal solid waste. Second-generation biofuels production is frequently reported, but is mainly constrained to the research laboratories or executed at pilot-scale studies only. Second-generation biofuels, especially ethanol, are not prevalent at commercial-scale production due to numerous challenges such as recalcitrance of LCB, crystallinity of cellulose, and the need for cost-intensive infrastructure. Deconstruction of LCB required a number to steps, and at each step, a number of specific enzymes are needed. The LCB required enzymatic hydrolysis, physical/chemical pretreatmentsteps that are also essential but usually cost intensive processes at large scale (Singh & Arya, 2021; Singh et al., 2023). However, reducing the capital and processing costs for the synthesis of biofuels from LCB may become possible in near future. Therefore, consolidated bioprocessing (CBP) is a novel strategy where all the phases needed for alcohol production ensue

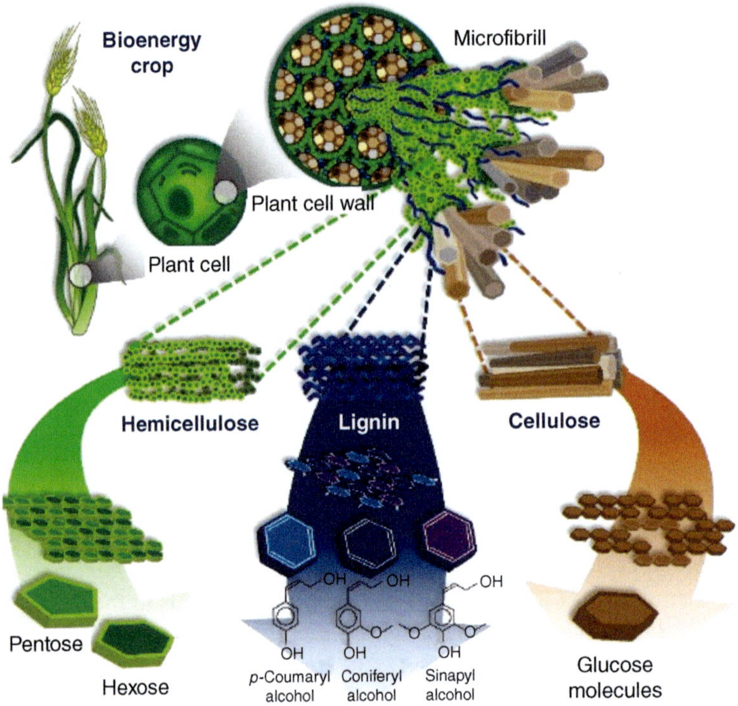

Figure 5.1 LCB structure consists of three main polymers: cellulose, hemicellulose, and lignin. Cellulose is the key polysaccharide, while lignin and hemicelluloses protect the cellulose from external degrading sources (Hernández-Beltrán et al., 2019).

in a single fermenter. A solitary microbe or a consortium of many microorganisms can convert the pretreated LCB to a value-added product like ethanol without the addition of cellulose-breaking enzymes, which would be presented as an inventive breakthrough for low-cost biomass processing (Li et al., 2024). Although the second-generation biofuels are not cost-effective in the current scenario, the technology still has a lot of potential because LCB is readily available as a raw material. That day may not be far when second-generation or cellulose-based ethanol will become cost-competitive with starch-based first-generation biofuels. There is a need for sincere research efforts and the preparation of strategies for the improvement of second-generation biofuel production at lower cost.

5.2 Important Aspects of CBP

5.2.1 CBP, A Single Step Technology and Economical Process

There are different and important steps involved in CBP: (1) production of enzymes such as cellulases and hemicellulases by a pure culture or a consortium of diverse microbes; (2, 3) hydrolysis of the cellulose and hemicelluloses (simultaneous saccharification and co-fermentation [SSCF]). The main sugar monomers such as glucose (hexose sugar) and pentose (xylose) are produced after the proper breakdown of the LCB. Most of the lignin

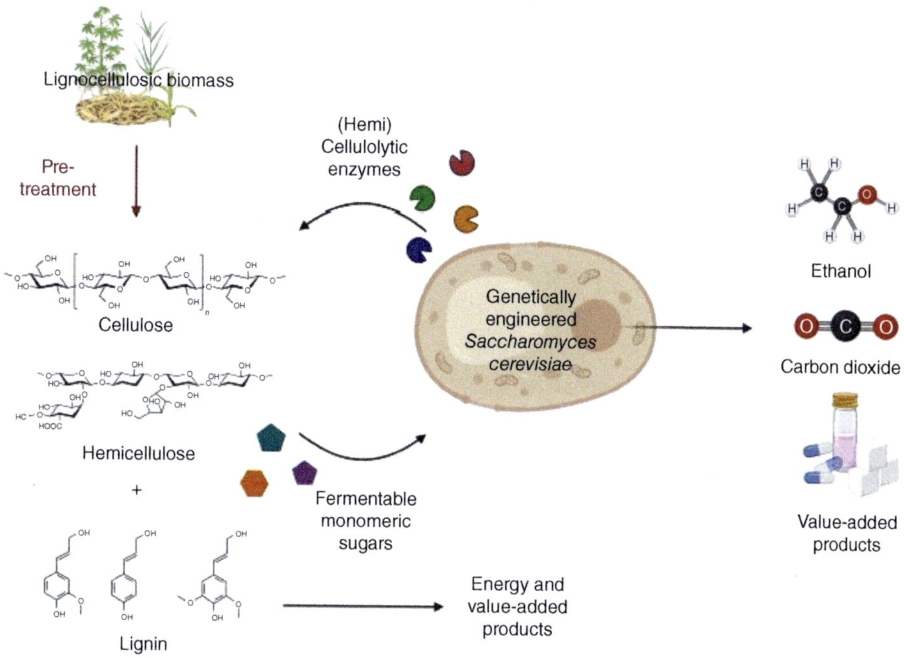

Figure 5.2 The basic principles of CBP, showing simultaneous breakdown of pretreated LCB by cellulase and hemicellulase degrading enzymes secreted by engineered yeast strain, along with SSCF and the production of alcohols and other value-added products (*Source:* Fortuin et al. (2025) / Springer Nature / CC BY 4.0).

is removed during the first step of pretreatment of LCB, such as heating at high temperature in the presence of extreme alkaline/acid conditions. The SSCF leads to the synthesis of ethanol and other bio-products in a single fermenter, which is the prime purpose of CBP (Figure 5.2). The concept and success of CBP depend on the proper functioning of the single or multiple microbes used in this process. Although in case of CBP, mostly genetically-modified yeasts are preferable, because in yeasts cellulases and hemicellulases producing genescould be inserted easily do the SSCF.

The cell surface engineering is indispensable because yeast can perform ethanol production from sugars but enzymes used for saccharification are not produced by yeast. The concept of cell surface engineering is a vital approach for the growth of such yeasts that may be successfully used in CBP (Sakuragi et al., 2011).

5.2.2 Importance of the Baker's Yeast (*Saccharomyces cerevisiae*) for CBP

Yeasts are the most successful eukaryotic alcohol-producing unicellular microbes that are exploited a lot at the industrial scale for the production of ethanol from hexose sugars. However, there is rare to rarest if any yeast especially *S. cerevisiae* that can break down the cellulose to glucose and simultaneously ferment to the ethanol. *S. cerevisiae* is a well-known, robust microbe that can easily be engineered to secrete enzymes required for the breakdown

Table 5.1 Some Desirable Characteristics of Yeast Strains Before Using in CBP.

1	Able to ferment hexose and pentose sugars
2	Able to tolerate high ethanol concentration
3	Able to tolerate the presence of fermentation inhibitors that originate from the pretreatment of LCB
4	Should be able to tolerate harsh industrial environmental conditions
5	Able to express the heterologous gene products like cellulases and xylanases
6	Able to concurrently ferment sugars
7	Should possess GRAS (Generally Recognized as Safe) status
8	Could be reusable like other industrial yeasts
9	Need minimal nutrient supplementation
10	Should be easily genetically modified

of cellulose and hemicelluloses. There are some essential traits of *S. cerevisiae* that are needed to be used in CBP (Table 5.1).

5.2.3 Strain Development/Improvement for the Successful Operation of CBP

The conversion of LCB to ethanol and other bioproducts is the best option and may server as a good substitute for petroleum-based fuels. As discussed earlier, CBP is considered more cost-effective than the conventional multistep conversion steps. Recent research has already concluded that not a single wild strain of microbe is sufficient to carry out CBP, because it is a cumbersome process that needs hydrolytic enzymes, and at the same time, saccharification and fermentation of sugars to biofuels is required. Therefore, researchers have developed many yeast strains over time with genetically modified diverse genes for the breakdown of cellulose and hemicellulose and the simultaneous fermentation of sugars. In this context, *S. cerevisiae* considered as the most robust microbe that can adapt the several environmental conditions due to its flexibility to the stress. The major challenges for the proper bio-conversion of LCB by engineered/modified *S. cerevisiae* include the heterologous expression of cellulose-degrading enzymes, the co-fermentation of glucose and pentose sugars, and resistance against fermentation inhibitors (Oh & Jin, 2020).

Designing or improving the microbial strain used in CBP requires sincere efforts of molecular engineering to manipulate the bio-machinery of yeast cells. CBP is a complex process in which breaking down the recalcitrance of LCB by use of hydrolytic enzymes, moreover SSCF is also involves. Further, the release of important enzymes by engineered yeast cells on their surface is also a metabolic challenge due to the limitation of endoplasmic reticulum in transport and folding of proteins. As a result of increased CBP burden on yeast cells and less production and transport of enzymes during heterologous expression of proteins are main concerns of yeasts used in the CBP. To overcome from all the hindrance of yeast cells there are lot of efforts has been done in the recent past (Figure 5.3). Emerging technologies such as OMICS, advances in computational biology, CRISPR-Cas9, and

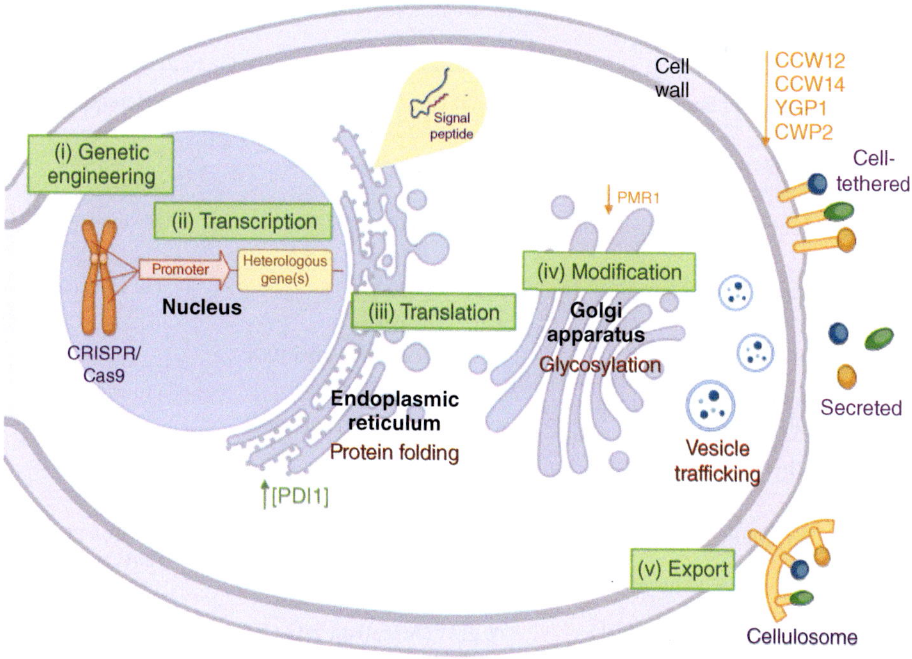

Figure 5.3 Strategies required to develop yeast strains for successful CBP of LCB. Several features of enzyme production and secretion pathways that should be manipulated through improved bioprocess along with the many gene targets that engineered for successful heterologous protein production. Important steps involved in the improvement of yeast strains are transcription, translation, modification and functioning of proteins, and export/secretion of LCB-degrading biocatalysts (*Source:* Minnaar et al. (2024) / with permission of Elsevier).

machine learning may also help the molecular biologists in future to design the more robust yeast strains that can useful for CBP at commercial scale. Inokuma et al. (2014) established the unique gene cassettes for the effective yeast cell-surface display of both endo and exo type of cellulolose degrading enzymes. This heterologous enzyme display on yeast cell surface was made possible through the applicability of glycosylphosphatidylinositol (GPI) anchoring system. The transformed yeast cells were able to hydrolyze rice straw and produce ethanol in a cost-effective CBP process. The *S. cerevisiae* YPH499 strain expressed the following enzymes such as xylose reductase, β-glucosidase, xylosidase, and xylanase as cell-surface-displayed biocatalysts. During CBP experiments, these enzymes successfully hydrolyzed rice straw and produced 5.8–37 gL^{-1} xylitol, a pharmaceutically important sweetener (Guirimand et al., 2016). Engineered *S. cerevisiae* with insertion of cellulase and xylanase genes as a result the breakdown of wheat straw was easy during CBP. Further production of ethanol was noticed that was 8.1 gL^{-1} as a 57.8 folds higher than the wild strain of *S. cerevisiae* in the process of CBP (Yang et al., 2016).

Different cellulases from various microbes were evaluated to increase the level of cellulase excretion in *S. cerevisiae* cells by screening through a translational fusion partner (TFP), which act as both a secretion signal and a fusion partner. Each recombinant yeast had the capacity to discharge up to 0.6–2.0 gL^{-1} of cellulases into the fermentation medium. Four

Table 5.2 Engineered *Saccharomyces cerevisiae* Strains Used in CBP of Different LCB Without the Inclusion of Exogenous Enzymes.

S. cerevisiae-Engineered Strains	Ethanol Titer/Yield	References
Expression of enzymes (EG, CBH, and BGL) through delta-integration	2.9 gL^{-1} (PASC) after 40 h	Fujita et al. (2004)
Expression of enzymes (EG, CBH, and BGL)	1.8 gL^{-1} (PASC) after 84 h	Mo et al. (2015)
Cell-attached expression of enzymes (EG, CBH1, CBH2, and BGL)	6.5 gL^{-1} (PASC) 1.4 gL^{-1} (Avicel) 1.3 gL^{-1} (rice straw) after 96 h	Liu et al. (2016)
Cocktail delta-integration to display EG, CBH1, CBH2, and BGL	2.9 gL^{-1} (Avicel) 0.8 gL^{-1} (rice straw) after 96 h	Liu et al. (2017)
Co-culture of (hemi)-cellulolytic Y5 strains	2.11 gL^{-1} (PASC) 1.6 gL^{-1} (corn stover) after 96 h	Chen et al. (2018)
Secretion of EG and BGL via natural isolate YI13	4.05 gL^{-1} (corn cob) after 168 h	Davison et al. (2019)
K. marxianus Demonstration of cellulosome (BGL, CBH, EG, LPMO, and CDH)	8.61 gL^{-1}(PASC) after 120 h 3.09 gL^{-1} (Avicel) after 144 h	Anandharaj et al. (2020)
CBH1, CBH2, and EG with cell-tethered BGL in natural *S. cerevisiae*	4–4.5 gL^{-1} (Avicel) after 120 h	Minnaar and Den Haan (2023)
Mini-cellulosomes display (CBH1, CBH2, EG, and BGL)	4.92 gL^{-1} (PASC) after 168 h	Ma et al. (2024)

PASC, phosphoric acid swollen cellulose; BGLs, β-glucosidases; CBH, cellobiohydrolase; EG, endo-glucanase; LPMO, lytic polysaccharide monooxygenase.
Source: Adapted from Fortuin et al. (2025).

yeast strains acted synergistically and hydrolyzed Avicel and other cellulosic biomasses. Further, co-fermentation of these yeast strains produced approximately 14 gL^{-1} ethanol from pretreated rice straw (Lee et al., 2017).

Modification of natural yeasts using CRISPR/Cas9 tools and increased expression of heterologous cellulases helps ethanol production in the presence of crystalline cellulose. It was observed that CBP was able to synthesize 4–4.5 gL^{-1} of ethanol in less oxygen supply (Minnaar & den Haan, 2023). Table 5.2 shows the CBP reports where there is no addition of LCB degrading enzyme from outside while cellulases produced by engineered yeasts and simultaneously synthesis of alcohols.

5.2.4 CBP in the Presence of More Than One Microbe or Consortium

The multi-species biofilm membrane (MBM) reactor has been used for CBP system, utilizing *Trichoderma reesei* for the secretion of cellulases and other enzymes. The MBM reactor facilitated both aerobic and anaerobic conditions for the production of ethanol by *S. cerevisiae*. Moreover, slurry made from dilute acid–treated wheat straw and without detoxification of this slurry was used for alcohol production up to 10 gL^{-1}. The authors claimed that may the ethanol production will be increased after the complete optimization of process parameters (Brethauer & Studer, 2014). Co-cultivation of *Trichoderma reesei* (aerobic fungus able to produce the cellulases) with facultative anaerobic lactic acid–producing prokaryotic strain *Lactobacillus pentosus* was successfully observed. This consortium produced 34 gL^{-1} lactic acid from 5% microcrystalline cellulose. Beechwood used in this study was pretreated and heated, resulting in the release of fermentation inhibitors such as acetic acid (aa) and carbon catabolite repression. In this case of CBP of synergistic action of bacteria and fungus hexoses as well as pentoses were easily consumed and leads to the degradation of aa (Shahab et al., 2018). The seeds of *Brosimum alicastrum*, which contain 61% of starch, were evaluated for the production of ethanol during CBP in the presence of *Trametes hirsuta* (Bm-2). The *T. hirsute* strain proved robust due to its ability to synthesize amylase (193 UmL^{-1}) and resist high concentration of ethanol in the environment. Finally, it was observed that *T. hirsute* was able to produce ethanol from *B. alicastrum* seed flour (10 gL^{-1}) with the yield of 123 $mLkg^{-1}$ flour. After the running of CBP, there was enough left over which contained the high percentage of proteins that can be used as animal feed (Olguin-Maciel et al., 2019). Multispecies biofilm membrane (MBM) reactor was used for the co-culturing of *Trichoderma reesei* as a cellulase-producing partner in aerobic chamber. A rumen microbiome was also cultivated in the same MBM reactor, and it was found that both of these microbes work synergistically and converted cellulose to short-chain fatty acids (7.3 gL^{-1}) when grown in 15% of crystalline cellulose at 30°C. Further, it was also observed that temperature played a major role for the production of short-chain fatty acids, and at lower temperature, fewer impurities were accumulated (Xiros et al., 2019). The cell-surface display of hemicellulolytic enzymes on an industrial *S. cerevisiae* engineered yeast strain enabled the release of hemicelluloses, mainly xylose, from corn cob. Simultaneous saccharification and fermentation resulted in the production of ethanol up to 11 gL^{-1}. The *S. cerevisiae* strain used in this study acted as a whole-cell biocatalyst, with robust traits as resistant to fermentation inhibitors and high temperature (Cunha et al., 2020). CBP of xylan was observed by hemicellulose-degrading, xylanase-producing bacteria *Thermoanaerobacterium thermosaccharolyticum* and succinic acid–producing prokaryote *Actinobacillus succinogenes.* About 32 gL^{-1} succinic acid (yield 0.39 gg^{-1}) was synthesized at the end of the process. On the other hand, 12 gL^{-1} succinic acid was produced from 80 gL^{-1} of corn cob (Lu et al., 2020). *Trichoderma reesei*, a cellulase-producing fungi, and *Ustilago maydis*, an itaconic acid–producing yeast were grown in the presence of cellulose in CBP; both acted synergistically and produced itaconic acid up to 34 gL^{-1} (Schlembach et al., 2020).

5.2.5 CBP at High Temperatures

The efficiency of the bioprocess system at higher temperature is considered advantageous due to several factors, such as less chances of contamination, less viscosity of the reactants, and final products then easy to down streaming. Further, the reaction system is fast when compared to mesophilic environments. Least pretreated or free from pretreatment process the LCB from populus tree was deconstructed by the *Clostridium thermocellum* in anaerobic conditions without addition of exogenous enzymes at higher temperature (Dumitrache et al., 2016). The *S. cerevisiae* MNII strain was engineered by the insertion of an artificial zinc finger protein that made the yeast able to function at 42°C. This modification of the microbe facilitates the enhanced breakdown of the swollen cellulose by 45%, which leads to the increased production of alcohol. Moreover, the fermentation time for Jerusalem artichoke stalk was reduced to 12 h at 42°C, and also resulted in improvements in ethanol synthesis during CBP (Khatun et al., 2017). The *Clostridium* sp. DBT was isolated from the Himalayan hot spring and is able to degrade a variety of cellulose- and hemicellulose-rich LCBs, converting them into ethanol, acetate, and lactate without the addition of exogenous enzymes during fermentation at a wide range of temperatures (55–65°C). It was noticed that the cellulose degradation potential was up to 10 gL^{-1} cellulose, and Avicel utilization reached 94% within 96 h of fermentation (Singh et al., 2017). The *Clostridium thermocellum* DSM showed the potential for ethanol production at a yield of 0.6 gL^{-1} in the presence of 0.5% cellobiose at 60°C after 72 h of fermentation. Moreover, the same bacteria generated ethanol at 0.68 gL^{-1} in the presence of alkali-treated sugarcane bagasse (SCB) within 120 h of CBP. The production of ethanol was increased to 0.82 gL^{-1} when fermentation was performed in a 3 L fermenter (Liu et al., 2020). After the acclimatization of the microbial community, it was able to produce butanol up to 13.8 gL^{-1}, which is 9 times higher than the parental microbial community. Further, it was also noticed that the adapted consortium also synthesized the fatty acid caproate (7.5 gL^{-1}). During the microbial community analysis, it was found that *Clostridium sensu stricto* and *Caproiciproducens* were the dominating bacteria (González-Tenorio et al., 2023).

5.3 Advantages and Disadvantages of CBP Technology

Although numerous reports are available on CBP, as well as several discussed in this chapter, CBP technology is still far from commercialization in biorefineries. Many factors

Table 5.3 Merits and Some Unresolved Drawbacks of CBP.

	Merits of CBP Technology
1	Entire process executed in a single bioreactor.
2	Breakdown of the LCB to sugars (hexoses and pentoses) occurs in the same vessel in which alcohol is produced.
3	Enzymes from outside are not required, so CBP is considered a cost-effective process.
4	Risk of contamination of CBP is lower because of less presence of carbohydrates.
5	Less energy is required from the initiation of CBP to end of the process.
	Drawbacks of CBP
1	Only engineered strains of microbes are preferred for the success of CBP.
2	Use of engineered microbes is restricted in many countries worldwide.
3	CBP requires too much time to produce alcohol/bioproducts.
4	In general, yields are lower.

stand against CBP, such as long incubation time, in general 3–12 days. Microbes required for CBP are genetically modified, so it is not fully ensured yet about their mechanism if they will contaminate the environment by chance. Parallel research is also ongoing on CBP by exploitation of wild strains of microbes, but their efficiency is not very high. Further, the use of genetically modified microbes consists of many genes; as a result, sometimes, the expression of such proteins also that may not require for the process. In Table 5.3, some important merits and demerits of CBP are listed.

5.4 Conclusions

Recent literature suggests CBP may be a highly useful concept for future biorefineries, but in existing situation, this technology is constrained to laboratories only, without effective commercialization. The backbone of successful CBP is a well-optimized engineered microbial strain, mainly yeast; however, yeasts inherently lack the ability to produce cellulases and other plant cell wall–degrading enzymes. After a lot of research on genetic engineering of yeast strains, still there is no such a strain in hand that can act as robust for the production of biofuels from LCB in the presence of fermentation inhibitors and without the addition of exogenous enzymes. The cell-tethering of cellulases seems to be a hopeful opportunity for further investigation; rational engineering of the yeast cell wall itself has been shown to further progress toward cell-tethering of cellulases, making it an attractive choice for CBP (Fortuin et al., 2025). The simultaneous breakdown of LCB, saccharification of cellulose, and fermentation to ethanol and other bioproducts is highly cost-effective and a single-step approach to achieve the industrial sustainability, which may accelerate the commercialization of LCB biofuels in the coming future.

References

Anandharaj, M., Lin, Y. J., Rani, R. P., Nadendla, E. K., Ho, M. C., Huang, C. C., Chang, J. F., Chang, J. J., & Li, W. H. (2020). Constructing a yeast to express the largest cellulosome complex on the cell surface. *Proceedings of the National Academy of Sciences*, *117*(5), 2385–2394.

Brethauer, S., & Studer, M. H. (2014). Consolidated bioprocessing of lignocellulose by a microbial consortium. *Energy & Environmental Science*, *7*(4), 1446–1453.

Chen, L., Du, J. L., Zhan, Y. J., Li, J. A., Zuo, R. R., & Tian, S. (2018). Consolidated bioprocessing for cellulosic ethanol conversion by cellulase-xylanase cell-surfaced yeast consortium. *Preparative Biochemistry & Biotechnology*, *48*(7), 653–661.

Cunha, J. T., Romaní, A., Inokuma, K., Johansson, B., Hasunuma, T., Kondo, A., & Domingues, L. (2020). Consolidated bioprocessing of corn cob-derived hemicellulose: Engineered industrial *Saccharomyces cerevisiae* as efficient whole cell biocatalysts. *Biotechnology for Biofuels*, *13*, 1–15.

Davison, S. A., Keller, N. T., van Zyl, W. H., & den Haan, R. (2019). Improved cellulase expression in diploid yeast strains enhanced consolidated bioprocessing of pretreated corn residues. *Enzyme and Microbial Technology*, *131*, 109382.

Dumitrache, A., Akinosho, H., Rodriguez, M., Meng, X., Yoo, C. G., Natzke, J., Engle, N. L., Sykes, R. W., Tschaplinski, T. J., Muchero, W., Ragauskas, A. J., Davison, B. H., & Brown, S. D. (2016). Consolidated bioprocessing of Populus using *Clostridium (Ruminiclostridium) thermocellum*: A case study on the impact of lignin composition and structure. *Biotechnology for Biofuels*, *9*, 1–14.

Fortuin, J., Hoffmeester, L. J., Minnaar, L. S., & den Haan, R. (2025). Advancing cellulose utilization and engineering consolidated bioprocessing yeasts: Current state and perspectives. *Applied Microbiology and Biotechnology*, *109*(1), 43.

Fujita, Y., Ito, J., Ueda, M., Fukuda, H., & Kondo, A. (2004). Synergistic saccharification, and direct fermentation to ethanol, of amorphous cellulose by use of an engineered yeast strain codisplaying three types of cellulolytic enzyme. *Applied and Environmental Microbiology*, *70*(2), 1207–1212.

González-Tenorio, D., Dudek, K., & Valdez-Vazquez, I. (2023). Butanol and caproate production by consolidated bioprocessing after adaptive evolution of a fermentative microbial community. *Biofuels, Bioproducts and Biorefining*, *17*(5), 1221–1235.

Guirimand, G., Sasaki, K., Inokuma, K., Bamba, T., Hasunuma, T., & Kondo, A. (2016). Cell surface engineering of *Saccharomyces cerevisiae* combined with membrane separation technology for xylitol production from rice straw hydrolysate. *Applied Microbiology and Biotechnology*, *100*, 3477–3487.

Hernández-Beltrán, J. U., Lira, H. D., Omar, I., Cruz-Santos, M. M., Saucedo-Luevanos, A., HernándezTerán, F., & Balagurusamy, N. (2019) Insight into pretreatment methods of lignocellulosic biomassto increase biogas yield: current state, challenges, and opportunities. *Appl. Sci.* *9*(18): 3721.

Inokuma, K., Hasunuma, T., & Kondo, A. (2014). Efficient yeast cell-surface display of exo- and endo-cellulase using the SED1 anchoring region and its original promoter. *Biotechnology for Biofuels*, *7*, 1–11.

Khatun, M. M., Yu, X., Kondo, A., Bai, F., & Zhao, X. (2017). Improved ethanol production at high temperature by consolidated bioprocessing using *Saccharomyces cerevisiae* strain engineered with artificial zinc finger protein. *Bioresource Technology*, *245*, 1447–1454.

Lee, C. R., Sung, B. H., Lim, K. M., Kim, M. J., Sohn, M. J., Bae, J. H., & Sohn, J. H. (2017). Co-fermentation using recombinant *Saccharomyces cerevisiae* yeast strains hyper-secreting different cellulases for the production of cellulosic bioethanol. *Scientific Reports*, *7*(1), 4428.

Li, Z., Waghmare, P. R., Dijkhuizen, L., Meng, X., & Liu, W. (2024). Research advances on the consolidated bioprocessing of lignocellulosic biomass. *Engineering Microbiology*, *4*(2), 100139.

Liu, Y., Xie, X., Liu, W., Xu, H., & Cao, Y. (2020). Consolidated bioprocess for bioethanol production from lignocellulosic biomass using *Clostridium thermocellum* DSM 1237. *BioResources*, *15*(4), 8355.

Liu, Z., Ho, S. H., Sasaki, K., Den Haan, R., Inokuma, K., Ogino, C., Van Zyl, W. H., Hasunuma, T., & Kondo, A. (2016). Engineering of a novel cellulose-adherent cellulolytic *Saccharomyces cerevisiae* for cellulosic biofuel production. *Scientific Reports*, *6*(1), 24550.

Liu, Z., Inokuma, K., Ho, S. H., den Haan, R., van Zyl, W. H., Hasunuma, T., & Kondo, A. (2017). Improvement of ethanol production from crystalline cellulose via optimizing cellulase ratios in cellulolytic *Saccharomyces cerevisiae*. *Biotechnology and Bioengineering*, *114*(6), 1201–1207.

Lu, J., Lv, Y., Jiang, Y., Wu, M., Xu, B., Zhang, W., Zhou, J., Dong, W., Xin, F., & Jiang, M. (2020). Consolidated bioprocessing of hemicellulose-enriched lignocellulose to succinic acid through a microbial cocultivation system. *ACS Sustainable Chemistry & Engineering*, *8*(24), 9035–9045.

Ma, X. Y., Coleman, B., Prabhu, P., Yang, M., & Wen, F. (2024). Engineering compositionally uniform yeast whole-cell biocatalysts with maximized surface enzyme density for cellulosic biofuel production. *ACS Synthetic Biology*, *13*(4), 1225–1236.

Minnaar, L., & den Haan, R. (2023). Engineering natural isolates of *Saccharomyces cerevisiae* for consolidated bioprocessing of cellulosic feedstocks. *Applied Microbiology and Biotechnology*, *107*(22), 7013–7028.

Minnaar, L. S., Kruger, F., Fortuin, J., Hoffmeester, L. J., & den Haan, R. (2024). Engineering *Saccharomyces cerevisiae* for application in integrated bioprocessing biorefineries. *Current Opinion in Biotechnology*, *85*, 103030.

Mo, C., Chen, N., Lv, T., Du, J., & Tian, S. (2015). Direct ethanol production from steam-exploded corn stover using a synthetic diploid cellulase-displaying yeast consortium. *BioResources*, *10*(3), 4460–4472.

Oh, E. J., & Jin, Y. S. (2020). Engineering of *Saccharomyces cerevisiae* for efficient fermentation of cellulose. *FEMS Yeast Research*, *20*(1), foz089.

Olguin-Maciel, E., Larqué-Saavedra, A., Lappe-Oliveras, P. E., Barahona-Pérez, L. F., Alzate-Gaviria, L., Chablé-Villacis, R., & Tapia-Tussell, R. (2019). Consolidated bioprocess for bioethanol production from raw flour of *Brosimum alicastrum* seeds using the native strain of *Trametes hirsuta* bm-2. *Microorganisms*, *7*(11), 483.

Sakuragi, H., Kuroda, K., & Ueda, M. (2011). Molecular breeding of advanced microorganisms for biofuel production. *BioMed Research International*, *2011*(1), 416931.

Schlembach, I., Tehrani, H. H., Blank, L. M., Büchs, J., Wierckx, N., Regestein, L., & Rosenbaum, M. A. (2020). Consolidated bioprocessing of cellulose to itaconic acid by a co-culture of *Trichoderma reesei* and *Ustilago maydis*. *Biotechnology for Biofuels*, *13*, 1–18.

Shahab, R. L., Luterbacher, J. S., Brethauer, S., & Studer, M. H. (2018). Consolidated bioprocessing of lignocellulosic biomass to lactic acid by a synthetic fungal-bacterial consortium. *Biotechnology and Bioengineering*, *115*(5), 1207–1215.

Singh, G., & Arya, S. K. (2019). Utility of laccase in pulp and paper industry: A progressive step towards the green technology. *International Journal of Biological Macromolecules*, *134*, 1070–1084.

Singh, G., & Arya, S. K. (2021). A review on management of rice straw by use of cleaner technologies: Abundant opportunities and expectations for Indian farming. *Journal of Cleaner Production*, *291*, 125278.

Singh, G., Kumar, S., Afreen, S., Bhalla, A., Khurana, J., Chandel, S., Aggarwal, A., & Arya, S. K. (2023). Laccase mediated delignification of wasted and non-food agricultural biomass: Recent developments and challenges. *International Journal of Biological Macromolecules*, *235*, 123840.

Singh, G., Sahu, S., Bharti, S., & Arya, S. K. (2024). Significance of enzymes for the recycling of wasted non-food biomass to value added products: A sustainable stewardship towards the cleaner environment. *Process Safety and Environment Protection*, *190*, 395–412.

Singh, N., Mathur, A. S., Tuli, D. K., Gupta, R. P., Barrow, C. J., & Puri, M. (2017). Cellulosic ethanol production via consolidated bioprocessing by a novel thermophilic anaerobic bacterium isolated from a Himalayan hot spring. *Biotechnology for Biofuels*, *10*, 1–18.

Xiros, C., Shahab, R. L., & Studer, M. H. P. (2019). A cellulolytic fungal biofilm enhances the consolidated bioconversion of cellulose to short chain fatty acids by the rumen microbiome. *Applied Microbiology and Biotechnology*, *103*, 3355–3365.

Yang, P., Zhang, H., & Jiang, S. (2016). Construction of recombinant *sestc Saccharomyces cerevisiae* for consolidated bioprocessing, cellulase characterization, and ethanol production by in situ fermentation. *3 Biotech*, *6*(2), 192.

6

Microbial Consortia in Consolidated Bioprocess Technology

Tehreem Maryam[1], Maryam Kainat[1], Maryam Maqsood[1], Yuliya Kulikova[2], Olga Babich[2], Afsana Huseynova Anvar[3], Hafiz Abdullah Shakir[4], Muhammad Khan[4], Titin Haryati[5], Marcelo Franco[6], and Muhammad Irfan[1]

[1]*Department of Biotechnology, University of Sargodha, Sargodha, Pakistan*
[2]*Research and Education Center "Industrial Biotechnologies", Immanuel Kant Baltic Federal University, Kaliningrad, Russia*
[3]*Department of Life Science, Khazar University, Baku, Azerbaijan*
[4]*Institute of Zoology, University of the Punjab, Lahore, Pakistan*
[5]*Research Center for Applied Microbiology, National Research and Innovation Agency, Bogor, West Java, Indonesia*
[6]*Department of Exact Science, State University of Santa Cruz, Ilheus, Brazil*

6.1 Introduction

World population is increasing significantly, and growing global energy demands have resulted in the exploitation of fossil fuel reservoirs. Continuous consumption of these fossil fuels is also a major cause of contamination in surrounding environment. However, the use of alternate energy sources can be an eco-friendly approach to tackle these challenges. Among the renewable energy sources, biofuels can serve as a better alternative. They have significantly reduced the reliance on non-renewable fossil fuels (Fatma et al., 2018). Bioprocessing is a key component of biotechnology that produces various types of products with the help of living organisms and their components (Show, 2021). Lignocellulose is a sustainable and abundant raw material for the production of valuable biomaterials like biofuels (Ojo, 2023), polymers, and bio-based chemicals (Fatma et al., 2018). Moreover, it does not interfere with food supply intended for human consumption (Deshavath et al., 2019).

Consolidated bioprocessing (CBP) is a next-generation bioconversion strategy. CBP enables the direct, one-step conversion of biomass into value-added compounds directly within a single bioreactor system (Rani Singhania et al., 2022). The conventional biorefinery process for lignocellulosic degradation and conversion occurs in three distinct stages: biomass pretreatment, enzymatic saccharification for sugar release, and microbial fermentation to produce bio-based products. In addition to this, CBP has a few limitations that make it a challenging task on large-scale operation. The high cost of cellulase enzyme synthesis and technical processing procedures are few obstacles in CBP. Moreover, the yield of fermentation strains is also not efficient. CBP technology endorses integrated operations like enzymatic hydrolysis, production of enzymes, and bioconversions using fermentation in a single bioreactor. This technology employs a pure culture of microorganisms or a mixed culture of microbes (Li et al., 2024).

CBP has an upper hand over standard bioprocessing techniques. Benefits include cost-effectiveness and minimized energy consumption (Rani Singhania et al., 2022). On account of lower production costs, CBP is economically feasible to be applied on industrial scale for lignocellulosic biofuel production (Olguin-Maciel et al., 2020). CBP has a few limitations regarding efficient biotransformations (Rani Singhania et al., 2022).

The prominent shortcomings of biotransformation are identifying microorganisms that are able to hydrolyze lignocellulose and produce valuable by-products (Li et al., 2024). Additionally, fabricating a single strain to make it capable of all these functions can result in physiological stress and instability (Chaves et al., 2019). Biotransformation of lignocellulosic material into useful bioproducts involves complicated metabolic pathways. These processes need varied environmental conditions (Cunha et al., 2019). On the other hand, it is not possible for a pure culture (single strain) to handle all optimal conditions at a time (Jawed et al., 2019). Consequently, single-microbe (pure culture) CBP systems are not compatible for industrial or large scale processes due to their lower efficiency (Olson et al., 2012; Rani Singhania et al., 2022). To overcome these limitations, mixed culture or microbial consortia play an important role in efficient biomass transformation (Shahab et al., 2018). In this technique, various microorganisms are operated synergistically (Zhang & Zhang, 2022). This strategy has been inspired from natural ecosystems (Scholz et al., 2018). Many studies highlight the applications of microbial consortia in consolidated bioprocessing technology.

Microbial consortia have an upper hand over single strains because they minimize the physiological stress and enhance efficiency by being capable of doing several different tasks simultaneously (Xu & Yu, 2021). Microbial consortia include a diverse group of microbial strains that are operated together to do difficult tasks. These microbes can cope with varied environmental conditions and stresses better than monocultures (Bhatia et al., 2018). There is a division of labor among microbes in microbial consortia, which leads to reduced metabolic stress and enhanced yield (Du et al., 2020; Shahab et al., 2020). For example, some species are capable of hydrolyzing lignocellulosic material into simpler fermentable sugars by releasing hydrolytic enzymes, while others are specialized to ferment these sugars into valuable products like biochemicals and biofuels.

Microbial species help each other to accomplish their tasks. They can be utilized in the production of natural products, valuable chemicals, and biofuels (Jiang et al., 2023). For example, different microorganisms can be used for the direct conversion of lignocellulosic biomass into lactic acid (Shahab et al., 2018). Notably, synthetic microbial consortia are designed for CBP. These mixed cultures play a significant role in the biochemical production (Gao et al., 2022). They are developed by the junction of microbiology and biotechnology (Jia et al., 2016). Two or more microorganisms are co-cultivated under certain environmental conditions in order to develop synthetic community systems. They have versatile biotechnological applications in many fields, including conservation of ecosystem (Liang et al., 2022), lignocellulose processing (Hu et al., 2017), human health, and bioremediation (Qian et al., 2019). Additionally, microbial consortia have applications in environmental remediation. They have augmented enzymatic activities and are able to degrade soil pollutants by being operationalized together (Zhang & Zhang, 2022).

Mixed cultures for lignocellulosic degradation are obtained either from natural sources or can be synthetically produced. Scientists have isolated four cellulose-degrading consortia from natural substrates. These consortia are able to degrade lignocellulose (Gladkov et al., 2022). Another well-studied example of natural microbial consortia is lichens. They are composed of fungi and cyanobacteria or algae. Lichens assist in efficient gas exchange and nutrient cycling, which id beneficial for biorefinery operations (Rodrigues Reis et al., 2018).

Engineered microbial co-cultures can be developed by using strategic methodologies. These strategies are based on interactions observed in natural consortia (Jia et al., 2016). These synthetic microbial consortia address challenges in scientific research and manufacturing processes. These engineered co-cultures are able to transform complex substrates into useful bioproducts. For example, fungal-bacterial co-cultures have been used in CBP for the synthesis of fumaric, succinic, and lactic acids from complex sugars like cellulose and hemicellulose (Scholz et al., 2018).

Further than using lignocellulosic feedstock, microbial consortia have also been used for the valorization of food waste and algal biomass to manufacture valuable resources. Raw glycerol, produced as a by-product of biodiesel synthesis from waste cooking oil, is difficult to utilize by pure cultures because of the presence of toxic impurities. In a study, researchers used four microbial consortia obtained from activated sludge to convert raw glycerol into 1,3-propanediol and lactate (Wang et al., 2019).

Microbes are carefully fabricated and optimized for the effective deployment of microbial co-cultures in CBP. However, maintaining stability during operation is crucial (Du et al., 2022). Additionally, strains in the co-culture system interact in various ways. To retain the stability and efficiency of microbes (mixed culture), these interactions are crucial (Che & Men, 2019). Positive interactions like mutualism, cooperation, and commensalism help microbial communities stay stable and productive. In contrast, negative interactions can disrupt the community instead (Che & Men, 2019; Ghosh et al., 2016).

There are tools used to design microbial consortia, but their stability and optimal functioning mainly depend on microbial interactions (Che & Men, 2019). To design a stable microbial consortium, key strategies include spatial or temporal segregation, separated utilization of nutrients, nutrient cross-feeding, and proper division of labor. However, novel tools such as quorum sensing systems and computational frameworks play a crucial role in regulating microbial communities. These technologies help maintain stability and enhance cooperative interactions within artificial microbial consortia (Xu & Yu, 2021). In this chapter, we have discussed the importance of microbial consortia in consolidated bioprocessing technology.

6.2 Microbial Consortia

Microbial consortia are communities of two or more microbial species that are organized together. These communities of microbial species coexist. They interact within a common ecological environment. These groups of microbes are characterized by various interactions. These interactions may be neutral, competitive, or cooperative. This results in a collective behavior and it is more efficient than the abilities of individual strains. Microbial consortia are abundant in nature. They play key roles in various processes. For example, nutrient cycling, organic matter degradation, and host-associated symbiosis (Kong et al., 2018).

It is a favorable approach to develop microbial consortia for various applications. These applications include synthesis of valuable product, lignocellulose utilization, human health care, bioremediation, and sustainable energy. There are various advantages to use microbial consortia. These include stability in the presence of environmental disturbances and reduced metabolic burden (Qian et al., 2019).

Microbial consortia are very helpful in the CBP. This is due to their capability to integrate multiple bioconversion steps in a single process. CBP intends to facilitate biomass conversion by excluding the separate steps for enzymatic hydrolysis and fermentation. At the same time, it is a major challenge to engineer a single microorganism that is able to efficiently degrade complex lignocellulosic biomass and ferment the resulting sugars. Microbial consortia present a solution to this problem. They do so by supporting a division of labor in which different species perform distinct tasks like some may be involved in enzyme production and saccharification and others for fermentation and detoxification (Brethauer & Studer, 2014).

The significant importance of microbial consortia is due to their potential to efficiently carry processes. They enhance product yields and substrate range. Consortia can overcome the limitations of monocultures by promoting synergistic interactions and combining complementary metabolic abilities. Therefore, they improve the viability of industrial CBP applications. Despite the benefits, artificial microbial consortia may confront the problems of low efficiency and instability due to metabolite incompatibility and growth competition. These issues need to be considered before employing microbial consortia in consolidated bioprocessing (Xu & Yu, 2021).

6.2.1 Ecological Basis of Microbial Consortia

The ecological interactions among members of microbial consortia determine their efficiency, stability, and functionality. These interactions can be classified into several types (Figure 6.1):

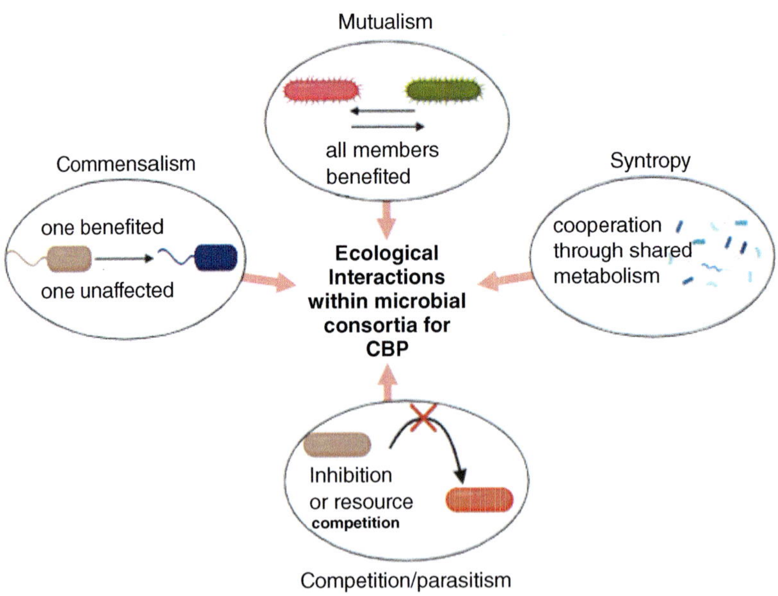

Figure 6.1 Ecological interactions among microbial consortia.

- **Mutualism:** In mutualistic relationships, all members benefit from the interaction. For example, a cellulose-degrading bacterium such as *Clostridium thermocellum* can break down cellulose into glucose and cellobiose. These sugars are then used by another organism like *Thermoanaerobacterium saccharolyticum*, which ferments them into ethanol (Hon et al., 2017). Another organism, *Saccharomyces cerevisiae* has also been utilized in mutualistic consortia. This yeast has a robust ethanol-producing ability. It is able to consume oxygen, thereby creating an anaerobic environment necessary for obligate anaerobes (Tabañag & Tsai, 2018).
- **Commensalism:** In this type of relationship, one organism benefits while the other is neither benefited nor harmed. For instance, in a consortium having *Bacillus subtilis* and *Zymomonas mobilis*, the first one can detoxify inhibitors like furfural during hydrolysis of lignocellulose. So, it indirectly supports the fermentative activity of *Z. mobilis* (Dong et al., 2013).
- **Syntrophy:** Syntrophy involves interdependence of metabolic activities. Here, the survival of one species relies on the metabolic activities of another. This is commonly observed in anaerobic digestion systems (Morris et al., 2013). For example, *Syntrophomonas wolfei* breaks down fatty acids to produce hydrogen. This hydrogen is instantly absorbed by hydrogenotrophic methanogens such as *Methanospirillum hungatei*. This immediate hydrogen consumption operates the energetically adverse reactions forward. So, it allows efficient biogas production (Sieber et al., 2010).
- **Competition and parasitism:** These types of interactions can destabilize consortia. They are usually undesirable in engineered systems. For instance, *Lactobacillus* species may subdue ethanol-producing microbes in some mixed cultures. They do so by consuming sugars quickly and producing lactic acid. This lactic acid can inhibit fermentative processes (García et al., 2019).

Understanding these ecological interactions is essential for developing stable and efficient microbial consortia. The central concept of this approach is the **division of labor**, where different microbial species carry out specific roles. So, they reduce the metabolic burden on individual strains and enhance the overall efficiency of the process.

6.2.2 Types of Microbial Consortia

Based on Origin

- **Natural consortia:** These microbial communities are obtained from natural environmental sources. These natural sources include soil, animal guts, or compost. These types of microbes are basically adapted to break down complex organic matter. They are resistant to environmental changes (Navarrete-Bolanos et al., 2007). Microbial consortia obtained from cow rumen have shown efficient degradation of lignocellulosic biomass. Key members of these consortia include *Fibrobacter succinogenes*, *Ruminococcus flavefaciens*, and *Methanobrevibacter ruminantium*. Among these, the first one is a cellulose degrader, the second is a hemicellulose degrader, and the third is a methanogen that consumes hydrogen and produces methane. Similarly, anaerobic sludge is obtained from biogas reactors. This sludge contains diverse consortia. They are able to convert organic waste into methane and CO_2 (Lazuka et al., 2015).

- **Synthetic consortia:** These are artificially designed and constructed consortia. They comprise certain microbial strains with specific functions. For example, a synthetic consortium of *Clostridium thermocellum* and *Thermoanaerobacter ethanolicus* has been used in CBP systems. The first one helps in cellulose hydrolysis, and the second in ethanol production (Mbaneme-Smith & Chinn, 2015). Another example is the co-culture of *Escherichia coli* and *S. cerevisiae*. The former is engineered for xylose fermentation and the latter for glucose fermentation. This allows simultaneous conversion of lignocellulose to mixed sugar substrates. Synthetic consortia can also include non-native pathways. These pathways are engineered via synthetic biology. For example, *Corynebacterium glutamicum* and *Pseudomonas putida* are used to produce bio-based polymers from lignin-derived compounds (Zhang et al., 2015).

Based on Culture Structure

- **Co-cultures:** Co-cultures include two or more defined microbial species that are grown together. These systems are easy to manage. They allow researchers to study specific interactions among microbes. It is also possible to optimize their metabolic interactions. For example, a co-culture of *Trichoderma reesei* and *S. cerevisiae*. The first one is a known cellulase producer, and the second one can be used for simultaneous saccharification and fermentation. It can convert cellulose to ethanol. Another co-culture system includes *Aspergillus niger* and *Z. mobilis*. It can be used for ethanol production from starch (Ado et al., 2010).
- **Mixed cultures:** These consist of a diverse group of microbial species. They are often undefined cultures. These can be derived from natural habitats. For example, mixed microbial cultures can be obtained from compost and wastewater treatment plants. They can be used for the production of volatile fatty acids from food waste. They offer robustness and diversity. But there are a few limitations as well. For example, it is difficult to control the consistency of products (Perez-Zabaleta et al., 2021).

Based on Interaction and Function

- **Metabolically cooperative consortia:** In this type of consortia, metabolic functions are distributed among different members of the community. It helps to optimize resource utilization and reduces the metabolic burden. One example is the co-culture of *Cellulomonas fimi* and *Clostridium acetobutylicum*. The former hydrolyzes cellulose to glucose and the latter ferments glucose into solvents such as acetone, ethanol, and butanol. Another example includes using *Bacillus subtilis* to produce cellulases and *Enterobacter cloacae* for hydrogen production (Roell et al., 2019).
- **Spatially structured consortia:** These consortia form well-organized physical structures. These structures include biofilms or encapsulated systems. This organization protects sensitive members from environmental stress. It also facilitates substrate channeling between species. For example, alginate beads have been used to immobilize *Bifidobacterium breve* and *Lactobacillus plantarum*. This enables stable co-cultivation and increased lactic acid production. *Shewanella oneidensis* and *Geobacter sulfurreducens* form biofilm-based consortia. They have been explored for improved electron transfer in microbial fuel cells (Said et al., 2020).
- **Temporally phased consortia:** In this type, the composition of community and their metabolic activities change over time. This is in response to the availability of substrate

or changes due to environmental conditions (Carlos et al., 2018). For instance, there is a two-stage fermentation process. First, thermophilic bacteria such as *Thermoanaerobacterium thermosaccharolyticum* can degrade hemicellulose at high temperatures. In the second stage, mesophilic fermentation occurs by *S. cerevisiae* at lower temperatures. This produces ethanol (Herring et al., 2016). Another example involves sequential degradation of lignocellulose. This is done by a fungal consortium. It is initially dominated by *Phanerochaete chrysosporium*, which breaks down lignin. It is then followed by cellulolytic bacteria such as *Cellulomonas uda* for carbohydrate hydrolysis (Peng et al., 2021).

6.3 Consolidated Bioprocessing

CBP technology combines the process of enzyme production, enzymatic breakdown, and microbial fermentation in a single integrated system for biomass conversion (Li et al., 2024). CBP is environmentally and economically benign as it has the capacity to reduce infrastructure and chemicals input for various steps (Rani Singhania et al., 2022; Figure 6.2).

Lignocellulosic materials are cost-effective and may not interfere with food crops for human use. Cellulolytic enzymes are typically needed to break down these materials into simpler sugars. The concept of CBP was introduced to potentially reduce the high cost of cellulase enzyme production for lignocellulosic biomass conversion (Gomes et al., 2021). As CBP is a highly integrated system, where enzyme production occurs simultaneously with the hydrolysis and fermentation of released sugars, without requiring the separate enzyme production step, CBP therefore offers the potential to save costs (Dempfle et al., 2021).

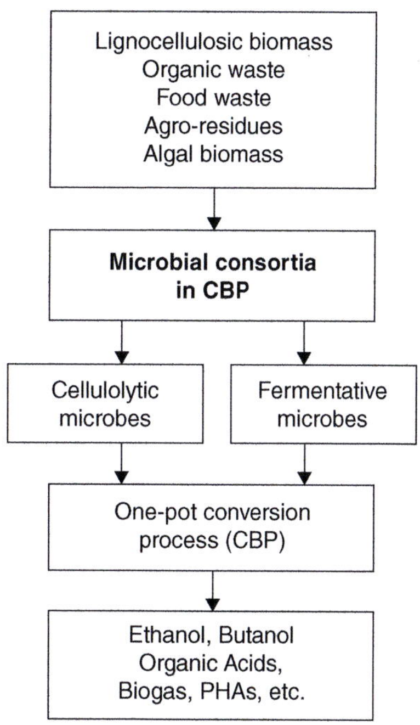

Figure 6.2 One-pot microbial conversion of biomass in CBP.

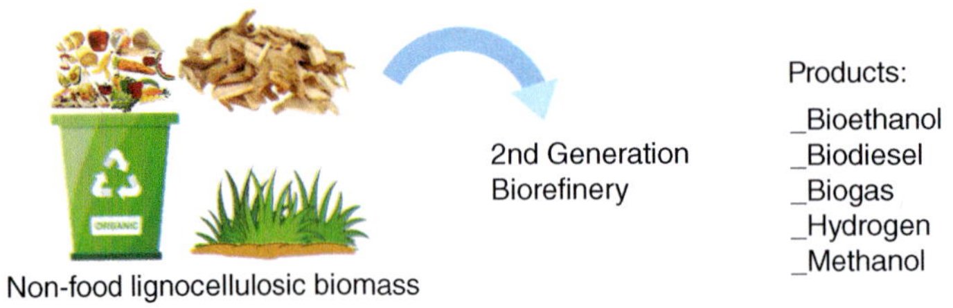

Figure 6.3 Advancing sustainability: non-food biomass in second-generation biorefineries. (*Source:* (a) Menikahnya / Adobe Stock Photos (b) Bert Folsom / Adobe Stock Photos (c) jihane37 / Adobe Stock Photos).

CBP technology utilizes lignocellulosic biomass as its primary feedstock (Li et al., 2024). Lignocellulosic biomass is the most widely available organic material on Earth, which is bio-renewable and can be utilized as a potential resource for the generation of sustainable fuels, chemicals, and valuable by-products with net-zero carbon emissions (Saravanan et al., 2023). Lignocellulose biomass include agricultural residues, forestry, and urban solid wastes (Inyang et al., 2022), and is composed of cellulose, hemicellulose, and lignin. The composition of biomass can vary. It depends on multiple factors; for example, according to species from which it is obtained and environmental factors under which it is grown. Lignocellulosic biomass can play an important role in renewable energy production as concerns about greenhouse gas emissions and environmental pollution grow (Ghislain et al., 2020).

Recently, second-generation biorefineries have gained much attention. The main objective of second-generation biorefineries is to control cost factor. They allow efficient and cost-effective valorization of lignocellulosic biomass. CBP is a main technique in second-generation biorefining (Olguin-Maciel et al., 2020; Figure 6.3).

6.4 Integration of Microbial Consortia in CBP

Microbial consortia refer to communities of multiple microbial species that are growing together. These communities are able to utilize diverse carbon sources. They can perform complex functions (Bhatia et al., 2018). In CBP systems, two or more microorganisms are cultured together. Each microorganism performs a specific task. One may hydrolyze biomass and others may be involved in biochemical production (Scholz et al., 2018). CBP can employ natural microorganisms that can degrade lignocellulose, engineered microorganisms, or microbial co-cultures (Li et al., 2024).

Various research studies have shown the applications of microbial consortia in consolidated bioprocess technology. For example, microbial co-culture composed of *Trichoderma reesei* and *Parageobacillus* species have been used for the conversion of cellulose to useful chemicals (Joshi et al., 2022). Another artificially synthesized microbial co-culture example is utilizing a blend of *Trichoderma reesei* and *Rhizopus delemar* in CBP. It is specialized to transform cellulose to fumaric acid efficiently. It does not require additional cellulase enzymes or costly chemicals (Scholz et al., 2018). In the same way, there is another mixed microbial culture composed of two strains. These are *Thermoanaerobacterium thermosaccharolyticum* and *Clostridium acetobutylicum*. They can produce butanol from xylan and untreated corncob efficiently (Jiang et al., 2020; Table 6.1).

Table 6.1 Representative Microbial Consortia for Consolidated Bioprocessing (CBP).

Consortium Composition	Substrate	Target Product	Key Features	References
Microbial consortium of *Thermoanaerobacterium thermosaccharolyticum* and *Actinobacillus succinogenes 130Z*	Hemicellulose-enriched lignocellulose	Succinic acid	Xylan to xylose conversion	Lu et al. (2020)
Engineering *Trichoderma reesei* and *Saccharomyces cerevisiae*	Lignocellulose	D-Glucaric acid	Synthetic consortium enabling direct glucaric acid production from biomass	Fang et al. (2022)
Clostridium thermocellum and *Thermoanaerobacterium thermosaccharolyticum*	Corn straw	Ethanol	Improved ethanol yields via co-culture	Pang et al. (2018)
Thermoanaerobacterium thermosaccharolyticum M5 and *Clostridium acetobutylicum* NJ4	Xylan, corn cob	Butanol	Efficient xylan degradation and butanol production through synergistic interaction	Jiang et al. (2020)
Clostridium thermocellum and *Clostridium beijerinckii*	Lignocellulosic corn cob	Butanol	Efficient degradation and fermentation of lignocellulosic biomass without butyrate addition	Lin et al. (2013)

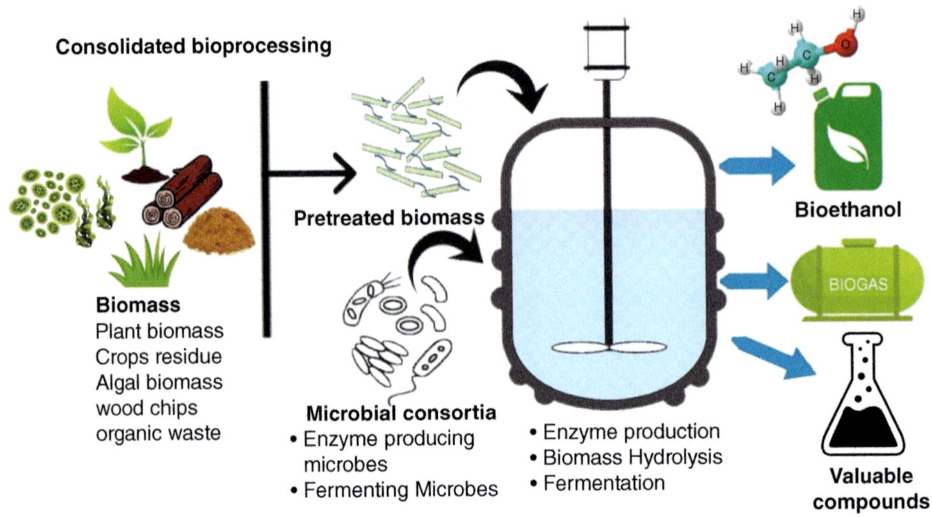

Figure 6.4 Integrated consolidated bioprocessing: microbial consortia for efficient biomass conversion. (*Source:* (b) Svitlana / Adobe Stock Photos (c) pathakdesigner / Adobe Stock Photos (d) teracreonte / Adobe Stock Photos (e) nsit0108 / Adobe Stock Photo) .

A comprehensive overview of CBP is shown in Figure 6.4. It is a technology that utilizes lignocellulosic biomass and produces sustainable fuels, chemicals, and other valuable products. The use of microbial consortia in this process allows the efficient conversion of biomass into various useful products.

6.5 Future Perspectives

Developing a microbial consortium is an advanced field that gives the scientists an avenue to explore the scope of biotechnology from a pure inoculum to mixed (consortia) cultures. Microbes can use a variety of carbon sources in mixed cultures. It enables microbes to respond to environmental factors effectively. In mixed cultures, microbes become capable of performing intricate functions that pure cultures can never perform. With technological developments, it has now become possible to develop mixed cultures and understand their interactions. Microbial consortia are categorized on the basis of mechanism of interaction, development of the consortium, and roles. Various patterns regarding the study of interaction modes and functions, such as microfluidics, membrane separation, Single_Cell genomics, and fluorescent imaging have been reviewed by Bhatia et al. (2018).

Current research trends in biotechnological methods have made valorization of food waste (FW) more feasible and efficient. Recently, consolidated models of biorefinery, construction of microbial transformation techniques, and deployment of circular economy tenets have been explored (Rajendran et al., 2021). The production of advanced microbial strains and recombinant enzymes has improved the transformation rate of FW into bioproducts (biochemical), biofuels, and other valuable items. These biological entities are engineered to degrade complex organic matter, making it possible to extract high-value products (Ankathi et al., 2024). Advanced biorefineries are implementing consolidated models that utilize the combination of several valorization pathways. This approach increases the yield and profitability of biotechnological refinery procedures. This also helps in the production of other by-products such as bioplastic and other bioenergy sources (Ubando et al., 2020).

By implementing circular economy principles in biorefineries, we can make processes more efficient, reduce waste, and recover valuable resources. This strategy is suitable for global ecosystem-friendly objectives and also helps in alleviating environmental aftermaths of biorefinery processes (Ubando et al., 2020).

The waste and emissions of conventional chemical processing are reduced by utilizing microbial mixed cultures in CBP, making this approach bio-economical and sustainable. In addition, the use of new (engineered) enzymes has increased the conversion of food waste into useful organic products due to their higher catalyzing ability and stability. Advancements in the development of microbial consortia for CBP are expected to drive innovations in sustainable manufacturing and contribute to the establishment of a circular bioeconomy. The environment surrounding the microbes during co-culturing has an impact on microbial interactions. One should consider a few technicalities of activation mechanisms while choosing the best co-culture such as adhesion, physical, gaseous, or diffusion (Padmaperuma et al., 2018). It is also important to consider the similarity of the media with natural environment for microbes because media components could affect the behavior of microbes. For example, fungi that thrive on surfaces like wood might be better cultured on agar rather than in liquid media. However, microorganisms can adapt to various environments, allowing flexibility in media selection (Kapoore et al., 2022).

Future research should focus on developing and understanding microbial consortia functionality and diversity. Additionally, creating comprehensive databases on nutrient requirements, metabolites, community interactions, and enzymes will be crucial. Designing synthetic microbial consortia for specific applications will also be a key area of study. Ultimately, applying microbial consortia in biotechnology sectors for large-scale pesticide degradation holds great promise for effective bioremediation solutions (Bhatt et al., 2021).

6.6 Conclusion

To summarize, microbial mixed cultures have shown a significant potential in CBP and can be used as a promising technique for sustainable production of energy. By utilizing the combined strength of different microbial strains, mixed cultures are able to convert lignocellulosic biomass into useful products. In this way, microbial consortia can overcome the limitations of pure culture and increase the efficacy of the process. The knowledge of the ecological behavior of microbial strains is important for developing stable and efficient consortia. Stability could also be achieved by lowering the metabolic stress and optimizing the utilization of substrates for better yield of products. Microbial consortia have a wide range of applications in CBP, including biofuel production and bioremediation. There are different classes of microbial consortia consisting of natural, engineered, and synthetic cultures. Natural consortia are acquired from environmental sources and they are capable of degrading complex organic matter. Synthetic consortia can be developed by using synthetic biology and making microbes able to utilize non-native pathways. Engineered cultures could be designed utilizing strategies based on interactions seen in natural consortia.

On the basis of morphological classification, there are two primary approaches including co-culture and mixed culture that can be used in CBP. On the whole, microbial mixed cultures are capable of transforming biomass into biofuels and bioproducts. This is an ecosystem-friendly and efficient approach that can combat with challenges of energy shortage and climate changes. By exploring the field of ecological interactions, we could develop more stable and efficient cultures and open new prospects of this technique for a more sustainable future.

References

Ado, S., Olukotun, G., Ameh, J., & Yabaya, A. (2010). Bioconversion of cassava starch to ethanol in a simultaneous saccharification and fermentation process by co-cultures of *Aspergillus niger* and *Saccharomyces cerevisiae*. *Journal of Pure and Applied Microbiology*, *4*(1), 55–61. https://doi.org/10.4314/swj.v4i1.51832.

Ankathi, S. K., Chaudhari, U. S., Handler, R. M., & Shonnard, D. R. (2024). Sustainability of biogas production from anaerobic digestion of food waste and animal manure. *Applied Microbiology*, *4*(1), 418–438. https://doi.org/10.3390/applmicrobiol4010029.

Bhatia, S. K., Bhatia, R. K., Choi, Y.-K., Kan, E., Kim, Y.-G., & Yang, Y.-H. (2018). Biotechnological potential of microbial consortia and future perspectives. *Critical Reviews in Biotechnology*, *38*, 1209–1229. https://doi.org/10.1080/07388551.2018.1471445.

Bhatt, P., Bhatt, K., Sharma, A., Zhang, W., Mishra, S., & Chen, S. (2021). Biotechnological basis of microbial consortia for the removal of pesticides from the environment. *Critical Reviews in Biotechnology*, *41*(3), 317–338. https://doi.org/10.1080/07388551.2020.1853032.

Brethauer, S., & Studer, M. H. (2014). Consolidated bioprocessing of lignocellulose by a microbial consortium. *Energy & Environmental Science*, *7*(4), 1446–1453. https://doi.org/10.1039/C3EE41753K.

Carlos, C., Fan, H., & Currie, C. R. (2018). Substrate shift reveals roles for members of bacterial consortia in degradation of plant cell wall polymers. *Frontiers in Microbiology*, *9*, 364. https://doi.org/10.3389/fmicb.2018.00364.

Chaves, J. E., Presley, G. N., & Michener, J. K. (2019). Modular engineering of biomass degradation pathways. *Processes*, *7*(4), 230. https://doi.org/10.3390/pr7040230.

Che, S., & Men, Y. (2019). Synthetic microbial consortia for biosynthesis and biodegradation: Promises and challenges. *Journal of Industrial Microbiology and Biotechnology*, *46*(9–10), 1343–1358. https://doi.org/10.1007/s10295-019-02211-4.

Cunha, J. T., Romaní, A., Costa, C. E., Sá-Correia, I., & Domingues, L. (2019). Molecular and physiological basis of *Saccharomyces cerevisiae* tolerance to adverse lignocellulose-based process conditions. *Applied Microbiology and Biotechnology*, *103*(1), 159–175. https://doi.org/10.1007/s00253-018-9478-3.

Dempfle, D., Kröcher, O., & Studer, M. H.-P. (2021). Techno-economic assessment of bioethanol production from lignocellulose by consortium-based consolidated bioprocessing at industrial scale. *New Biotechnology*, *65*, 53–60. https://doi.org/10.1016/j.nbt.2021.07.005.

Deshavath, N. N., Veeranki, V. D., & Goud, V. V. (2019). Lignocellulosic feedstocks for the production of bioethanol: Availability, structure, and composition. In M. Rai & A. P. Ingle (Eds.), *Sustainable Bioenergy* (pp. 1–19). Elsevier. https://doi.org/10.1016/B978-0-12-817654-2.00001-0.

Dong, H. W., Fan, L. Q., Luo, Z., Zhong, J. J., Ryu, D. D., & Bao, J. (2013). Improvement of ethanol productivity and energy efficiency by degradation of inhibitors using recombinant *Zymomonas mobilis* (pHW20a-fdh). *Biotechnology and Bioengineering, 110*(9), 2395–2404. https://doi.org/10.1002/bit.24897.

Du, R., Li, C., Lin, W., Ki Lin, C. S., & Yan, J. (2022). Domesticating a bacterial consortium for efficient lignocellulosic biomass conversion. *Renewable Energy, 189*, 359–368. https://doi.org/10.1016/j.renene.2022.02.093.

Du, Y., Yang, B., Yi, Z., Hu, L., & Li, M. (2020). Engineering *Saccharomyces cerevisiae* coculture platform for the production of flavonoids. *Journal of Agricultural and Food Chemistry, 68*(7), 2146–2154. https://doi.org/10.1021/acs.jafc.9b07916.

Fang, H., Deng, Y., Pan, Y., Li, C., & Yu, L. (2022). Distributive and collaborative push-and-pull in an artificial microbial consortium for improved consolidated bioprocessing. *AIChE Journal, 68*(11), e17844. https://doi.org/10.1002/aic.17844.

Fatma, S., Hameed, A., Noman, M., Ahmed, T., Shahid, M., Tariq, M., Sohail, I., & Tabassum, R. (2018). Lignocellulosic biomass: A sustainable bioenergy source for the future. *Protein and Peptide Letters, 25*(2), 148–163. https://doi.org/10.2174/0929866525666180122144504.

Gao, H., Li, M., Yang, L., Jiang, Y., Jiang, W., Yu, Z., Zhang, W., Xin, F., & Jiang, M. (2022). Spatial niche construction of a consortium-based consolidated bioprocessing system. *Green Chemistry, 24*(20), 7941–7950. https://doi.org/10.1039/D2GC01850K.

García, C., Rendueles, M., & Díaz, M. (2019). Liquid-phase food fermentations with microbial consortia involving lactic acid bacteria: A review. *Food Research International, 119*, 207–220. https://doi.org/10.1016/j.foodres.2019.01.043.

Ghislain, T., Duret, X., Diouf, P. N., & Lavoie, J.-M. (2020). Lignocellulosic biomass. In A. Nzihou (Ed.), *Handbook on characterization of biomass, biowaste and related by-products* (pp. 499–535). Springer. https://doi.org/10.1007/978-3-030-35020-8.

Ghosh, S., Chowdhury, R., & Bhattacharya, P. (2016). Mixed consortia in bioprocesses: Role of microbial interactions. *Applied Microbiology and Biotechnology, 100*, 4283–4295. https://doi.org/10.1007/s00253-016-7448-1.

Gladkov, G. V., Kimeklis, A. K., Afonin, A. M., Lisina, T. O., Orlova, O. V., Aksenova, T. S., Kichko, A. A., Pinaev, A. G., & Andronov, E. E. (2022). The structure of stable cellulolytic consortia isolated from natural lignocellulosic substrates. *International Journal of Molecular Sciences, 23*(18), 10779. https://doi.org/10.3390/ijms231810779.

Gomes, D. G., Cunha, J. T., Zanuso, E., Teixeira, J. A. C., & Domingues, L. (2021). Strategies towards reduction of cellulases consumption: Debottlenecking the economics of lignocellulosics valorization processes. *Polysaccharides, 2*, 287–310. https://doi.org/10.3390/polysaccharides2020020.

Herring, C. D., Kenealy, W. R., Joe Shaw, A., Covalla, S. F., Olson, D. G., Zhang, J., Ryan Sillers, W., Tsakraklides, V., Bardsley, J. S., & Rogers, S. R. (2016). Strain and bioprocess improvement of a thermophilic anaerobe for the production of ethanol from wood. *Biotechnology for Biofuels, 9*, 1–16. https://doi.org/10.1186/s13068-016-0536-8.

Hon, S., Olson, D. G., Holwerda, E. K., Lanahan, A. A., Murphy, S. J., Maloney, M. I., Zheng, T., Papanek, B., Guss, A. M., & Lynd, L. R. (2017). The ethanol pathway from *Thermoanaerobacterium saccharolyticum* improves ethanol production in *Clostridium thermocellum*. *Metabolic Engineering, 42*, 175–184. https://doi.org/10.1016/j.ymben.2017.06.011.

Hu, J., Xue, Y., Guo, H., Gao, M. T., Li, J., Zhang, S., & Tsang, Y. F. (2017). Design and composition of synthetic fungal-bacterial microbial consortia that improve lignocellulolytic enzyme activity. *Bioresource Technology, 227*, 247–255. https://doi.org/10.1016/j.biortech.2016.12.058.

Inyang, V., Laseinde, O., & Kanakana, G. M. (2022). Techniques and applications of lignocellulose biomass sources as transport fuels and other bioproducts. *International Journal of Low-Carbon Technologies, 17*, 900–909. https://doi.org/10.1093/ijlct/ctac068.

Jawed, K., Yazdani, S. S., & Koffas, M. A. (2019). Advances in the development and application of microbial consortia for metabolic engineering. *Metabolic Engineering Communications, 9*, e00095. https://doi.org/10.1016/j.mec.2019.e00095.

Jia, X., Liu, C., Song, H., Ding, M., Du, J., Ma, Q., & Yuan, Y. (2016). Design, analysis and application of synthetic microbial consortia. *Synthetic and Systems Biotechnology, 1*(2), 109–117. https://doi.org/10.1016/j.synbio.2016.02.001.

Jiang, Y., Lv, Y., Wu, R., Lu, J., Dong, W., Zhou, J., Zhang, W., Xin, F., & Jiang, M. (2020). Consolidated bioprocessing performance of a two-species microbial consortium for butanol production from lignocellulosic biomass. *Biotechnology and Bioengineering, 117*(10), 2985–2995. https://doi.org/10.1002/bit.27464.

Jiang, Y., Wu, R., Zhang, W., Xin, F., & Jiang, M. (2023). Construction of stable microbial consortia for effective biochemical synthesis. *Trends in Biotechnology, 41*(11), 1430–1441. https://doi.org/10.1016/j.tibtech.2023.05.008.

Joshi, C., Kumar, M., Bennett, M., Thakur, J., Leak, D. J., Sharma, S., MacKinnon, N., & Masakapalli, S. K. (2022). Integration of synthetic microbial consortia based bioprocessing with pyrolysis for efficient conversion of cellulose to valuables. *BioRxiv*. https://doi.org/10.1101/2022.06.26.497679.

Kapoore, R. V., Padmaperuma, G., Maneein, S., & Vaidyanathan, S. (2022). Co-culturing microbial consortia: approaches for applications in biomanufacturing and bioprocessing. *Critical Reviews in Biotechnology, 42*(1), 46–72. https://doi.org/10.1080/07388551.2021.1921691.

Kong, W., Meldgin, D. R., Collins, J. J., & Lu, T. (2018). Designing microbial consortia with defined social interactions. *Nature Chemical Biology, 14*(8), 821–829. https://doi.org/10.1038/s41589-018-0091-7.

Lazuka, A., Auer, L., Bozonnet, S., Morgavi, D. P., O'Donohue, M., & Hernandez-Raquet, G. (2015). Efficient anaerobic transformation of raw wheat straw by a robust cow rumen-derived microbial consortium. *Bioresource Technology, 196*, 241–249. https://doi.org/10.1016/j.biortech.2015.07.084.

Li, Z., Waghmare, P. R., Dijkhuizen, L., Meng, X., & Liu, W. (2024). Research advances on the consolidated bioprocessing of lignocellulosic biomass. *Engineering Microbiology, 4*(2), 100139. https://doi.org/10.1016/j.engmic.2024.100139.

Liang, Y., Ma, A., & Zhuang, G. (2022). Construction of environmental synthetic microbial consortia: Based on engineering and ecological principles. *Frontiers in Microbiology, 13*, 829717. https://doi.org/10.3389/fmicb.2022.829717.

Lin, Y., Wen, Z., Zhu, L., Lin, J., & Cen, P. (2013). Butanol production from corncob in the sequential co-culture of *Clostridium thermocellum* and *Clostridium beijerinckii*. *Journal of Chemical Engineering of Chinese Universities, 3*, 444–449.

Lu, J., Lv, Y., Jiang, Y., Wu, M., Xu, B., Zhang, W., Zhou, J., Dong, W., Xin, F., & Jiang, M. (2020). Consolidated bioprocessing of hemicellulose-enriched lignocellulose to succinic acid through a microbial cocultivation system. *ACS Sustainable Chemistry & Engineering, 8*(24), 9035–9045. https://doi.org/10.1021/acssuschemeng.0c01865.

Mbaneme-Smith, V., & Chinn, M. S. (2015). Consolidated bioprocessing for biofuel production: Recent advances. *Energy and Emission Control Technologies, 3*, 23–44. https://doi.org/10.2147/EECT.S63000.

Morris, B. E., Henneberger, R., Huber, H., & Moissl-Eichinger, C. (2013). Microbial syntrophy: Interaction for the common good. *FEMS Microbiology Reviews, 37*(3), 384–406. https://doi.org/10.1111/1574-6976.12019.

Navarrete-Bolanos, J., Serrato-Joya, O., Botello-Alvarez, E., Jimenez-Islas, H., Cardenas-Manriquez, M., Conde-Barajas, E., & Rico-Martinez, R. (2007). Analyzing microbial consortia for biotechnological processes design. Méndez-Vilas A, *Communicating current research and educational topics and trends in applied microbiology, 1*, 437–449 Formatex. https://www.researchgate.net/publication/278667261_Analyzing_microbial_consortia_for_biotechnological_processes_design.

Ojo, A. O. (2023). An overview of lignocellulose and its biotechnological importance in high-value product production. *Fermentation, 9*(11), 990. https://doi.org/10.3390/fermentation9110990.

Olguin-Maciel, E., Singh, A., Chable-Villacis, R., Tapia-Tussell, R., & Ruiz, H. A. (2020). Consolidated bioprocessing, an innovative strategy towards sustainability for biofuels production from crop residues: An overview. *Agronomy, 10*(11), 1834. https://doi.org/10.3390/agronomy10111834.

Olson, D. G., Mcbride, J. E., Shaw, A., Lynd, L. R., Liao, J., & Messing, J. (2012). Recent progress in consolidated bioprocessing. *Current Opinion in Biotechnology, 23*(3), 396–405. https://doi.org/10.1016/j.copbio.2011.11.026.

Padmaperuma, G., Kapoore, R. V., Gilmour, D. J., & Vaidyanathan, S. (2018). Microbial consortia: A critical look at microalgae co-cultures for enhanced biomanufacturing. *Critical Reviews in Biotechnology, 38*(5), 690–703. https://doi.org/10.1080/07388551.2017.1390728.

Pang, J., Hao, M., Li, Y., Liu, J., Lan, H., Zhang, Y., Zhang, Q., & Liu, Z. (2018). Consolidated bioprocessing using *Clostridium thermocellum* and *Thermoanaerobacterium thermosaccharolyticum* co-culture for enhancing ethanol production from corn straw. *BioResources, 13*(4), 8209–8221.

Peng, X., Wilken, S. E., Lankiewicz, T. S., Gilmore, S. P., Brown, J. L., Henske, J. K., Swift, C. L., Salamov, A., Barry, K., & Grigoriev, I. V. (2021). Genomic and functional analyses of fungal and bacterial consortia that enable lignocellulose breakdown in goat gut microbiomes. *Nature Microbiology, 6*(4), 499–511. https://doi.org/10.1038/s41564-020-00861-0.

Perez-Zabaleta, M., Atasoy, M., Khatami, K., Eriksson, E., & Cetecioglu, Z. (2021). Bio-based conversion of volatile fatty acids from waste streams to polyhydroxyalkanoates using mixed microbial cultures. *Bioresource Technology*, *323*, 124604. https://doi.org/10.1016/j.biortech.2020.124604.

Qian, X., Chen, L., Sui, Y., Chen, C., Zhang, W., Zhou, J., Dong, W., Jiang, M., Xin, F., & Ochsenreither, K. (2019). Biotechnological potential and applications of microbial consortia. *Biotechnology Advances*, *40*, 107500. https://doi.org/10.1016/j.biotechadv.2019.107500.

Rajendran, N., Gurunathan, B., Han, J., Krishna, S., Ananth, A., Venugopal, K., & Priyanka, R. S. (2021). Recent advances in valorization of organic municipal waste into energy using biorefinery approach, environment and economic analysis. *Bioresource Technology*, *337*, 125498. https://doi.org/10.1016/j.biortech.2021.125498.

Rani Singhania, R., Kumar Patel, A., Singh, A., Haldar, D., Soam, S., Chen, C. W., Tsai, M. L., & Dong, C.-D. (2022). Consolidated bioprocessing of lignocellulosic biomass: Technological advances and challenges. *Bioresource Technology*, *354*, 127153. https://doi.org/10.1016/j.biortech.2022.127153.

Rodrigues Reis, C. E., Rajendran, A., Silva, M. B., Hu, B., & De Castro, H. F. (2018). The application of microbial consortia in a biorefinery context: Understanding the importance of artificial lichens. In O. Singh & A. Chandel (Eds.), *Sustainable biotechnology-enzymatic resources of renewable energy* (pp. 423–437). Springer. https://doi.org/10.1007/978-3-319-95480-6_16.

Roell, G. W., Zha, J., Carr, R. R., Koffas, M. A., Fong, S. S., & Tang, Y. J. (2019). Engineering microbial consortia by division of labor. *Microbial Cell Factories*, *18*, 1–11. https://doi.org/10.1186/s12934-019-1083-3.

Said, S. B., Tecon, R., Borer, B., & Or, D. (2020). The engineering of spatially linked microbial consortia–potential and perspectives. *Current Opinion in Biotechnology*, *62*, 137–145. https://doi.org/10.1016/j.copbio.2019.09.015.

Saravanan, A., Yaashikaa, P., Kumar, P. S., Thamarai, P., Deivayanai, V., & Rangasamy, G. (2023). A comprehensive review on techno-economic analysis of biomass valorization and conversional technologies of lignocellulosic residues. *Industrial Crops and Products*, *200*, 116822. https://doi.org/10.1016/j.indcrop.2023.116822.

Scholz, S. A., Graves, I., Minty, J. J., & Lin, X. N. (2018). Production of cellulosic organic acids via synthetic fungal consortia. *Biotechnology and Bioengineering*, *115*(4), 1096–1100. https://doi.org/10.1002/bit.26509.

Shahab, R. L., Brethauer, S., Davey, M. P., Smith, A. G., Vignolini, S., Luterbacher, J. S., & Studer, M. H. (2020). A heterogeneous microbial consortium producing short-chain fatty acids from lignocellulose. *Science*, *369*(6507). https://doi.org/10.1126/science.abb1214.

Shahab, R. L., Luterbacher, J. S., Brethauer, S., & Studer, M. H. (2018). Consolidated bioprocessing of lignocellulosic biomass to lactic acid by a synthetic fungal-bacterial consortium. *Biotechnology and Bioengineering*, *115*(5), 1207–1215. https://doi.org/10.1002/bit.26541.

Show, P. L. (2021). Special Issue on "Biotechnology for sustainability and social well being". *Processes*, *9*(2), 216. https://doi.org/10.3390/pr9020216.

Sieber, J. R., Sims, D. R., Han, C., Kim, E., Lykidis, A., Lapidus, A. L., McDonnald, E., Rohlin, L., Culley, D. E., & Gunsalus, R. (2010). The genome of *Syntrophomonas wolfei*: New insights into syntrophic metabolism and biohydrogen production. *Environmental Microbiology*, *12*(8), 2289–2301. https://doi.org/10.1111/j.1462-2920.2010.02237.x.

Tabañag, I. D. F., & Tsai, S.-L. (2018). Hemicellulose degradation and utilization by a synthetic *Saccharomyces cerevisiae* consortium. *BioRxiv*, 244244. https://doi.org/10.1101/244244.

Ubando, A. T., Felix, C. B., & Chen, W.-H. (2020). Biorefineries in circular bioeconomy: A comprehensive review. *Bioresource Technology*, *299*, 122585. https://doi.org/10.1016/j.biortech.2019.122585.

Wang, X. L., Zhou, J. J., Sun, Y. Q., & Xiu, Z. L. (2019). Bioconversion of raw glycerol from waste cooking-oil-based biodiesel production to 1,3-propanediol and lactate by a microbial consortium. *Frontiers in Bioengineering and Biotechnology*, *7*, 14. https://doi.org/10.3389/fbioe.2019.00014.

Xu, C., & Yu, H. (2021). Insights into constructing a stable and efficient microbial consortium. *Chinese Journal of Chemical Engineering*, *30*, 112–120. https://doi.org/10.1016/j.cjche.2020.12.012.

Zhang, H., Pereira, B., Li, Z., & Stephanopoulos, G. (2015). Engineering *Escherichia coli* coculture systems for the production of biochemical products. *Proceedings of the National Academy of Sciences*, *112*(27), 8266–8271. https://doi.org/10.1073/pnas.1506781112.

Zhang, T., & Zhang, H. (2022). Microbial consortia are needed to degrade soil pollutants. *Microorganisms*, *10*(2). https://doi.org/10.3390/microorganisms10020261.

7

Challenges and Solutions of Consolidated Bioprocessing: A Case Study of Biogas Production

Muhammad Anas[1], Waseem Ahmed Khattak[2], Erdogan E. Hakki[1], Javed Iqbal[3], Banzeer Ahsan Abbasi[4], Ziaur Rehman[5], and Akhtar Munir[5]

[1]*Department of Soil Science & Plant Nutrition, Selçuk University, Konya, Türkiye*
[2]*Department of Plant Sciences, Faculty of Biological Sciences, Quaid-I-Azam University, Islamabad, Pakistan*
[3]*Department of Botany, Bacha Khan University, Charsadda, Pakistan*
[4]*Department of Botany, Rawalpindi Women University, Rawalpindi, Pakistan*
[5]*Department of Chemistry, Quaid-i-Azam University, Islamabad, Pakistan*

7.1 Introduction

7.1.1 Overview of Consolidated Bioprocessing (CBP)

Consolidated Bioprocessing (CBP) is an advanced biotechnological multistep process that involves the transformation of the lignocellulosic bio-background into biofuels or biochemicals as a one-step integrated process. Unlike traditional multistep bioconversion, where enzyme production, biomass hydrolysis, and fermentation are done using separate steps, CBP combines them in a single microbial platform or microbial consortium (Kukreti et al., 2024). Such concentration significantly reduces the complexity of operation, energy demand, and the cost of production, making CBP an interesting and sustainable solution to large-scale bioconversion of renewable biomass. This conceptual core of CBP depends on special or designed microorganisms, which could synthesize the required hydrolytic enzymes (cellulases and hemicellulases) at the same time, break down the complex plant polymer into fermentable sugars, and transform those sugars into products of interest, such as ethanol, organic acids, and biogas. There are two major approaches to the current CBP research: (i) an attempt to engineer one versatile microbial strain that could perform all functions effectively and (ii) the design of synergistic microbial consortia where various microorganisms contribute different functions to work in a synergistic way. Remarkable entities used in relation to CBP are *Clostridium thermocellum*, *Trichoderma reesei*, *Caldicellulosiruptor* spp., as well as thermophilic bacteria that can make the direct biomass-to-fuel conversion (Olaganathan, 2025). Two factors have supported the development of CBP, and reflecting the global interest in making biofuel production both economically viable and environmentally sustainable. Since energy generated using biomass is a very important aspect of the renewable-energy domain, CBP is an innovative way of thinking of balancing biomass exploitation and reducing its preprocessing. It does not involve enzymatic pretreatment, which is costly, and it capitalizes on the metabolism by the microbes to save cost. Instead, establishing a CBP transforms a fragmented, resource-intensive pathway into a unified, scalable, and environmentally friendly process. (Vijayan et al., 2023).

7.1.2 Importance in Biogas Production

Biogas production is one of the major pillars of renewable energy plans, especially with respect to the world's challenges of managing wastes, achieving carbon neutrality, and sustaining fuel sources (Kabeyi & Olanrewaju, 2022). The most common biogas is a combination of methane (CH_4) and carbon dioxide (CO_2), produced through anaerobic digestion (AD) of organic wastes, including agricultural residues, food waste, municipal sewage sludge, and resources made up of lignocellulosic biomass. The recalcitrance of lignocellulose, however, is a major constraint to efficient biogas production, since lignocellulose is the most important component that constitutes the plant biomass. Its dense cellulose, hemicellulose, and lignin are a serious hindrance to the digestibility of microbes, thus reducing methane production (Li et al., 2018). In this context, one of the promising solutions appears to be consolidated biological processing (CBP), which could improve the efficiency of biogas systems. CBP integrates both enzymatic hydrolysis process and microbial fermentation into one operation by disintegrating lignocellulosic biomass more efficiently, avoiding extrinsic addition of enzymes and the need for energy-consuming chemical pretreatments. The feature is particularly desirable in rural and industrial contexts where sources of biomass are abundant, but capital and operational costs are limited. Through CBP, straw, corn stover, sugarcane bagasse, and forestry residues are transformed into biogas with improved methane content and an increase in the conversion rates (Wang et al., 2023). Moreover, CBP allows broadening the investigation scope to the thermophilic and extremophile-based AD approaches, in which specially evolved microorganisms are shown to be compatible with the high-temperature environments, thus encouraging the development of more rapid reaction rates and alleviating pathogenic issues. Thermophilic CBP strains have the capacity to ferment and hydrolyze biomass under conditions that can accommodate both mesophilic and thermophilic AD infrastructures, thus making the installation of CBP modules synonymous with anaerobic biogas plants (Althuri & Mohan, 2020).

A new method of biogas production is Consolidated Bioprocessing (CBP), which can take advantage of microbial consortia that simulate natural ecosystems, known as gateways to biogas formation, like cow rumen or termite guts, in which prolonged degradation of cellulose and production of methane are attained by means of synergistic interactions of various microbes. Engineered or enriched consortia constituted by the union of hydrolytic fungi or bacteria with methanogenic archaea have been demonstrated to be capable of not only slowing the breakdown of complex biological substrates but also reducing the associated methane production, thus reducing hydraulic retention time (HRT) and increasing the amount of energy hoarded per volume of medium. Also remarkable is the ability of CBP to drive decentralized biogas energy systems, especially in developing countries and rural areas where advanced pretreatment technology is restricted (Singh et al., 2024). The ease and the unproblematic incorporation of CBP eliminate significant technological adverse factors, making it possible to design small and less-expensive digesters to treat multicomponent organic wastes directly. These aspects allow receiving local energy independence and moving toward waste valorization and a circular economy (Mishra et al., 2022).

Economically, at the investment level, CBP enhances the feasibility of biogas through cost minimization (in factors such as procurement of commercial cellulases) and enhancement in overall process yield, bringing an increase in the rate of return on investment (ROI) by biogas plants (Kubis, 2023). In an environmental aspect, the technology enables carbon neutrality, whereby it encourages the use of agricultural waste materials and other wastes as a source of energy feedstock, reduces dependence on fossil fuels, and reduces greenhouse gases. In addition, products of CBP also present a simplified route for the conversion of lignocellulosic feedstock to methane, a potent complement to traditional biogas practices, and a significant potential to increase levels of energy and economic biogas potential in search of scalable, sustainable, bio-energy materials (Srivastava, 2019).

7.1.3 Challenges in Consolidated Bioprocessing

One of the new technologies that has come up to completely transform the process of production of biofuels, particularly biogas, using lignocellulosic biomass is termed consolidated bioprocessing (CBP). CBP removes the requirement of multiple reactors that are being conventionally utilized in bioconversion by incorporating enzyme production, hydrolysis of biomass, and fermentation in a single reactor and operating under either simplex or multiplex conditions using one or several microbial species. However, there exist some scientific, technical, and economic obstacles that hinder its industrialization. The given analysis examines four key obstacles to CBP development, which are microbial strain constraints, complexities and predication needs of the substrates, process efficiency and yield maximization, and scale-up problems (Figure 7.1).

7.1.3.1 Limitations of Microbial Strain

This is probably one of the daunting challenges, considering the identification or genetic engineering of strong microbial strains that can perform all three functions of CBP with high efficiency in an industrial setup, namely, enzyme production, hydrolysis, and fermentation.

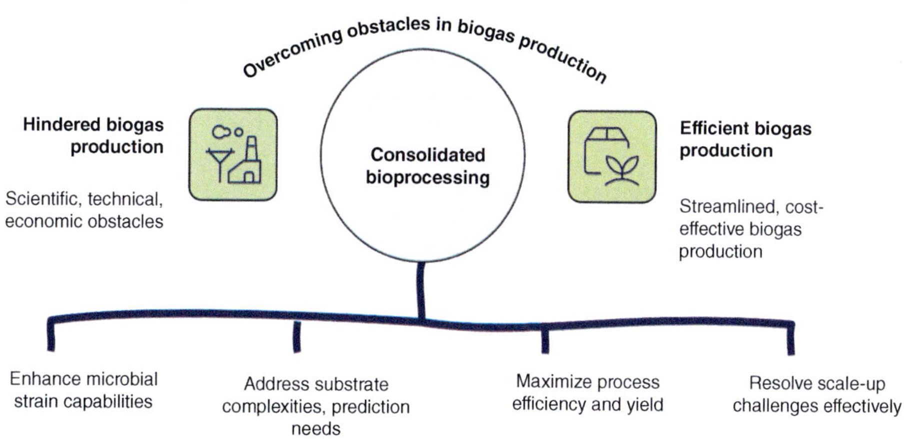

Figure 7.1 Key challenges and solutions in biogas production, highlighting obstacles in substrate processing, microbial efficiency, and process optimization.

7.1.3.2 Metabolic Limitations and Genetic Problems

Microorganisms that naturally occur usually excel in doing one or two of such functions, but in rare cases, they perform all three. To illustrate the above, *Clostridium thermocellum* has the magnificent capacity to break down cellulose due to its cellulosome complex but has low ethanol titers and is prone to ethanol poisoning. Genetic engineering to boost metabolic production commonly causes metabolic loads or strain instabilities (Wu et al., 2016).

7.1.3.3 Trade-Offs of Enzyme Production

Hydrolytic enzyme production is energetically costly, leaving few resources for the host cell biomass conversion and creation of the products. Cells tend to reuse energy even in scenarios when the metabolic pathways have been engineered, making the quantity of biogas or ethanol produced to be lower (Straathof, 2014).

7.1.3.4 Tolerance Problems and Stability Problems

Genetic modification gives a poor tolerance to industrial stressful conditions, like sudden alteration of temperature, pH alteration, and high concentration of products, among others. In many of the thermophilic or acidophilic types employed in CBP, there is inadequate tolerance to ethanol or acid, which consequently threatens process inhibition or failure (Salehi Jouzani & Taherzadeh, 2015).

7.2 Substrate Complexity and Pretreatment Needs

Pretreatment and enzymatic hydrolysis steps are both long and expensive processes that must be followed during the commercial conversion of biomass. To avoid these burdens, the idea of continuous-flow Consolidated Bioprocessing (CBP) was devised. However, the recalcitrant nature of lignocellulose remains a most daunting challenge (Hassan et al., 2019).

7.2.1 Lignocellulosic Heterogeneity

Lignocellulosic biomass is made up of closely bonded cellulose, hemicellulose, and lignin. The fact that cellulose is in a crystalline structure and lignin has an amorphous aromatic nature makes the substrate impenetrable to microbial degradation. The cellulose cannot be fermented by the microorganisms that can ferment it unless the lignin structure is broken through (Chang et al., 2012).

7.2.1.1 Pretreatment Is Nonetheless Required

Albeit CBP vows to forego pretreatment, some mechanical, chemical, or biological pretreatment will be a prerequisite to reinforce the availability of substrates. Steam explosion, dilute acid hydrolysis, or alkaline treatment adds an additional cost and environmental implications to its methods (Singh et al., 2015).

7.2.1.2 Degradation By-Product Inhibition

The inhibitory substances, such as furfural, 5-hydroxymethylfurfural HMF, and phenolics, produced often during pretreatment processes, inhibit the behavior of microbes. These toxins may also cause a decrease in hydrolytic activity even in consortia indigenous to CBP (Bhatia et al., 2018).

7.2.2 Process Efficiency and Yield Optimization

Achieving an optimum balance between maximum biogas production and economic viability is also another complex aspect of CBP technology. During consolidated reactors, the process of enzymatic hydrolysis and microbial fermentation does not occur simultaneously, which limits the yield of products. The formation of co-products further complicates the process, including the conversion of the substrate into byproducts such as acetic acid, lactate, or hydrogen. Low yields of methane are typical in cases of accumulation of volatile fatty acids, changes in pH of the water, and insufficiency of complex substance conversion. Energy savings in multimodal systems may be compromised by balancing issues in terms of energy saving, since many of the units involved in the production tend to integrate (Mauser, 2017).

7.3 Case Study: Biogas Production via CBP

7.3.1 System Design and Operational Parameters

Consolidated bioprocessing (CBP) represents a breakthrough in the working efficiency of biogas systems, since it integrates the production of enzymes, biomass hydrolysis, and the process of microbial fermentation into a single stage (Singhania et al., 2022). A notable hallmark of CBP-based technology applied to biogas production is the cotreatment-CBP consolidated bioprocessing (C-CBP) platform, a novel ethanol production system developed by. This platform utilizes *Clostridium thermocellum* to perform cellulose hydrolysis and ethanol production from corn stover. The system is specifically designed to consume the untreated or slightly pretreated lignocellulosic feedstocks in a form of thermophilic fermentation (55 °C) and a neutral pH of around 6.8, and later digest the remaining solids in an anaerobic system to produce biogas that can meet all the internal process heat demands.

It consists of two main modules: a CBP reactor, where the process of hydrolysis and fermentation takes place, and an anaerobic digester, where the rest of the solids are further converted into methane. The CBP reactor has high solid loading rates (15–20% total solids), hence maximizing the substrate density at minimal water use. Calcium hydroxide acts as a stabilizer, ensures pH balance, and enhances the efficiency of the enzymes. Solids further undergo separation and are taken to a mesophilic digester operating at 37 °C and possessing an HRT of 25 days. This architecture makes compatibility with methanogenic archaea involved in the breakdown of organic residues after the process of fermenting ethanol (Amin et al., 2021).

In parallel, researchers at École Polytechnique Fédérale de Lausanne (EPFL) EPFL proposed a scalable CBP model based on consortia combining cellulolytic and methanogenic microorganisms, which is a fusion of microbial communities in a semi-continuous stirred reactor. The best performance was achieved at an organic loading rate (OLR) of 2.5 g VS/L d and a solids concentration of 10% total solids. The studies mentioned above together prove that the successful design of CBP needs to reconcile the enzyme kinetics with microbial fermentation kinetics in a way that would maximize the biomass conversion. A study by Bioverfahrenstechnik and Kaya (2022) also demonstrated an effective CBP configuration using a fungal–bacterial system to convert cotton stalks and olive mill waste.

The system employed *Trichoderma reesei* for enzymatic hydrolysis and *Methanosarcina barkeri* for methanogenesis. The batch fermentation was operated at 45 °C with an 8% solids concentration, highlighting flexibility in operational parameters depending on feedstock type (Flevaris & Chatzidoukas, 2021).

7.3.2 Key Performance Metrics

The demand for continuous CBP-based biogas production requires one to take into consideration several indicators of performance, namely the yield of methane, reduction of solids, and recovery of energy and greenhouse-gas mitigation (Kabeyi et al., 2024). A methane yield of 4.6 GJ per dry metric ton (DMT) of corn stover was reported by the C-CBP model created by Kubis and Lynd (2023), which is enough to support the whole biorefinery. When ethanol production was added to the scheme, the overall energy recovery reached about 79% of the theoretical maximum that can be obtained from carbohydrates (Díaz & Willis, 2019). Also, this method produced net-negative greenhouse-gas emissions, decreasing by up to 140 g CO_2-eq/MJ of ethanol when the avoided fossil-fuel use was considered, and the carbon was captured during combustion of biogas. With the system, it was possible to launch 70–75% VS degrading steady-state conditions with reactor productivities of 0.9–1.3 m^3 CH_4/m^3 day. Notably, the research applied automated feedback loops that preserved pH and buffer dose, thus countering the rise and fall of microbial activity as well as gradual rise of volatile fatty acids, a problem that is typical of mixed CBP systems. In their investigation, Kaya found the production of methane under the yields of 0.26 m^3 CH^4/kg VS and achieved 73% hydrolysis output. These findings were associated with high COD removal and destruction of organic matter. The research also revealed that the final biogas had a methane content of 64–68%, which falls within the desired range for energy applications (Kruczek et al., 2019).

These results put an emphasis on maximizing the inoculum-to-substrate ratio and maintaining the coordinated action of the hydrolytic and methanogenic organisms. The definition of the product stream produced by CBP technology should be noted as an important measure of its performance (Tseng et al., 2009). Empirical research has proved that CBP provides digestates of higher quality compared to traditional AD. The major process behind this betterment is enhanced hydrolysis of cellulose and hemicellulose during the pre-treatment phase, and there are no high levels of fiber and phenolic molecules in the resultant digestates as compared to traditional digestates. Once they managed to limit the initial content of lignin in the feedstock to a maximum of about 20%, the authors noted major improvements concerning both the production of biogas (about 25–35%) and the quality of the digestate, allowing its re-use in the agricultural sector (Anas et al., 2025).

Simultaneously, life cycle and techno-economic analyses predict that implementing CBP systems could reduce the levelized cost of biogas (LCOB) by up to 30% by 2030. This reduction occurs through three main pathways: (i) lower enzyme consumption, (ii) upgrades to processing facilities, and (iii) increased conversion rates. To evaluate the performance of a ruminant-inspired CBP configuration, the anaerobic digestion facility of a pig farm was adapted. In this setup, goats were maintained at the interface of the two anaerobic digestion stages to prevent the direct action of enzymes and organic acids secreted by rumen microbes and other microorganisms. The researchers monitored the biogas release rate to maximize biogas extraction. Without focusing on a specific conversion, they concluded that the model achieved a significant net energy gain, with remarkable chemical oxygen demand reduction (about 78%) and favorable process thermodynamics (Yan et al., 2020).

7.4 Solutions and Innovations

7.4.1 Genetic and Metabolic Engineering of Microbial Consortia

Genetic and metabolic engineering of microbial consortia is currently changing the face of CBP by increasing the synergy of hydrolytic and methanogenic organisms (Re & Mazzoli, 2023). The key issue of the CBP is the design of microbes that are effective in the destruction of lignocellulosic waste, along with the production of biogas or second-generation biofuels. The three main approaches currently being developed to address this problem are: (i) enhancing enzyme production in cellulolytic microorganisms, (ii) redirecting metabolic pathways to generate higher concentrations of methane precursors, and (iii) improving tolerance to lignin-derived phenolic inhibitors. The recent breakthrough in the field of synthetic biology has empowered the idea of driving *Clostridium thermocellum* to coproduce ethanol and acetate while being resilient to its thermophilic status (Sitara et al., 2024). The co-culture systems exhibit enhanced production of methane in engineered consortia amid *Methanosarcina* species and *Clostridium thermocellum* due to the clear flow of carbon within the hydrolysis and methanogenesis steps. At the same time, CRISPR-Cas9 technology is implemented to regulate the genes associated with the hydrogenase and cellulase functions in *Caldicellulosiruptor* and other anaerobic thermophiles. The division of labor is also achieved by the development of a microbial consortium that is also synthetic. Amplification of the methane outputs refers to large-scale augmentation with constituted methanogens that use formate and acetate optimally. *Methanobacterium formicicum* has been metabolically engineered to become more tolerant to volatile fatty acids and to perform well under increased loading (Wainaina et al., 2019). This can be seen recently and states that the cooperation of metabolism in multi-species consortia can be additionally optimized using omics-based profiling of communities, thus allowing the fine-tuning of interspecies nutrient transfer and by-product management. In conjunction with pathway engineering, stability optimization, and systems-level metabolic modeling, all of these are converting CBP into a modular, biologically integrated reality (Fong, 2014).

7.4.2 Advanced Pretreatment Techniques

An efficient substrate pretreatment is an unaltered requirement of CBP, especially in connection with lignocellulose feedstocks, corn stover, wheat straw, and woody biomass (Singhania et al., 2022). CBP aims at reducing reliance on financially demanding pre-treatments, but recent technologies can provide low-energy, enzyme-friendly, and bio-friendly processes that support the availability of the substrates to the CBP-related microorganisms. An interesting approach is biological pretreatment using microbial consortia or fungi such as *Phanerochaete chrysosporium* to selectively remove lignin and leave intact the cellulose (Ramarajan & Manohar, 2017). Such systems reduce the demand for chemicals and can still be applied to the thermophilic CBP microbes. It has been indicated that by the use of biodelignification, it is possible to reduce the lignin content by 40% and at the same time increase the cellulose digestibility by 60%. Another topic of investigation under development is microbial consortia-assisted pretreatment, where cellulolytic and lignin-degrading microorganisms are added in co-culture or successively. Incorporating hydrolytic microbes into the CBP system leads to the fact that a separate enzymatic saccharification stage is not needed. The consequent microbial communities produce an amalgam of endoglucanases, xylanases, and ligninases and hence a perfect utilization of biomass *in situ* (Rasool & Irfan, 2024). Further, alkaline oxidation procedures using dilute levels of hydrogen peroxide or ozone have also been standardized to be acceptable to downstream CBP microbes. The methods have proven to reduce the interferences of lignin by 70% and cut down the energy requirement of the traditional steam explosion by 30%. The existing pretreatment methods are increasingly combined with real-time monitoring systems, which probe enzymatic activity and adjust the pH or add buffer reagents. This kind of feedback control has been seen to be useful in semi-continuous CBP systems where variations in composition of feedstock can lead to loss of uniformity of hydrolysis. New solutions are the usage of microbe-loaded biochar, which is used as a pretreatment matrix and facilitates microbial carriers, thereby increasing the accessibility of the substrates as well as processing stability concurrently (Saravanan et al., 2023).

7.4.3 Process Integration and Intensification

The commercialization of continuous industrial-scope biogas production needs process combinations and intensification as non-negotiable preconditions. The conventional multi-stage system through hydrolysis, saccharification, fermentation, and AD is currently being reduced to low-scale bioreactor sets and co-processing platforms that maximize incoming energy, metabolic activities, and product extraction. A prominent example of process intensification would be simultaneous saccharification and methanogenesis (SSM), where hydrolytic microorganisms and methanogens are kept together and are co-cultivated under optimized conditions (Bothma, 2020). Instead, two-stage thermophilic–mesophilic systems are extensively performed. In the first stage, hydrolysis and fermentation are carried out at 55–60 °C, whereas the second stage is carried out at 35–40 °C, which is for methanogenesis. This sequential order allows that the respective microbial communities to function within their most preferred temperature ranges (Westerholm et al., 2018). These systems are reported to reduce energy losses by approximately 20% while simultaneously improving volumetric productivity. Process intensification has relied heavily on the use of digitalization and automation of processes. In real time, sensors monitor pH, gas mixture, and biomass degradation, thus producing data that is processed by AI-based control systems and is used to optimize the feeding, buffering, and temperature records. The value of closed-loop control in ensuring the stable output conditions of the various situations of the feedstock. Also, new membrane bioreactor (MBR) designs and integrated electrochemical-biological systems (IEBS) are being tested at the pilot scale. Such systems not only increase the level of conversion but also allow the process of nutrient recovery and purification of biogas at the same time, thus making the operation of CBP more economical (Figure 7.2) (Lin et al., 2021).

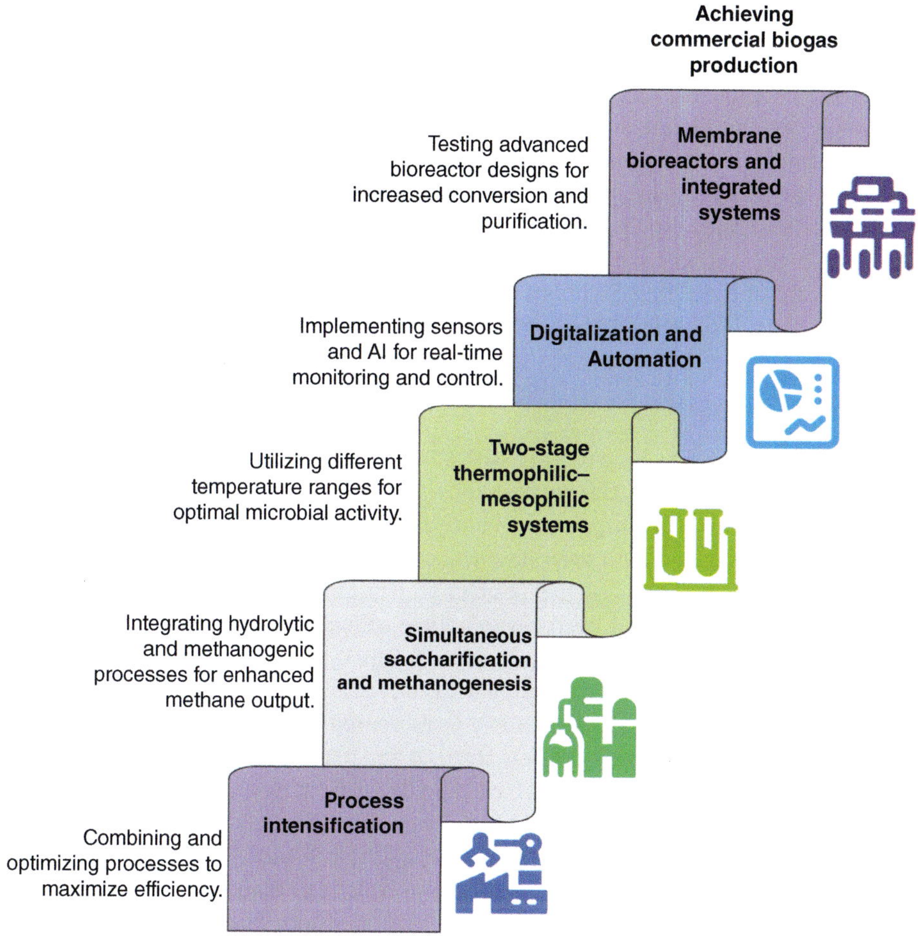

Figure 7.2 Advanced bioreactor technologies for commercial biogas production, including membrane bioreactors, AI-driven monitoring, and process-intensification strategies.

7.5 Future Perspectives

7.5.1 Technological Advancements

The evolution of CBP is tightly coupled with technological innovations in genetic engineering, process control, and bioreactor design. Recent years have seen marked progress in these domains, pushing CBP from experimental proof-of-concept toward semi-industrial implementation. At the core of these advances is metabolic engineering of microbial strains such as *Clostridium thermocellum and Caldicellulosiruptor bescii*, which can simultaneously produce hydrolytic enzymes and convert sugars into ethanol or methane precursors (Zhang et al., 2021). Emerging technologies like adaptive laboratory evolution and CRISPR-Cas9 multiplex editing allow tailored modification of microbial genomes for enhanced resistance to inhibitors and faster substrate utilization. Coupled with omics-guided strain selection, these strategies are making it feasible to construct synthetic consortia optimized for different feedstocks and operational environments. In terms of physical systems, modular bioreactor designs incorporating temperature zoning (thermophilic–mesophilic stages) and real-time analytics have been proposed. Additionally, AI and digital twins are being applied to simulate metabolic and process kinetics for predictive control and fault detection (Mosquera-Lopez & Jacobs, 2024). Digitalization plays a key role in enhancing system reliability and enabling remote operation for decentralized rural installations. Despite these advances, scale-up remains a technological bottleneck, especially due to the shear sensitivity of microbes and uneven mass transfer at high solids loading. Solutions like biofilm-based CBP and membrane-integrated bioreactors are being explored as next-generation platforms.

7.5.2 Commercial Viability

Commercial viability is a major challenge with CBP systems, although some laboratory and pilot-scale tests are promising. Capital costs, enzyme and inoculum reuse, efficient energy use, and policy incentives are important viability determinations (de Lima et al., 2022). This approach eliminates the need for separate enzyme production, which typically accounts for 20–30% of the cost of cellulosic biofuel production. The solution to this presented by CBP is the utilization of engineered microbes to secrete cellulases and hemicellulases locally and thereby substantially minimize input charges. CBP systems can be made to diversify revenue streams when used in combination with co-product recovery (e.g., ethanol, hydrogen, and organic acids) (Hafyan et al., 2024). As measured in the Levelized Cost of Biogas (LCOB), estimated to be in a range of between 0.04 and 0.07 per kilowatt hour under optimal conditions, CBP-integrated systems have been deemed competitive with that of natural gas on European markets, especially where carbon credits or subsidies are available or applicable. Terragia Bioenergy and Mascoma (formerly involved in the research of CBP) are companies that have tested CBP-related systems based on ethanol/biogas. Commercial roll-out, however, has fixated on the large scale, with feedstock heterogeneity and microbial instability in continuous processes acting as a hindrance. The working of biorefineries from an economic aspect reveals that economies of scale can only be achieved when biorefineries process biomass >500 dry tons/day, or CBP modules are attached to existing AD or pulp and paper plants (Chaturvedi, 2017). Co-location can lower the costs of logistics and sharing thermal energy systems. In terms of financing, there is an increasing trend that hybrid business models, consisting of the combination of biogas, biochar, and nutrient recovery, would be economically sustainable. Such multiproduct plants have the ability of tapping into various subsidy streams, and they are not as susceptible to market changes. Wise techno-economic formulation, together with strong microbial and pre-treating technologies, is needed to move CBP out of the lab-scale and commercially into biorefineries.

7.5.3 Sustainability Outlook

The technology of CBP holds great promise to meet the concern on sustainability that comes with the traditional production of bioenergy. This efficiency allows CBP to consume less money, chemicals, and energy in general by combining a number of steps and operating them within one system, which categorizes it as a microbial system (Agbor et al., 2014; Parisutham et al., 2014). One of the most important environmental advantages is the minimization of greenhouse gases. Lifecycle analyses of CBP-based systems indicate that CBP has the potential to produce negative net emissions, particularly when combined with sequestration of flue gas carbon or digestate. As an example, Kubis (2023) listed net-negative emissions of −140 g CO_2-eq/MJ associated with AD and biomethane harvest in the context of CBP-based ethanol production. Besides, CBP fosters the valorization of agricultural wastes, that is, straw, stover, and pruning residues, which remain underused or are burnt, thus playing a role in the air pollution (Thakur et al., 2024). CBP can decouple food and energy because it enables the use of raw or low-prepared biomass, which is not to be grown at the expense of agricultural land unlike maize or sugarcane. Sustainability is also concerned with resource efficiency, particularly on water and nutrient use. Higher circularity may be achieved, as ammonium, phosphate, and potassium can be recovered for reuse as biofertilizers by CBP bioreactors with closed digestate loops. Energy balances can also be further optimized by integration with renewable hydrogen or solar-thermal units and enable the upgrading of methane into biomethane or syngas (Agrafiotis et al., 2014). Social sustainability will rely on decentralized deployment models, especially in rural areas and agro-industrial clusters. Modular CBP units may be used to meet local waste management needs and produce energy and organic fertilizers to encourage regional bioeconomies. However, there is the fear of biosafety, particularly when it comes to genetically modified organisms (GMOs) in CBP. Regulatory frameworks must change to encourage safe and responsible practices involving synthetic biology, especially in open or semi-open digestion settings. Risk assessments, as well as risk containment measures, will be critical in ensuring that these issues are accepted by wider society (Aven, 2016).

7.6 Conclusion

Consolidated bioprocessing (CBP) stands at the forefront of sustainable biogas production, offering a transformative approach by unifying enzyme production, substrate hydrolysis, and fermentation within a single, streamlined process. By integrating traditionally separate stages, CBP eliminates the need for external enzymatic inputs and reduces the energy and cost burden associated with conventional biorefinery operations. This innovation is especially crucial for the effective valorization of lignocellulosic biomass—such as agricultural residues and organic wastes—that would otherwise remain underutilized or contribute to environmental degradation through open-field burning or landfilling. Recent advancements in microbial strain engineering, particularly through synthetic biology and CRISPR-based tools, have significantly enhanced the metabolic efficiency and stability of cellulolytic and methanogenic consortia. These developments are complemented by emerging pretreatment technologies that are both eco-friendly and compatible with microbial health, further improving substrate accessibility and process yields. Additionally, intensified process designs—such as thermophilic–mesophilic bioreactor configurations, feedback-automated systems, and membrane-assisted digestion—are demonstrating potential for industrial-scale applications. Despite these innovations, challenges remain regarding scale-up, microbial robustness under variable feedstock conditions, and achieving consistent economic viability across diverse geographies. However, techno-economic assessments suggest that with proper integration and co-product recovery strategies, CBP systems can achieve competitive biogas production costs, reduce greenhouse gas emissions, and support circular bioeconomies. Looking forward, the future of CBP in biogas production lies in modularity, digital integration, and policy-driven incentives that support decentralized renewable energy systems. When supported by responsible biotechnology regulations and stakeholder collaboration, CBP can not only revolutionize bioenergy production but also play a pivotal role in achieving global climate and sustainability targets through low-carbon, resource-efficient, and resilient bioprocessing solutions.

References

Agbor, V., Carere, C., Cicek, N., Sparling, R., & Levin, D. (2014). Biomass pretreatment for consolidated bioprocessing (CBP). *Advances in Biorefineries*, 234–258. https://doi.org/10.1533/9780857097385.1.234.

Agrafiotis, C., Von Storch, H., Roeb, M., & Sattler, C. (2014). Solar thermal reforming of methane feedstocks for hydrogen and syngas production—A review. *Renewable and Sustainable Energy Reviews*, *29*, 656–682.

Althuri, A., & Mohan, S. V. (2020). Sequential and consolidated bioprocessing of biogenic municipal solid waste: A strategic pairing of thermophilic anaerobe and mesophilic microaerobe for ethanol production. *Bioresource Technology*, *308*, 123260.

Amin, F. R., Khalid, H., El-Mashad, H. M., Chen, C., Liu, G., & Zhang, R. (2021). Functions of bacteria and archaea participating in the bioconversion of organic waste for methane production. *Science of the Total Environment*, *763*, 143007.

Anas, M., Khalid, A., Saleem, M. H., Ali Khan, K., Ahmed Khattak, W., & Fahad, S. (2025). Symbiotic synergy: Unveiling plant-microbe interactions in stress adaptation. *Journal of Crop Health*, *77*(1), 1–21.

Aven, T. (2016). Risk assessment and risk management: Review of recent advances on their foundation. *European Journal of Operational Research*, *253*(1), 1–13.

Bhatia, S. K., Bhatia, R. K., Choi, Y.-K., Kan, E., Kim, Y.-G., & Yang, Y.-H. (2018). Biotechnological potential of microbial consortia and future perspectives. *Critical Reviews in Biotechnology*, *38*(8), 1209–1229.

Bioverfahrenstechnik, M. S., & Kaya, G. (2022) Evaluation of Bioenergy Recovery from Olive Mill Waste and Cotton Stalks in Turkey: Case Study for Biogas Production.

Bothma, L. (2020). Study on the microbial community shift during co-fermentation of substrates. (Doctoral dissertation, North-West University (South Africa)).

Chang, J., Cheng, W., Yin, Q., Zuo, R., Song, A., Zheng, Q., Wang, P., Wang, X., & Liu, J. (2012). Effect of steam explosion and microbial fermentation on cellulose and lignin degradation of corn stover. *Bioresource Technology*, *104*, 587–592.

Chaturvedi, T. (2017). *Process design, economics and policies for biorefineries*. (Doctoral dissertation, Masdar Institute of Science and Technology).

de Lima, E. A., Mandelli, F., Kolling, D., Souza, J. M., de Oliveira Filho, C. A., da Silva, M. R., de Mesquita Sampaio, I. L., Junqueira, T. L., Chagas, M. F., & Teodoro, J. C. (2022).

Development of an economically competitive Trichoderma-based platform for enzyme production: Bioprocess optimization, pilot plant scale-up, techno-economic analysis and life cycle assessment. *Bioresource Technology*, *364*, 128019.

Díaz, V. H. G., & Willis, M. J. (2019). On the economic optimisation of ethanol production using corn stover feedstock: A new kinetic model, a green recovery system and a de-acetylation step. *Energy Conversion and Management*, *202*, 112200.

Flevaris, K., & Chatzidoukas, C. (2021). Optimal fed-batch bioreactor operating strategies for the microbial production of lignocellulosic bioethanol and exploration of their economic implications: A step forward towards sustainability and commercialization. *Journal of Cleaner Production*, *295*, 126384.

Fong, S. S. (2014). Computational approaches to metabolic engineering utilizing systems biology and synthetic biology. *Computational and Structural Biotechnology Journal*, *11*(18), 28–34.

Hafyan, R. H., Mohanarajan, J., Uppal, M., Kumar, V., Narisetty, V., Maity, S. K., Sadhukhan, J., & Gadkari, S. (2024). Integrated biorefinery for bioethanol and succinic acid co-production from bread waste: Techno-economic feasibility and life cycle assessment. *Energy Conversion and Management*, *301*, 118033.

Hassan, S. S., Williams, G. A., & Jaiswal, A. K. (2019). Moving towards the second generation of lignocellulosic biorefineries in the EU: Drivers, challenges, and opportunities. *Renewable and Sustainable Energy Reviews*, *101*, 590–599.

Kabeyi, M. J. B., & Olanrewaju, O. A. (2022). Biogas production and applications in the sustainable energy transition. *Journal of Energy*, *2022*(1), 8750221.

Kabeyi, M. J. B., Olanrewaju, O. A., & Akpan, J. (2024). Biogas production and process control improvements. In *From biomass to biobased products*. IntechOpen. DOI: 10.5772/intechopen.113061.

Kruczek, G., Przybyła, G., Ziółkowski, Ł., & Adamczyk, W. P. (2019). Comparative assessment of the application of methane and biogas in energy production: An experimental and numerical investigation. *Renewable Energy*, *143*, 1519–1530.

Kubis, M. R. (2023). *Lignocellulose conversion via consolidated bioprocessing: High solid loadings, bioreactor development, and technoeconomic analysis*. (Doctoral dissertation, Dartmouth College).

Kukreti, N., Kumar, P., & Kataria, R. (2024). Sustainable biotransformation of lignocellulosic biomass to microbial enzymes: An overview and update. *Industrial Crops and Products*, *222*, 119432.

Li, W., Khalid, H., Zhu, Z., Zhang, R., Liu, G., Chen, C., & Thorin, E. (2018). Methane production through anaerobic digestion: Participation and digestion characteristics of cellulose, hemicellulose and lignin. *Applied Energy*, *226*, 1219–1228.

Lin, R., O'Shea, R., Deng, C., Wu, B., & Murphy, J. D. (2021). A perspective on the efficacy of green gas production via integration of technologies in novel cascading circular bio-systems. *Renewable and Sustainable Energy Reviews*, *150*, 111427.

Mauser, I. (2017). *Multi-modal building energy management* (Doctoral dissertation, Dissertation, Karlsruhe, Karlsruher Institut für Technologie (KIT), 2017).

Mishra, R., Naik, B. K. R., Raut, R. D., & Paul, S. K. (2022). Circular economy principles in community energy initiatives through stakeholder perspectives. *Sustainable Production and Consumption*, *33*, 256–270.

Mosquera-Lopez, C., & Jacobs, P. G. (2024). Digital twins and artificial intelligence in metabolic disease research. *Trends in Endocrinology & Metabolism*, *35*(6), 549–557.

Olaganathan, R. (2025). 7 jet fuel production. In S. K. Bhatia, P. S. Panesar, & R. Gurav (Eds.), *Microbial biofuel: A sustainable source of renewable energy* (p. 173). CRC Press.

Parisutham, V., Kim, T. H., & Lee, S. K. (2014). Feasibilities of consolidated bioprocessing microbes: From pretreatment to biofuel production. *Bioresource Technology*, *161*, 431–440.

Ramarajan, R., & Manohar, C. S. (2017). Biological pretreatment and bioconversion of agricultural wastes, using ligninolytic and cellulolytic fungal consortia. *Bioremediation Journal*, *21*(2), 89–99.

Rasool, G., & Irfan, M. (2024). The role of microbial diversity in lignocellulosic biomass degradation: A biotechnological perspective. *ChemBioEng Reviews*, *11*(3), 613–635.

Re, A., & Mazzoli, R. (2023). Current progress on engineering microbial strains and consortia for production of cellulosic butanol through consolidated bioprocessing. *Microbial Biotechnology*, *16*(2), 238–261.

Salehi Jouzani, G., & Taherzadeh, M. J. (2015). Advances in consolidated bioprocessing systems for bioethanol and butanol production from biomass: A comprehensive review. *Biofuel Research Journal*, *2*(1), 152–195.

Saravanan, A., Swaminaathan, P., Kumar, P. S., Yaashikaa, P., Kamalesh, R., & Rangasamy, G. (2023). A comprehensive review on immobilized microbes-biochar and their environmental remediation: Mechanism, challenges and future perspectives. *Environmental Research*, *236*, 116723.

Singh, D., Tembhare, M., Dikshit, A. K., Dangi, M. B., & Kumar, S. (2024). Technical and operational challenges in setting up a decentralized biogas plant: Opportunities and future perspective toward sustainable nation. *Process Safety and Environmental Protection*, *185*, 392–407. https://doi.org/10.1016/j.psep.2024.02.079.

Singh, J., Suhag, M., & Dhaka, A. (2015). Augmented digestion of lignocellulose by steam explosion, acid and alkaline pretreatment methods: A review. *Carbohydrate Polymers*, *117*, 624–631.

Singhania, R. R., Patel, A. K., Singh, A., Haldar, D., Soam, S., Chen, C.-W., Tsai, M.-L., & Dong, C.-D. (2022). Consolidated bioprocessing of lignocellulosic biomass: Technological advances and challenges. *Bioresource Technology*, *354*, 127153.

Sitara, A., Hocq, R., Horvath, J., & Pflügl, S. (2024). Industrial biotechnology goes thermophilic: Thermoanaerobes as promising hosts in the circular carbon economy. *Bioresource Technology*, *408*, 131164.

Srivastava, R. K. (2019). Bio-energy production by contribution of effective and suitable microbial system. *Materials Science for Energy Technologies*, *2*(2), 308–318.

Straathof, A. J. (2014). Transformation of biomass into commodity chemicals using enzymes or cells. *Chemical Reviews*, *114*(3), 1871–1908.

Thakur, A., Kumar, A., & Somya, A. (2024). Forestry and agricultural residues-based wastes. In C. Verma & S. Dubey (Eds.), *Biomass wastes for sustainable industrial applications: A waste-to-wealth approach* (p. 95). CRC Press.

Tseng, F.-M., Chiu, Y.-J., & Chen, J.-S. (2009). Measuring business performance in the high-tech manufacturing industry: A case study of Taiwan's large-sized TFT-LCD panel companies. *Omega*, *37*(3), 686–697.

Vijayan, D. S., Devarajan, P., Sivasuriyan, A., Stefańska, A., Koda, E., Jakimiuk, A., Vaverková, M. D., Winkler, J., Duarte, C. C., & Corticos, N. D. (2023). A state of review on instigating resources and technological sustainable approaches in green construction. *Sustainability, 15*(8), 6751.

Wainaina, S., Lukitawesa, Kumar Awasthi, M., & Taherzadeh, M. J. (2019). Bioengineering of anaerobic digestion for volatile fatty acids, hydrogen or methane production: A critical review. *Bioengineered, 10*(1), 437–458.

Wang, J., Ma, D., Lou, Y., Ma, J., & Xing, D. (2023). Optimization of biogas production from straw wastes by different pretreatments: Progress, challenges, and prospects. *Science of the Total Environment, 905*, 166992.

Westerholm, M., Isaksson, S., Lindsjö, O. K., & Schnürer, A. (2018). Microbial community adaptability to altered temperature conditions determines the potential for process optimisation in biogas production. *Applied Energy, 226*, 838–848.

Wu, G., Yan, Q., Jones, J. A., Tang, Y. J., Fong, S. S., & Koffas, M. A. (2016). Metabolic burden: Cornerstones in synthetic biology and metabolic engineering applications. *Trends in Biotechnology, 34*(8), 652–664.

Yan, P., Shi, H.-X., Chen, Y.-P., Gao, X., Fang, F., & Guo, J.-S. (2020). Optimization of recovery and utilization pathway of chemical energy from wastewater pollutants by a net-zero energy wastewater treatment model. *Renewable and Sustainable Energy Reviews, 133*, 110160.

Zhang, K., Zhao, W., Rodionov, D. A., Rubinstein, G. M., Nguyen, D. N., Tanwee, T. N., Crosby, J., Bing, R. G., Kelly, R. M., & Adams, M. W. (2021). Genome-scale metabolic model of *Caldicellulosiruptor bescii* reveals optimal metabolic engineering strategies for bio-based chemical production. *MSystems, 6*(3), 10–1128.

8

Technological Assessment of Bioprocess and Consolidated Process: A Way Forward

Waseem Ahmed Khattak[1], Erdogan E. Hakki[2], Muhammad Anas[2], Javed Iqbal[3], and Banzeer Ahsan Abbasi[4]

[1]*Department of Plant Sciences, Faculty of Biological Sciences, Quaid-I-Azam University, Islamabad, Pakistan*
[2]*Department of Soil Science & Plant Nutrition, Selçuk University, Konya, Türkiye*
[3]*Department of Botany, Bacha Khan University, Charsadda, Pakistan*
[4]*Department of Botany, Rawalpindi Women University, Rawalpindi, Pakistan*

8.1 Introduction

Bioprocess-based industries are the leading sustainable industries of our time: pharmaceuticals, biofuels, bioplastics, and food (Ramchuran et al., 2023). With the increasing demand for cleaner, greener technology, the use of biological systems for remarkable advances in technology is fast becoming popular. Consolidated bioprocessing (CBP) is one of these strategies, and it promises to make both the bioprocess and the CBP system simpler and more direct by combining (consolidating) the production of a product, along with its secreted enzymes, into a single process. The technological evaluation of bioprocessing systems, specifically geared toward consolidated processes, their assessment, and future routes toward the industrially applicable stage, are addressed in this chapter. To achieve a successful transfer of these technologies into commercial and environmental practices, it is imperative to comprehend their advantages, drawbacks, and testing techniques (Shrivastava, 2018).

8.1.1 Overview of Bioprocess Technologies

Bioprocess technologies are related to the use of living organisms or their substances (such as enzymes or metabolites) for any industrial, bio-based products. These processes are central to many industries, such as pharmaceuticals, agriculture, energy production, and waste treatment. A conventional bioprocess mainly includes three units: upstream (substrate preparation and inoculum build-up), bioconversion (bioreaction), and downstream (product recovery and purification) (Kalaichelvan & Pandi, 2019). Typically, bioprocesses are broken down into different phases, and each is individually optimized for efficiency or output. The compartmentalized steps, however, while effective, may require more capital and operational costs, longer production times, and larger energy requirements. Advances in recent years have expanded the horizons of bioprocesses

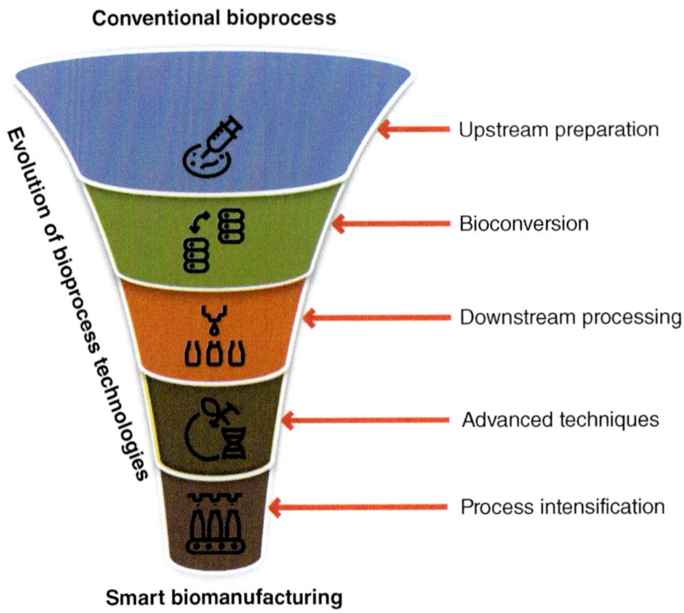

Figure 8.1 Evolution of bioprocess technologies, highlighting key components such as smart biomanufacturing, upstream preparation, and downstream processing. The diagram outlines the progression and integration of advanced techniques in modern bioprocessing (*Source:* Macrovector / Adobe Stock Photos).

by developments in metabolic engineering, synthetic biology, systems biology, and process intensification. These tools enable a more personalized development of microbial strains and bioreactor designs, with positive prospects for higher yields, shorter processing, and more sustainable processes (Figure 8.1). The challenge is the development of damage-tolerant and high-temperature systems, which can operate under industrial conditions for a short period with limited environmental impact. In addition, more recent designs like fed-batch, continuous, and perfusion strategies offer more control over nutrient concentration, oxygen transfer, and by-product inhibition. To this purpose, the application of monitoring and controlling online and in real time has improved the accuracy and speed of response dynamics in the bioprocesses. Together, these advances suggest a departure from conventional fermentation approaches to smarter, more automated, and more integrated biomanufacturing platforms (Gargalo et al., 2020).

8.1.2 Importance of Technological Assessment in Bioprocessing

Due to the increased global awareness of climate change, energy security, and sustainable industrial practice, technological assessment of bioprocesses is inevitable (Melendez et al., 2022). In short, bioprocesses are now evaluated not only in terms of yield/productivity, but also by a number of other dimensions such as economic viability, environmental sustainability, scale-up potential, and social acceptance. Technological assessment involves the formal analysis of the functionality, impact, efficiency, and other aspects of a technology or technological practice. In the field of bioprocessing, it aids in the evaluation of the industrial readiness of a process, bottlenecks that need to be addressed, environmental hazards, and costs involved. This is especially true for downstream processing of second-generation biorefinery systems such as CBP, in which most of the processes are combined in a single operation, opening new opportunities as well as challenges (Hassan et al., 2019). For example, a fermentation process that is high yielding might be non-scalable if it relies on a rare or expensive feedstock or if it necessitates unusually aseptic conditions that are difficult to replicate on a large scale. Second, if the production of valuable products requires too much water or the creation of toxic by-products, the resulting process would similarly not be classified as sustainable. Therefore, life cycle assessment (LCA), techno-economic quantity (TEA), and risk analysis are indispensable parts of technology validation. Such exercises also provide decision-makers—that is, industrial partners and researchers alike—with the good and not-so-good about different technologies. They add transparency based on economics, regulation, market, and environmental costs and benefits. The use of digital tools such as process simulation, digital twins, and artificial intelligence (AI)-enabled control systems would be able to further improve the accuracy, decrease time-to-market, and provide support to predictive scaling strategies included in the technological assessment (Holland & Shapira, 2024). Technological assessment also contributes to policy coherence at a time when countries need to prioritize bio-based technologies that align with their energy and environmental goals (Holland & Shapira, 2024).

8.1.3 Scope and Objectives of the Chapter

This chapter attempts to offer a holistic analysis of new and current bioprocessing technologies, particularly focusing on CBP, which is the process of biologically converting the components of biomass and its processing intermediates into products in a single integrated fermentation process. CBP has been identified as a potentially transformative concept for producing second-generation biofuels as well as other biorefinery applications by reducing costs and simplifying the process. The main aim of this chapter is to close the knowledge gap between a conceptual breakthrough and industrial application by evaluating the techno-economic, scalable, and sustainable potential for CBP and associated bioprocesses. It is intended to provide a comprehensive summary of the principles, design, technology, facilities, and factors that influence the successful development of processes and the advancement of global innovation theory for future bio-based technologies, as well as the knowledge required to accelerate the development of biological host systems. ; to review new technologies in synthetic biology, enzyme engineering, reactor design, and process integration that are leading the way in the development of the bioprocessing industry; to evaluate technological assessment criteria, based on performance measurements, economic feasibility, environmental performance (LCA), and regulation; to determine obstacles and bottlenecks like substrate deficiency, enzymes cost, process inhibition, and downstream processing complexity; to provide industrial best-practice case studies and industrial viewpoints, demonstrating examples of successful implementation of novel bioprocesses, with a focus on bioenergy, bioplastics, and waste-to-resource industries. The Phenomena of World Resource and Ecology initiative invited discussions on future directions, including the integration of AI and digital technologies, the advancement of the circular bioeconomy, and contributions to the Sustainable Development Goals (SDGs). In the end, the chapter aims to pave the way for researchers, engineers, and policymakers who are now pursuing the use of bioprocessing for sustainable development.

It also highlights the importance of a comprehensive understanding of what technology is capable of, considering real-world limits, and being guided by a "forward-looking" approach. By integrating scientific understanding and experiential knowledge, the chapter seeks to encourage the uptake of robust, resource-effective, and economically competitive bioprocessing technologies that will be the key to the bio-based industries of the future.

8.2 Fundamentals of Bioprocessing

Bioprocessing involves the use of living organisms or their cells to transform materials and perform chemical synthesis and is used to manufacture end products like fuels, food additives, and biopharmaceuticals. It is the bedrock of industrial biotechnology and is involved in various industries (pharmaceutics, agronomy, energy production, food management, and environmental control). Knowledge of the fundamentals of the building blocks, process types, and biological pathways becomes crucial for performance optimization and further technological development. This chapter presents the main bases of bioprocessing, the distinction between various process modes, and the relevance of microorganisms and enzymes in such systems (Moser, 2012).

8.2.1 Key Components and Stages of Bioprocesses

A generic bioprocess includes three major steps: upstream, bioconversion, and downstream processing. All these stages are interdependent, and each is necessary; they all together contribute to the overall system effectiveness and productivity. The term "upstream" encompasses everything that happens before the actual conversion process takes place. This involves choosing and modifying the microbe or enzyme system to be used, culture media formulation, sterilization, and development of the inoculum (Mould et al., 2005). The objective at this stage is to find the best environment for the biological media to grow and to do its work as well as it can.

Stage of bioconversion or fermentation: This is the core of the bioprocess. In this process, a microorganism (or an enzyme) produces valuable products (often from biomass raw materials or substrates) such as organic acids, alcohols, proteins, enzymes, and antibiotics. This usually takes place in bioreactors, in which pH, temperature, aeration, agitation, and substrate concentration are very tightly regulated.

Downstream stage: This involves the isolation and purification of the product from the fermentation broth. This can include cell separation, filtration, centrifugation, solvent extraction, precipitation, chromatography, and drying. Downstream processing is, for the greater part, product- and purity-dependent, and thus complex and expensive. For instance, the pharmaceutical-grade proteins require stricter and more expensive techniques to purify than the case where the end product is ethanol or organic acids. Other supporting factors are bioreactor design, process control systems, and automation solutions allowing for upscaling of the bioprocess from laboratory to industrial scale while ensuring reproducibility, safety, and functionality (Mitra & Murthy, 2022).

8.2.2 Classification of Bioprocesses: Batch, Fed-Batch, and Continuous

Bioprocesses can be classified based on how substrates are supplied and how products are harvested. The three main types are batch, fed-batch, and continuous processes. Each has its own advantages, limitations, and applications, depending on the nature of the product, organism, and market demand (Carus & Dammer, 2018).

8.2.2.1 Batch Bioprocessing

In batch culture, all medium components, including microorganisms, are introduced at once into the bioreactor, and the system is closed and proceeds for a predetermined time. There is no further feed addition, and the products are collected only when the run is over. The operation of this method is easy, it is easy to control, and it is applicable for small batch or odd-lot production. The method has productivity restrictions, however, as the organism could go into nutrient starvation, or the presence of inhibitory by-products could be increased during the cultivation (Rajpurohit & Eiteman, 2022).

8.2.2.2 Fed-Batch Bioprocessing

Fed-batch designs are better than batch systems. Nutrients are fed slowly or subject to control in the bioconversion step to keep the culture in a growth phase for a longer duration. This approach minimizes substrate inhibition, prolongs the production phase, and improves both yield and productivity. Fed-batch operation is a popular downstream processing option in the pharmaceutical and enzyme industries, where high-value biomolecules, including insulin, monoclonal antibodies, and therapeutic enzymes, are being manufactured (Matanguihan & Wu, 2022).

8.2.2.3 Continuous Bioprocessing

Fresh substrate is also continuously fed to the bioreactor, and at the same time, the product-containing broth is taken out from it, in the case of continuous systems. As a result, a continuous process becomes possible and permits manufacturing at the largest possible reactant volume. Continuous operations are more efficient and have better Free Process Economics, but are more difficult to monitor and control to achieve the desired steady state. These represent ideal process types for commodities such as ethanol, organic acids, or waste treatment, where the economy of scale and low-cost operation are critical factors. Hybrid approaches such as perfusion-culture systems and semi-continuous modes are also being developed, taking advantage of the positive aspects of each of these classical methods, especially for fragile or high-yielding processes (Romann, 2023).

8.2.3 Role of Microorganisms and Enzymes

Microorganisms and enzymes form the biological core of any bioprocess. Their selection, engineering, and optimization directly affect process efficiency, product yield, and system resilience.

8.2.3.1 Microorganisms in Bioprocessing

Bacteria, yeast, fungi, and algae are commonly used microorganisms, as they offer the capacity to convert a variety of substrates into a variety of bioproducts. For instance:

- To produce recombinant proteins, bacteria such as *Escherichia coli* and *Bacillus subtilis* are used.
- Except for *Saccharomyces cerevisiae*, yeast is used in ethanol production and baking processes.
- Fungi such as *Aspergillus niger* are effective organic acid or industrial enzyme producers.
- Algae are promising resources in biofuel, pigment, and nutraceutical production.

The activity of these microbes is based on their metabolism, ability to thrive under stress, growth properties, etc. Breakthroughs in synthetic biology, systems biology, and adaptive evolution made it possible to develop engineered strains that yield, are robust, and utilize substrate more efficiently than wild type (Tang et al., 2021).

8.2.3.2 Enzymes in Bioprocessing

Enzymes, the biological catalysts, can promote specific biochemical reactions effectively with high selectivity. They are of high interest in applications such as biomass hydrolysis, sugar fermentation, and biotransformation in pharmaceutical production. Industrial enzymes such as amylases, proteases, lipases, cellulases, and laccases are applied in the food, textile, and energy sectors. Enzymes have several advantages: they can function under moderate conditions (pH, temperature), and the chemical consumption and non-specific reactions are lower. BBP aims to produce in a single reactor (through "one pot," or CBP) the enzymes that are needed from one or more microbial strains, which means that no other microorganism or facility is necessary to produce enzymes or hydrolyze polysaccharides. The recent advances are focusing on enzyme immobilization for reusability, directed evolution for better activity and stability, and fusion enzymes with two or more catalytic activities into a single protein (Figure 8.2; Porter et al., 2016).

8.3 Consolidated Bioprocessing (CBP): Concept and Evolution

8.3.1 Definition and Core Principles

CBP is an innovative bioconversion process that combines several stages of biomass utilization, including enzyme formation, hydrolysis, and fermentation, into a single process (Singhania et al., 2022). CBP is based on the concept of doing away with independent enzyme production and hydrolysis phases by developing microorganisms engineered to produce hydrolytic enzymes and convert the resulting sugars into target products in the same bioreactor. This combined process greatly simplifies the process design, lowers capital and operational costs, and improves conversion efficiency. Conventional bioprocessing of lignocellulosic biomass, for example, often comprises (i) pretreatment of the biomass, (ii) production of hydrolytic enzymes, (iii) enzymatic saccharification, and (iv) microbial fermentation (Abo et al., 2019). CBP integrates steps (ii), (iii), and (iv) in a single reactor with a single microorganism or a compatible microbial consortium. For CBP purposes, the perfect microbial host would include the entire complement of cellulases and would efficiently ferment hexoses and pentoses to produce biofuels or high-value chemicals. Several

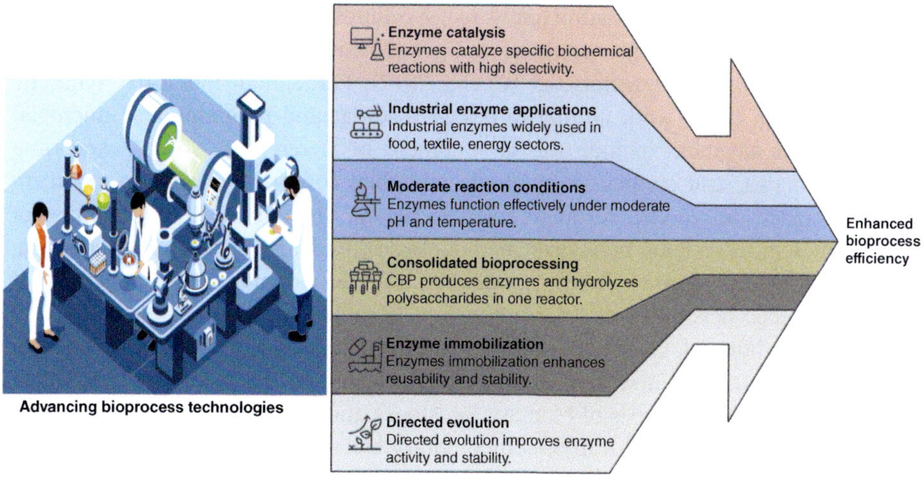

Figure 8.2 Key aspects of enzyme catalysis, including industrial applications, reaction conditions, and advanced techniques like directed evolution and immobilization. The diagram highlights modern bioprocess technologies for efficient biochemical transformations.

aspects are important for the success of CBP, such as the creation of microorganisms with wide substrate utilization, high product tolerance, balanced metabolic pathways, and effective expression of cellulolytic systems (den Haan et al., 2015). The economic attractiveness of CBP is based on the potential to decrease the cost of enzyme production (which represents a large fraction of process cost in lignocellulosic biorefineries). In addition, CBP enables process intensification, which results in higher reactor throughput and reduced energy demand. CBP represents a transformative paradigm in bioprocessing that aligns with the principles of green chemistry and the circular bioeconomy. Emerging themes in this field emphasize the need for efficient biomass valorization, reduction of environmental footprints, and the advancement of renewable bio-based chemicals and fuels. (Pandey & Sharma, 2024).

8.3.2 Historical Development of CBP

CBP was developed as an alternative to complex multi-step biomass conversion processes, specifically for lignocellulosic biofuel production (Sarangi & Nanda, 2019). Research organizations and funding institutions, such as the US Department of Energy (DOE), realized since the early 2000s that the high cost of enzyme production and the complexity of processes were the main bottlenecks for the commercialization of second-generation biofuels. This observation initiated the exploration of CBP as an inexpensive and scalable option. Early attempts involved the screening and manipulation of model native cellulolytic microorganisms such as *Clostridium thermocellum*, *Caldicellulosiruptor saccharolyticus*, and *Trichoderma reesei*, which have the requisite enzymatic machinery to break down cellulose (Rabby et al., 2022). But many of these strains did not have adequate routes to ferment the sugars that were generated into valuable products such as alcohol and organic acids. On the other hand, the industrial fermenters like *S. cerevisiae* and *Zymomonas mobilis* seemed to be efficient in fermentation but were not cellulolytic. This led to the engineering of recombinant strains through metabolic and genetic strategies to co-couple saccharification and fermentation characteristics (Cheng, 2017). Throughout the 2010s, synthetic biology was taken up in work aiming to design modular and programmable microbial systems. Recent advances include the development of *S. cerevisiae* strains expressing bacterial cellulases, and studies on synthetic microbial consortia where one organism degrades biomass and another ferments released sugars. These works took advantage of omics technologies, systems biological methodologies, and high-throughput screening, which have provided improved and streamlined strain development. The application of CBP has expanded beyond bioethanol production to include a wide range of bio-based products such as butanol, lactic acid, succinic acid, and bioplastics. CBP has evolved from its original incarnation to new versions, including technologies that are being modified to accommodate different types of biomass feedstocks and to combine pretreatment and downstream recovery into the hub of contemporary biorefinery design. Its development is being coalesced by an increasing intersection between the fields of engineering biology, materials science, and process engineering (National Research Council et al., 2014).

8.3.3 Comparison with SHF and SSF

Consolidated bioprocessing (CBP) is a development of previous concepts for the hydrolysis of biomass to sugars and the fermentation of sugars to products, such as Separate Hydrolysis and Fermentation (SHF) and Simultaneous Saccharification and Fermentation (SSF) (Joshi et al., 2021). All these approaches have their process setups, operating conditions, and economic considerations. SHF has two separate stages: enzyme hydrolysis of the pretreated biomass and yeast fermentation of the released sugars. This system enables each process to be optimized separately, for example, temperature and pH. For example, enzymes, such as cellulases, work best at 50°C, while fermenting organisms, such as *S. cerevisiae*, work best at 30°C. However, SHF has the disadvantage of longer residence times and greater capital and operating costs due to the requirement for two distinct reactors and enzyme production plants (Mithra et al., 2018). SSF, however, can combine hydrolysis and fermentation in one bioreactor, and thus perform saccharification and sugar fermentation simultaneously. This minimizes equipment requirements and processing time and reduces end-product inhibition of enzymes by continuously removing sugars via fermentation. However, SSF still depends on exogenously produced enzymes, and is typically limited by a temperature discrepancy between the hydrolysis and the fermentation stages. CBP surpasses both SHF and SSF by consolidating enzyme production, biomass saccharification, and sugar fermentation into a single process. This eliminates the requirement to produce enzymes externally or to increase the number of reactors, which reduces costs. CBP is also a process that is easier to control and provides greater productivity per volume unit. Indeed, its usefulness is restricted by the difficulty in developing organisms that are both cellulolytic and efficient at fermentation at present. Finally, while SHF is beneficial because it is flexible, and SSF is advantageous because of its higher conversions, CBP has the potential to offer the most cost reduction and process intensification perhaps if the biological challenges and technical obstacles can be overcome. The future of CBP will be in the production of strong microbial systems that integrate the greatest features of these classic formulations in a streamlined and industrially practical format (Meyer & Schmidhalter, 2012).

8.4 Technological Advancements in Bioprocess and CBP

8.4.1 Synthetic Biology and Metabolic Engineering

Synthetic biology and metabolic engineering have become disruptive strategies to improve bioprocesses for CBP application (Banner et al., 2021; Yan & Fong, 2017). These methods enable targeted engineering of microbial genomes for imparting desired properties such as improved substrate utilization and inhibitor tolerance, as well as co-production of multiple enzymes and target chemicals. Metabolic engineering refers to the targeted modification of metabolic pathways in microbes to channel carbon toward the production of desirable products such as biofuels, bioplastics, and organic acids (Lee et al., 2025). For these CBP applications, organisms are typically engineered to co-express cellulases and hemicellulases together with the pathways for ethanol or lactic acid formation. One of the difficulties here is to reconcile the three aspects: cell growth, expression of enzymes, and product formation. Integrating higher, synthetic levels, synthetic biology is aimed at defining typical biological building blocks (such as promoters, ribosome binding sites s, and operons) and abstracting them to construct functional transcriptional devices. This enables the programmable regulation of gene expression and the incorporation of complicated functions. For instance, recombinant *S. cerevisiae* strains capable of hydrolyzing cellulose and fermenting hexose and pentose sugars contaminated in lignocellulose biomass are now available, those that are natural yeast strains (Madhavan et al., 2012). Recent developments include genome editing by CRISPR-Cas9, modular cloning systems such as Golden Gate and BioBrick, and biosensors for dynamic control of metabolic pathways. By using these technologies, the strain development cycles have been drastically shortened, and high-throughput screening of the engineered variants has become possible. Synthetic biology and metabolic engineering are advancing the engineering of "super strains" specifically customized for CBP, which are able to provide high titers, yields, and productivity in a consolidated bioprocess. Their combination with computational modeling and machine learning is increasing the predictability of native and optimized cell performance in an industrial setting (Krzywanski et al., 2024).

8.4.2 Enzyme Engineering and Cocktail Optimization

Enzyme processing is a key aspect of bioprocessing, and this is especially true for CBP, where biomass hydrolysis must be efficient at process-compatible conditions. Most naturally occurring cellulases, hemicellulases, and ligninases are not stable and active, or not combined in the right synergism to an extent required for industrial applications. Enzyme engineering can overcome this issue through the manipulation of enzymes to increase their thermostability, substrate specificity, resistance to inhibitors, and high expression in host organisms (Sharma et al., 2021). Directed evolution and rational design are the two favored breeding tactics for enhancing enzyme efficiency. Directed evolution emulates a form of natural selection by generating mutant libraries and selecting mutants with a certain desirable phenotype; the idea behind rational design is that it uses the principles of structural biology and computational modeling to engineer specific amino acids (Gollihar Jr, 2017). Together, these strategies have resulted in the production of thermoactive cellulases, highly stable β-glucosidase, and enzyme fusions with increased catalytic activity. Cocktail optimization, on the other hand, refers to the identification and combination of enzymes that act in a coordinated manner to degrade recalcitrant lignocellulose (Adsul et al., 2020). Cocktails showing the highest potency are those having a combination of endoglucanases, exoglycanases, and β-glucosidases for cellulose hydrolysis, as well as xylanases and ligninases for hemicellulose and lignin depolymerization. Optimization should be based on the type of substrate, pre-treatment applied, and process conditions. For CBP, the ideal endpoint would be the direct production of these enzyme cocktails in plant. Approaches range from the construction of microbes for producing multi-functional enzymes to the construction of synthetic operons to express them together. A few attempts also deal with cell surface display systems, where enzymes are immobilized onto the microbial membrane to act directly on solid substrates. By enhancing the efficiency of enzymes and fine-tuning cocktails according to biomass, enzyme engineering expedites the speed of hydrolysis, diminishes the need to load enzyme, and therefore makes CBP technologies economically feasible (Mukherjee et al., 2024).

8.4.3 Reactor Design and Process Integration

The reactor design is another important factor for scaling up bioprocesses and attaining the maximum performance with CBP. The bioreactor is the core of functioning, where microbial growth, enzyme production, biomass hydrolysis, and product formation occur in parallel (Pino et al., 2018). Developing a system in which these several roles can be met under the best possible physiological circumstances and engineering parameters is a daunting task, yet it is of critical importance. Stirred-tank reactors (STRs) still represent the most popular design, thanks to their flexibility and control simplicity. For CBP, however, STRs typically require modifications to accommodate the viscosity and heterogeneity of lignocellulose slurries (Knutsen & Liberatore, 2010). These include improved impeller design, gas sparging, and heat exchangers to ensure homogeneity and efficient mass and heat transfer. Packed-bed and fluidized-bed reactors have been considered for processes with immobilized enzymes or biofilms, and the former can also involve cells for which a separate immobilization step is not needed, while membrane bioreactors maintain cells as well as enzymes, generating in all cases a continuous process. Solid-state fermentation systems are also being designed for CBP in those areas where water is scarce or where low energetic demand is required (Karimi et al., 2025). Process integration deals with connecting upstream pretreatment, CBP, and downstream product recovery into a continuous process. Features that include in-line biomass feed, pH, and dissolved oxygen control, as well as continuous extraction modules, minimize downtime, limit contamination, and optimize overall productivity. A notable development has been the shift toward modular bioreactors that may be quickly reconfigured for different feedstocks and products. Also, process intensification with high cell density systems or co-cultures leads to improved reactor performance. In the end, fully integrated reactor designs and process designs would not only eliminate the need for high operating stability but also would prevent the economics of CBP processes from being disadvantaged vs. those derived from fossil feedstocks (Kubis & Lynd, 2023).

8.4.4 Process Monitoring and Automation

Effective monitoring and automation are crucial for upholding the optimum environment in complex bioprocesses such as CBP, where various biological and chemical effects must be controlled on the fly. The use of state-of-the-art process analytical technologies (PATs) and automation platforms allows real-time monitoring of key parameters such as temperature, pH, the concentration of dissolved oxygen, nutrients, enzyme activity, and metabolites (Moretti et al., 2024). Current biopharmaceutical production processes incorporate *in situ* sensors for online monitoring of data and computer-controlled loops for regulation of aeration, agitation, feed rates, and temperature. Spectroscopic techniques, such as Raman and Near-Infrared (NIR) spectroscopy, enable the non-invasive *in situ* detection of substrate conversion and product formation. Concurrently, biosensors—designer proteins or cellular systems that read out on given metabolites—are being installed in microbial hosts to alert to process changes or induce adaptive gene expression. For bioprocesses, model users could have a look at the Supervisory Control and Data Acquisition (SCADA), digital twins, and how they enable operators to remotely model, simulate, and optimize bioprocesses (Mu'azzam et al., 2024). Such systems provide features for predictive analytics, which could help in early warning of process deviations, contamination incidents, or system failures, thereby lowering risk and enhancing reliability. CBP is also incorporating AI and machine learning (ML) into CBP platforms to optimize efficiencies and accelerate decision support (Ahmed et al., 2020). For example, ML models can compare historical process logs with final performance measures, detect patterns, and suggest process adjustments to improve yield. The integration of bioprocess automation and Industry 4.0 technologies is enabling smart, self-optimizing systems with reduced labor, minimized waste, and consistent product quality. While CBPs are transitioning to the production scale, automated and digitized control systems are expected to become critical for reproducibility, scalability, and regulatory approval, as well as for wide industrial application (Subramanian, 2021).

8.5 Assessment Parameters for Bioprocess Technologies

8.5.1 Technical Feasibility and Efficiency Metrics

Technical feasibility is the capability of a bioprocess to perform to its target specifications under known operating conditions. It includes the assessment of whether the biological system (microorganisms or enzymes) and the reactor configuration, as well as the process operation, are adequate for a cost-effective conversion of substrates into products. Such an assessment is especially important for CBP, where stages such as enzyme production, biomass hydrolysis, and fermentation are taking place at once. Critical operational performance measures comprise product yield, volumetric productivity, substrate conversion efficiency, enzyme activity, and residence time. Yield is defined as the amount of product generated per amount of substrate consumed, and productivity represents the amount of product formed per unit reactor mass or volume per unit of time. Such parameters are useful to compare performances between different microbial strains, substrates, and process designs. Another important technical feature is process robustness, which is the ability of various factors (such as temperature variations, pH shifts, or substrate inhibition) to affect the stability of the process. Controllability, the possibility of contamination occurring, and integration into the process must inevitably be considered for any bioprocess. For CBP systems, an additional concern is the efficiency of enzyme expression, that is, if the organism could produce enough of the active enzymes without impinging on its growth or fermentation performance. Technical feasibility testing is usually done by lab testing and simulation and then very early-stage pilot testing for scale-up validation. An evaluation of technical feasibility forms the basis for subsequent economic and environmental analyses, enabling researchers and industries to decide whether a certain bioprocess is industrially feasible to be developed and implemented (Pollock et al., 2017).

8.5.2 Economic Evaluation and Cost–Benefit Analysis

The economic assessment is crucial to assess the commercial potential of a bioprocess technology. Regardless of technical efficacy, a process must be financially viable to pass from laboratory to industrial level. A full cost–benefit analysis (CBA) would consider capital expenditure (CAPEX), operational elements (OPEX), ROI, and break-even over a period of time (Sartori et al., 2014). CAPEX costs (investment costs) are typically involved with equipment, facility, and process erection and installation. The combination of enzyme production and fermentation in one process reduces the necessity of equipment and investment in CBP (Wilkinson et al., 2017). The use of special bioreactors or downstream purification systems can cause a rise in costs. OPEX includes recurring expenses, including raw material (e.g., biomass feedstock), utilities (power and water), labor, maintenance, and waste disposal. Feedstock feeding is a determining factor in economic feasibility, and low-cost and easily available feedstock seems to be preferable to make the process more profitable. For this reason, agricultural or industrial waste as a substrate may enhance profit. The cost associated with enzyme production or purchase is another important issue, which CBP addresses through *in vivo* enzyme synthesis. Process economics can be modeled using economic modeling software such as SuperPro Designer or BioSTEAM to screen and evaluate critical processes and drivers of cost. Sensitivity analysis enables evaluation of the sensitiveness of the profitability to variations of some specific input such as feedstock cost, yield, and product prices. Value can also be created by-product valorization, carbon credits, and integration to circular economy approaches (Hagman & Feiz, 2021). The structure of costs may also be affected by subsidies, fiscal incentives, and policies. The economic assessment guides technology prioritization as well as investment decisions through securing resources in technologies promising both a market potential and a financial value.

8.5.3 Environmental Impact and Life Cycle Assessment (LCA)

Assessment of environmental impacts is essential in ensuring that bioprocess technologies are beneficial in terms of sustainability. LCA is the most widely used methodological tool to assess the environmental performance of a system, process, product, or service through its life cycle—from raw material extraction through production, use, and end-of-life management. LCA according to ISO 14040/44 comprises four sequential phases: goal and scope definition, inventory analysis, impact assessment, and interpretation (Rosenbaum, 2017). For bioprocesses, it considers inputs and outputs at every stage, including upstream processing, fermentation, downstream purification, and transportation (energy, water, raw materials, emissions, waste, and by-products). Typical environmental indicators evaluated in LCA are Global Warming Potential (GWP), eutrophication, acidification, energy demand, and water footprint. In the case of CBP, the environmental benefits frequently come from less usage of chemicals (as a result of enzymes produced *in situ*), decreased water and energy inputs, and less waste generation. The utilization of renewable biomass or waste substrates provides additional improvement to the sustainability profile. Comparative LCAs can be carried out for conventional chemical and bioprocess-based alternatives to pinpoint environmental trade-offs (Hasnain et al., 2025). For instance, bioethanol production from corn by the SHF route could emit higher GWP and have higher eutrophication values than CBP from lignocellulosic wastes. Sophisticated tools, including SimaPro, GaBi, and OpenLCA, are commonly employed to model environmental flows and make impact assessments. LCA is not only a tool for decision support in Research and Development (R&D), but also for regulatory demands, environmental labeling, and communication toward stakeholders. It offers a clear, science-based approach to the process of evaluating and reducing the environmental impacts generated by bioprocess technologies (Figure 8.3; Gargalo et al., 2016).

8.5.4 Scalability and Industrial Applicability

The scalability of a bioprocess to industrial production can have a significant impact on the successful application of a lab-scale procedure in large-scale manufacturing

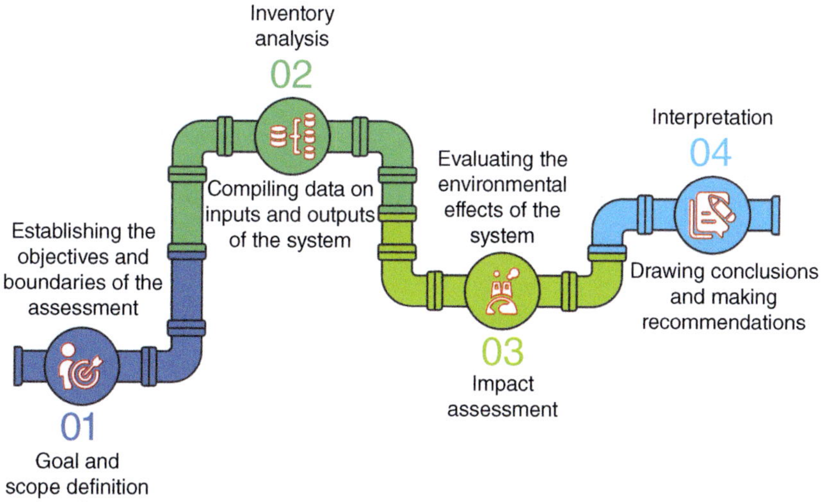

Figure 8.3 Phases of inventory analysis, including goal definition, data compilation, impact assessment, and interpretation.

(Wang, Haringa, et al., 2020). It is for evaluating a process performance/efficiency/stability for scaling up to pilot and industrial reactors from bench-top model reactors. Bioreactor design, process control, mass and heat transfer, and mixing efficiency are among the key factors that affect scalability. Reactor homogeneity and the spatial distribution of temperature and substrate become harder to achieve as the size of the reactor increases, especially in solid–liquid systems such as those for CBP (Wang, Shao, et al., 2020). Foam production, shear sensitivity of microorganisms, and limitations of oxygen transfer are also frequent scaling and design challenges. Feasibility at the industrial scale-up will also depend upon the adaptability of the reaction to existing facility infrastructure, safety implications, regulatory clearance, and alignment with feedstock supply chains. Bioprocessing capabilities for non-sterile operations, robustness to contaminants, and use of raw, inexpensive feedstocks improve the industrial relevance of a bioprocess. Before commercialization, pilot-scale trials and techno-economic assessments are required. These trials are used to optimize process conditions, to determine capital costs, and to solve operational problems in an environment that simulates industrial practice closely (Georgiadis et al., 2019). Moreover, it must be compatible with the industrial standard and regulation (e.g., Food and Drug Administration (FDA), Environmental Protection Agency (EPA) and Registration, Evaluation, Authorization and Restriction of Chemicals (REACH)) as well as criteria like the safety and environmental emission of products. Technologies that do so are more likely to be embraced by stakeholders and investors. The appearance of modular bioprocess kits and local biorefineries is helping to facilitate scalability and localization of the bioprocesses. Scalability and compatibility considerations with other unit operations should be made at the process design stage. Bioprocesses can play a leading role in the bio-based economy and longer-term sustainability goals when designed for scalability and compatibility with other operations (Gerssen-Gondelach et al., 2014).

8.6 Challenges and Bottlenecks

Although impressive advancements were made in the field of bioprocessing and CBP, a number of obstacles have precluded the commercialization of lactate for large-scale applications. These limitations pose challenges to the technical, economic, and regulatory feasibility of bio-transformation systems. By understanding and solving these issues, process performance may be improved, costs may be reduced, and competitiveness compared with fossil-derived counterparts may be increased (Karka et al., 2021).

8.6.1 Substrate Limitations and Pretreatment Issues

Complexity and recalcitrance of lignocellulosic biomass are among the most challenging obstacles in bioprocesses, particularly in CBP. Cellulose, hemicellulose, and lignin are abundant in agricultural by-products, forestry waste, and energy crops (Anas, Falak, et al., 2025). Bacteria form aggregated chains with a high degree of organization, which makes them resistant to biodegradation and difficult to access without pretreatment. Pretreatments like dilute acid traces, steam explosion, ammonia fiber expansion, and alkaline treatment cause disruption of hemicellulose and lignin, thereby making cellulose susceptible to enzymatic action. However, these are typically high energy-consuming processes, and they often generate inhibitory compounds (e.g., furfural and hydroxymethyl furfural) for the subsequent bioconversions that inhibit microbial growth and enzyme activity (Anas, Khalid, et al., 2025). Furthermore, variability in feedstock, such as moisture levels, composition, and seasonal availability, can affect process reliability and scalability. Construction of tolerant microbes, though, is still a challenge by allowing them to tolerate a vast number of feedstocks and inhibitors. There are attempts to develop more benign, cheaper, and efficient pretreatment options (to include biological methods with organisms such as white-rot fungi or ionic liquids). This latter process needs to match CBP-compatible conditions to be suitable in real-life use (Cauvain & Cauvain, 2015).

8.6.2 Product Inhibition and Yield Losses

Product inhibition is a major limitation in bioprocesses. During fermentation, with the increase in the products of interest (ethanol, solvents, and organic acids), microbial growth may be inhibited as well as its metabolism. For example, the denaturation of enzymes and the loss of membrane integrity due to ethanol above the concentration limit will reduce cell viability and decrease the final cell density. In CBP processes, the problem is aggravated because the saccharification and fermentation processes are performed concomitantly, and by-products from enzymatic hydrolysis can inhibit the fermentative microorganisms. In addition, sugars would tend to accumulate due to inefficient hydrolysis, leading to catabolite repression, which would impact both enzyme production and microbial activity. To address this, tolerant microbial strains are being developed that can tolerate high amounts of product and remain metabolically active (Anas, Falak, et al., 2025). Among the strategies, the overexpression of stress response genes, modification of membrane composition, and instillation of efflux pumps for the elimination of toxic compounds (extrusion) are described. Another strategy is *in situ* product removal (ISPR), in which products are continuously taken out of the system (for instance, by pervaporation, gas stripping, and adsorption). This reduces accumulation of product and extends microbial productivity. Finally, reduction in product inhibition is a core element for increased volumetric productivity and feasibility of continuous, or semi-continuous, operation of bioprocesses (Sriraman et al., 2024).

8.6.3 Enzyme Cost and Recycling

The production and cost of enzymes are significant components of overall bioprocess cost, particularly for conventional processes requiring separate hydrolysis. Although this issue is tackled by producing enzymes in the microorganism host, the efficiency and stability of the enzymes produced are still bottlenecks. The development of more efficient and more stable enzyme cocktails depends on the appropriate expression systems, fermentation process, and posttranslational modifications. Recombinant expression in *E. coli, Pichia pastoris*, and *B. subtilis* is typical but usually plagued by problems such as misfolding or low production (Yang et al., 2021). In addition, even in CBP, the enzymes should not overload the host cell through *in situ* enzyme production, which also should ferment. Such a metabolic load may give rise to suboptimal growth rates and product titers. Enzyme recycling and immobilization are attempts to enhance enzyme usage. Immobilization on supports, such as silica, alginate beads, or biochar, enables the reuse of enzymes through several cycles, having economic considerations. It is, however, possible that immobilization leads to either a decrease in enzyme activity or mass transfer limitation. An alternate approach is to overproduce the enzymes, employ cell-surface display techniques, and co-culture systems in which one strain yields the enzymes and a second strain ferments the sugars. Although significant progress has been made in these respects, sustainable enzyme performance in industrial settings (high solids loading and presence of inhibitors) is still a challenge. Elimination of enzyme-related bottlenecks is a prerequisite for economically viable and high-throughput bioprocess systems.

8.6.4 Regulatory and Market Constraints

Bioprocessing technologies, especially those based on genetically modified organisms (GMOs), enzymes, or new bio-based products, are highly scrutinized and subject to market risk. Regulations vary greatly by jurisdiction, and the approval of new cells, microbial strains, or bioproducts can be time-consuming and expensive. Intensive toxicological, environmental, and performance testing should be a part of the bioproducts' safety assessment process for human health and the environment in most countries. This should consider regulations like the EPA (USA), EFSA (EU), REACH, and national bioeconomy schemes. Biosafety and containment of genetically engineered strains are additional challenges for the products. Marketwise, consumer attitude, price competitiveness, and supply chain preparedness affect adoption as well. However, bio-based products have a hard time competing in the market with their well-established fossil-based equivalents unless they are supported by subsidies and incentives or by regulations (such as carbon taxation or plastics bans). Moreover, non-harmonized global standards for bio-based certification, labeling, and sustainability metrics exist, which in turn complicates the international marketing of products. Barriers of knowledge and proprietary IP, as well as lack of access to large-scale demonstration units, also tend to retard innovation. Issues, such as engaging with regulators, providing safety data, and considering Stage I compliance early, need to be dealt with and developed. To further drive the market acceptance, public–private partnerships (PPPs), green financing schemes, and supportive policy environments could also have a significant impact. Addressing regulatory and market bottlenecks is key to realizing the potential of bioprocesses and investments in next-generation bio-industrial platforms (Dietz et al., 2024).

8.7 Future Trends and Opportunities

8.7.1 Integration with Artificial Intelligence and Digital Twins

The combination of AI and digital twins for bioprocessing is transforming how biomanufacturing systems are monitored, optimized, and scaled. These products deliver predictive real-time applications resulting in more effective process control, less waste, and greater efficiency (Saez et al., 2018). AI and ML algorithms can learn from huge datasets produced by sensors, process logs, and lab experiments to look for patterns and provide predictions on performance, and provide recommended best process parameters (Malik & Khattak, 2023). In CBP, AI still can tune for better enzyme expression, substrate conversion, and product generation conditions and, in many cases, save dollars, lives, and time by avoiding expensive trial-and-error experiments. Digital twins are represented as virtual counterparts of physical bioprocesses. They employ up-to-date data to model, monitor, and anticipate system behavior so as to make decisions under various circumstances. This allows for a proactive approach to decision-making and risk reduction. In industrial CBP operations, digital twins can let operators model different feedstock strategies, identify anomalies, or even predict equipment failure, all to maintain uptime and process consistency. The tools also enable model predictive control, where operating parameters are tuned online with feedback from AI to keep the microbial activity and the yield in the optimal state. Given increasing computational power and precision of data collection, the use of AI and digital twins is likely the future standard in biomanufacturing plants. The convergence between biotechnology and digital technologies is a major trend in terms of Industry 4.0 for bioprocessing and a game-changer, offering higher sustainability, precision, and scalability (Betz et al., 2023).

8.7.2 Circular Bioeconomy and Sustainable Development Goals

A circular bioeconomy represents an economy in which biological resources are reused, recycled, and managed in a sustainable manner in the development of food, materials, energy systems, and chemicals. Bioprocesses, and particularly bioprocesses using waste biomass and renewable feedstock, form the core of this concept. In this regard, CBP, among other next-generation bioprocess technologies, entails low resource consumption, low environmental burden, and the valorization of waste by-products, very much in line with a circular approach. For example, agricultural waste can be made to yield biofuels, enzymes, bioplastics, and soil amendments: a circular economy model in which producing and consuming are circular, not linear. These developments also make important contributions to the UN's SDG. This is facilitated by ensuring that CBP and biorefineries contribute to SDG 7 (Affordable and Clean Energy), through bioenergy provision; SDG 12 (Ensure Sustainable Consumption and Production Patterns) by decreasing dependence on fossil resources; SDG 13 (Action to Combat Climate Change) by reducing greenhouse gases; and SDG 9 (Industry, Innovation, and Infrastructure) by green industrialization. Bio-based innovations can also contribute to rural development and job creation, which contributes to supporting SDG 8 (Decent Work and Economic Growth) and SDG 1 (No Poverty). Through policy instruments, LCA tools, and eco-labeling, incentives toward a more circular, sustainable, and competitive industrial transformation path are being established. In the future, sustainability should be an integrated part of every stage of bioprocess development so that these technologies can deliver on their promise to regenerate the economy (Aguilar et al., 2019).

8.7.3 Policy Support and Public–Private Partnerships

A strong regulatory framework and dynamic PPP are necessary to drive the industrial implementation of bioprocess technologies. Although scientific and technical progress has enabled CBP and advanced bioprocessing, upscaling will need supportive regulation, funding, and collaboration among the players. Governments also have a role to play by formulating bioeconomy strategies and providing both subsidies and grants, as well as tax incentives and regulatory fast lanes for bio-based innovations. For instance, demand may be cultivated using carbon pricing, requirements for renewable fuels, and bio-based product purchasing. Bio-based R&D is getting increasing focus by international organizations and national research councils in mission-driven programs. Programs like the European Union's Bio-Based Industries Joint Undertaking (BBI-JU) or India's National Bio-Energy Mission demonstrate how leveraging public investment can help advance industrial biotechnology. PPPs between universities, technology developers, feedstock suppliers, and end-users collaborate to co-develop and commercialize effective bioprocess solutions. These collaborations help spread out the financial risk, provide necessary common infrastructure (such as pilot plants and testing labs), and make sure ideas that come out of the partnership are needed in the real world. Additionally, these partnerships stimulate brain drain, support human capital development, and facilitate international cooperation, the former of which will be essential for mitigating climate change and resource depletion. Going forward, industrial biotechnology needs to be harmonized with the policy roadmaps, green finance instruments, and multi-stakeholder platforms for transforming from a fossil economy to a bioeconomy (Iriarte et al., 2021).

8.8 Conclusion

The recent development of bioprocess technologies (especially CBP) is a major step toward sustainable Industrial Biotechnology (IBB). This chapter addressed the decades-old basics, technical innovations, and meaningful criteria for evaluating systematic synthesis and development of modern bioprocesses. CBP, which combines the production of enzymes with their hydrolysis and fermentation into one step, allows the possibility to overcome the economic and operational difficulties of multistage biomass conversion processes. Advances in synthetic biology, metabolic engineering, enzyme cocktail design, and smart bioreactor systems have established the means for making CBP viable at scale. At the same time, AI and digital twins are ushering in an age of precision bioprocessing, offering real-time monitoring, predictive control, and optimization. Nevertheless, challenges such as feedstock diversity, product inhibition, expensive enzyme production, and a complex regulatory environment still exist. As has been repeatedly verified in the past, the success of bioprocesses to come will depend not only on technical but also on overall technological progress.

Prospects and Challenges: Detailed LCA, cost–benefit analysis, upscaling trials, and environmental analysis are required to substantiate environmental claims and commercial potential. By embedding these assessments in the early stages of the technology development pipeline, new technologies will be underpinned by circular economy and global sustainability requirements. Prospects: the integration of biotechnology with digital infrastructure, supportive policies, and cross-sectoral partnerships will be important to overcome the current bottlenecks. We can also shorten the time to market of effective, low-cost, and green bioprocess technologies by encouraging close cooperation between academia, industry, and the state. In summary, the way forward is a multi-faceted one, which includes a continuum of rigorous science, innovation in technology and practice, and critical assessments of sustainability. With targeted support, bioprocesses and CBP can play a central role in the bioeconomy by providing cleaner production systems and value-added resource recovery and creating a sustainable environment for the future in the face of global challenges.

References

Abo, B. O., Gao, M., Wang, Y., Wu, C., Ma, H., & Wang, Q. (2019). Lignocellulosic biomass for bioethanol: An overview on pretreatment, hydrolysis and fermentation processes. *Reviews on Environmental Health*, *34*(1), 57–68. https://doi.org/10.1515/reveh-2018-0054.

Adsul, M., Sandhu, S. K., Singhania, R. R., Gupta, R., Puri, S. K., & Mathur, A. (2020). Designing a cellulolytic enzyme cocktail for the efficient and economical conversion of lignocellulosic biomass to biofuels. *Enzyme and Microbial Technology*, *133*, 109442. https://doi.org/10.1016/j.enzmictec.2019.109442.

Aguilar, A., Twardowski, T., & Wohlgemuth, R. (2019). Bioeconomy for sustainable development. *Biotechnology Journal*, *14*(8), 1800638. https://doi.org/10.1002/biot.201800638.

Ahmed, Z., Mohamed, K., Zeeshan, S., & Dong, X. (2020). Artificial intelligence with multi-functional machine learning platform development for better healthcare and precision medicine. *Database*, baaa010. https://doi.org/10.1093/database/baaa010.

Anas, M., Falak, A., Hassan, S., Khattak, W. A., Saleem, M. H., Khan, K. A., Khalid, A., & Fahad, S. (2025). Microbial interactions and bacterial responses to metal stress in plants: Mechanisms, adaptations, and applications for sustainable agriculture. *Journal of Crop Health*, *77*(1), 36. https://doi.org/10.1007/s10343-025-01108-w.

Anas, M., Khalid, A., Saleem, M. H., Ali Khan, K., Ahmed Khattak, W., & Fahad, S. (2025). Symbiotic synergy: Unveiling plant-microbe interactions in stress adaptation. *Journal of Crop Health*, *77*(1), 1–21. https://doi.org/10.1007/s10343-024-01070-z.

Banner, A., Toogood, H. S., & Scrutton, N. S. (2021). Consolidated Bioprocessing: Synthetic Biology Routes to Fuels and Fine Chemicals. *Microorganisms*, *9*(5), 1079. https://doi.org/10.3390/microorganisms9051079.

Betz, U. A., Arora, L., Assal, R. A., Azevedo, H., Baldwin, J., Becker, M. S., Bostock, S., Cheng, V., Egle, T., & Ferrari, N. (2023). Game changers in science and technology—Now and beyond. *Technological Forecasting and Social Change*, *193*, 122588. https://doi.org/10.1016/j.techfore.2023.122588.

Carus, M., & Dammer, L. (2018). The circular bioeconomy—Concepts, opportunities, and limitations. *Industrial Biotechnology*, *14*(2), 83–91. https://doi.org/10.1089/ind.2018.29121.mca.

Cauvain, S. (2015). *Breadmaking Processes*. Springer, Cham: Technology of Breadmaking. https://doi.org/10.1007/978-3-319-14687-4_2.

Cheng, Cheng, C. (2017). Chemicals and Biofuels Production from CO2 and Biomass by Fermentation [Doctoral dissertation, Ohio State University]. OhioLINK Electronic Theses and Dissertations Center. http://rave.ohiolink.edu/etdc/view?acc_num=osu1511827507089631.

den Haan, R., van Rensburg, E., Rose, S. H., Görgens, J. F., & van Zyl, W. H. (2015). Progress and challenges in the engineering of non-cellulolytic microorganisms for consolidated bioprocessing. *Current Opinion in Biotechnology*, *33*, 32–38.

Dietz, T., Bogdanski, A., Boldt, C., Börner, J., von Braun, J.,Choncubhair, O. N., Durham, B., Ecuru, J., & Lang, C., Li, Y., Lund, M., MacRae, E., Maxon, M., Miranda, H. C., Mizunashi, W., Mungeyi, P., O'Hara, I. M., Fonseca, L. P. (2024). *Bioeconomy Globalization: Recent Trends and Drivers of National Programs and Policies* (November 16, 2024). Available at SSRN: https://ssrn.com/abstract=5023374 or http://dx.doi.org/10.2139/ssrn.5023374.

Gargalo, C. L., Cheali, P., Posada, J. A., Carvalho, A., Gernaey, K. V., & Sin, G. (2016). Assessing the environmental sustainability of early stage design for bioprocesses under uncertainties: An analysis of glycerol bioconversion. *Journal of Cleaner Production*, *139*, 1245–1260.

Gargalo, C. L., Udugama, I., Pontius, K., Lopez, P. C., Nielsen, R. F., Hasanzadeh, A., Mansouri, S. S., Bayer, C., Junicke, H., & Gernaey, K. V. (2020). Towards smart biomanufacturing: A perspective on recent developments in industrial measurement and monitoring technologies for bio-based production processes. *Journal of Industrial Microbiology & Biotechnology: Official Journal of the Society for Industrial Microbiology and Biotechnology*, *47*(11), 947–964.

Georgiadis, G. P., Elekidis, A. P., & Georgiadis, M. C. (2019). Optimization-based scheduling for the process industries: From theory to real-life industrial applications. *Processes*, *7*(7), 438.

Gerssen-Gondelach, S., Saygin, D., Wicke, B., Patel, M., & Faaij, A. (2014). Competing uses of biomass: Assessment and comparison of the performance of bio-based heat, power, fuels and materials. *Renewable and Sustainable Energy Reviews*, *40*, 964–998.

Gollihar Jr, J. D. (2017). *Methods in protein engineering and screening: From rational design to directed evolution and beyond* (Doctoral dissertation).

Hagman, L., & Feiz, R. (2021). Advancing the circular economy through organic by-product valorisation: A multi-criteria assessment of a wheat-based biorefinery. *Waste and Biomass Valorization*, *12*(11), 6205–6217.

Hasnain, M., Zainab, R., Ali, F., & Abideen, Z. (2025). An overview of biosorption, remediation, and desalination potential of algae to remove wastewater toxic pollutants for environmental safety. *Environmental Monitoring and Assessment*, *197*(8), 853.

Hassan, S. S., Williams, G. A., & Jaiswal, A. K. (2019). Moving towards the second generation of lignocellulosic biorefineries in the EU: Drivers, challenges, and opportunities. *Renewable and Sustainable Energy Reviews*, *101*, 590–599.

Holland, C., & Shapira, P. (2024). Building the bioeconomy: A targeted assessment approach to identifying biobased technologies, challenges and opportunities. *Engineering Biology*, *8*(1), 1–15.

Iriarte, L., Fritsche, U., & van Dam, J. (2021). *Sustainability governance of bioenergy and the broader bioeconomy*. Technical paper prepared for IEA Bioenergy Task, 45.

Joshi, A., Kanthaliya, B., Meena, S., Khan, F., & Arora, J. (2021). Process consolidation approaches for cellulosic ethanol production. In R. C. Ray (Ed.), *Sustainable biofuels* (pp. 43–72). Elsevier.

Kalaichelvan, P., & Pandi, I. A. (2019). *Bioprocess technology*. MJP Publisher.

Karimi, F., Mazaheri, D., Saei Moghaddam, M., Mataei Moghaddam, A., Sanati, A. L., & Orooji, Y. (2025). Solid-state fermentation as an alternative technology for cost-effective production of bioethanol as useful renewable energy: A review. *Biomass Conversion and Biorefinery*, *15*, 29523–29539.

Karka, P., Johnsson, F., & Papadokonstantakis, S. (2021). Perspectives for greening European fossil-fuel infrastructures through use of biomass: The case of liquid biofuels based on lignocellulosic resources. *Frontiers in Energy Research*, *9*, 636782.

Knutsen, J. S., & Liberatore, M. W. (2010). Rheology modification and enzyme kinetics of high solids cellulosic slurries. *Energy & Fuels*, *24*(5), 3267–3274.

Krzywanski, J., Sosnowski, M., Grabowska, K., Zylka, A., Lasek, L., & Kijo-Kleczkowska, A. (2024). Advanced computational methods for modeling, prediction and optimization—A review. *Materials*, *17*(14), 3521.

Kubis, M. R., & Lynd, L. R. (2023). Carbon capture from corn stover ethanol production via mature consolidated bioprocessing enables large negative biorefinery GHG emissions and fossil fuel-competitive economics. *Sustainable Energy & Fuels*, *7*(16), 3842–3852.

Lee, J., Yu, H. E., & Lee, S. Y. (2025). Metabolic engineering of microorganisms for carbon dioxide utilization. *Current Opinion in Biotechnology*, *91*, 103244.

Madhavan, A., Srivastava, A., Kondo, A., & Bisaria, V. S. (2012). Bioconversion of lignocellulose-derived sugars to ethanol by engineered *Saccharomyces cerevisiae. Critical Reviews in Biotechnology*, *32*(1), 22–48.

Malik, I., & Khattak, W. A. (2023). Enhancing urban health: Machine learning applications in environmental managemen. *Journal of Sustainable Infrastructure for Cities and Societies*, *8*, 1–21.

Matanguihan, C., & Wu, P. (2022). Upstream continuous processing: Recent advances in production of biopharmaceuticals and challenges in manufacturing. *Current Opinion in Biotechnology*, *78*, 102828.

Melendez, J. R., Mátyás, B., Hena, S., Lowy, D. A., & El Salous, A. (2022). Perspectives in the production of bioethanol: A review of sustainable methods, technologies, and bioprocesses. *Renewable and Sustainable Energy Reviews*, *160*, 112260.

Meyer, H.-P., & Schmidhalter, D. R. (2012). Microbial expression systems and manufacturing from a market and economic-perspective. In E. C. Agbo (Ed.), *Innovations in biotechnology* (pp. 211–250). IntechOpen.

Mithra, M., Jeeva, M., Sajeev, M., & Padmaja, G. (2018). Comparison of ethanol yield from pretreated lignocellulo-starch biomass under fed-batch SHF or SSF modes. *Heliyon*, *4*(10), e00885.

Mitra, S., & Murthy, G. S. (2022). Bioreactor control systems in the biopharmaceutical industry: A critical perspective. *Systems Microbiology and Biomanufacturing*, *2*(1), 91–112.

Moretti, A., Ivan, H. L., & Skvaril, J. (2024). A review of the state-of-the-art wastewater quality characterization and measurement technologies. Is the shift to real-time monitoring nowadays feasible? *Journal of Water Process Engineering*, *60*, 105061.

Moser, A. (2012). *Bioprocess technology: Kinetics and reactors*. Springer Science & Business Media.

Mould, F., Kliem, K., Morgan, R., & Mauricio, R. (2005). In vitro microbial inoculum: A review of its function and properties. *Animal Feed Science and Technology*, *123*, 31–50.

Mu'azzam, K., da Silva, F. V. S., Murtagh, J., & Gallagher, M. J. S. (2024). A roadmap for model-based bioprocess development. *Biotechnology Advances*, *73*, 108378.

Mukherjee, P., Pal, S., & Sivaprakasam, S. (2024). Process parameter controls for efficient enzymatic hydrolysis of cellulosic biomass. In *Handbook of Biorefinery Research and Technology: Biomass Logistics to Saccharification* (pp. 1–29). Dordrecht: Springer.

National Research Council, Division on Earth, Life Studies, Board on Life Sciences, & Committee on Key Challenge Areas for Convergence. (2014). *Convergence: Facilitating transdisciplinary integration of life sciences, physical sciences, engineering, and beyond.*

Pandey, A., & Sharma, Y. C. (2024). Advancements in biomass valorization in integrated biorefinery systems. *Biofuels, Bioproducts and Biorefining*, *18*(6), 2078–2090.

Pino, M. S., Rodríguez-Jasso, R. M., Michelin, M., Flores-Gallegos, A. C., Morales-Rodriguez, R., Teixeira, J. A., & Ruiz, H. A. (2018). Bioreactor design for enzymatic hydrolysis of biomass under the biorefinery concept. *Chemical Engineering Journal*, *347*, 119–136.

Pollock, J., Coffman, J., Ho, S. V., & Farid, S. S. (2017). Integrated continuous bioprocessing: Economic, operational, and environmental feasibility for clinical and commercial antibody manufacture. *Biotechnology Progress*, *33*(4), 854–866.

Porter, J. L., Rusli, R. A., & Ollis, D. L. (2016). Directed evolution of enzymes for industrial biocatalysis. *ChemBioChem*, *17*(3), 197–203.

Rabby, M. R. I., Ahmed, Z. B., Paul, G. K., Chowdhury, N. N., Akter, F., Razu, M. H., Karmaker, P., & Khan, M. (2022). A combined study on optimization, in silico modeling, and genetic modification of large scale microbial cellulase production. *Biochemistry Research International*, *2022*(1), 4598937.

Rajpurohit, H., & Eiteman, M. A. (2022). Nutrient-limited operational strategies for the microbial production of biochemicals. *Microorganisms*, *10*(11), 2226.

Ramchuran, S. O., O'Brien, F., Dube, N., & Ramdas, V. (2023). An overview of green processes and technologies, biobased chemicals and products for industrial applications. *Current Opinion in Green and Sustainable Chemistry*, *41*, 100832.

Romann, P. (2023). *Advancing continuous biomanufacturing: Quality control and process intensification strategies for perfusion cell culture* [Dissertation, Technische Universität Wien]. reposiTUm. https://doi.org/10.34726/hss.2023.102142.

Rosenbaum, R. K. (2017). Selection of impact categories, category indicators and characterization models in goal and scope definition. In M. Curran (Ed.), *Goal and scope definition in life cycle assessment* (pp. 63–122). Springer.

Saez, M., Maturana, F. P., Barton, K., & Tilbury, D. M. (2018). Real-time manufacturing machine and system performance monitoring using internet of things. *IEEE Transactions on Automation Science and Engineering*, *15*(4), 1735–1748.

Sarangi, P. K., & Nanda, S. (2019). Recent advances in consolidated bioprocessing for microbe-assisted biofuel production. In S. Nanda, P. K. Sarangi, & D.-V. N. Vo (Eds.), *Fuel processing and energy utilization* (pp. 141–157). Chapman and Hall.

Sartori, D., Catalano, G., Genco, M., Pancotti, C., Sirtori, E., Vignetti, S., & Del Bo, C. (2014). *Guide to cost-benefit analysis of investment projects. Economic appraisal tool for cohesion policy, 2020.*

Sharma, A., Gupta, G., Ahmad, T., Mansoor, S., & Kaur, B. (2021). Enzyme engineering: Current trends and future perspectives. *Food Reviews International*, *37*(2), 121–154.

Shrivastava, P. (2018). Environmental technologies and competitive advantage. In A. E. Singer (Ed.), *Business ethics and strategy* (Vol. I and II, pp. 317–334). Routledge.

Singhania, R. R., Patel, A. K., Singh, A., Haldar, D., Soam, S., Chen, C.-W., Tsai, M.-L., & Dong, C.-D. (2022). Consolidated bioprocessing of lignocellulosic biomass: Technological advances and challenges. *Bioresource Technology*, *354*, 127153.

Sriraman, V., Johnrajan, J., Yazhini, K., & Rathinasabapathi, P. (2024). Bioprocess engineering essentials: Cultivation strategies and mathematical modeling techniques. In S. Khanal (Ed.), *Industrial microbiology and biotechnology: A new horizon of the microbial world* (pp. 247–276). Springer.

Subramanian, G. (2021). *Process control, intensification, and digitalisation in continuous biomanufacturing*. John Wiley & Sons.

Tang, T.-C., An, B., Huang, Y., Vasikaran, S., Wang, Y., Jiang, X., Lu, T. K., & Zhong, C. (2021). Materials design by synthetic biology. *Nature Reviews Materials*, *6*(4), 332–350.

Wang, G., Haringa, C., Noorman, H., Chu, J., & Zhuang, Y. (2020). Developing a computational framework to advance bioprocess scale-up. *Trends in Biotechnology*, *38*(8), 846–856.

Wang, J., Shao, Y., Yan, X., & Zhu, J. (2020). Review of (gas)-liquid-solid circulating fluidized beds as biochemical and environmental reactors. *Chemical Engineering Journal*, *386*, 121951.

Wilkinson, S., Smart, K. A., James, S., & Cook, D. J. (2017). Bioethanol production from brewers spent grains using a fungal consolidated bioprocessing (CBP) approach. *Bioenergy Research*, *10*, 146–157.

Yan, Q., & Fong, S. S. (2017). Challenges and advances for genetic engineering of non-model bacteria and uses in consolidated bioprocessing. *Frontiers in Microbiology*, *8*, 2060.

Yang, H., Qu, J., Zou, W., Shen, W., & Chen, X. (2021). An overview and future prospects of recombinant protein production in *Bacillus subtilis*. *Applied Microbiology and Biotechnology*, *105*(18), 6607–6626.

9

Reactor Design and Consolidated Bioprocess Technology: Practical Use and Impact

Navodita Maurice

Prophyl Limited, Bár, Hungary

Abbreviations

ABC	ATP binding cassette
ADMET	absorption, distribution, metabolism, excretion, and toxicology
AI	artificial intelligence
ARLs	airlift bioreactors
ATA	atorvastatin acid
BAK	bioartificial kidney
BAL	bioartificial liver device
BBB	blood–brain barrier
BCRs	bubble column bioreactors
BFB	biological fluidized bed
BLSS	Excorp medical bioartificial liver support system
Bo	Bodenstein number
BOD	biological oxygen demand
CFB	cytophaga–flavobacterium–bacteroides
CFD	computational fluid dynamics
CHO'	Chinese hamster ovary
CKD	chronic kidney disease
CO_2	carbon dioxide
COD	chemical oxygen demand
CYP	cytochrome P450
CSTR	continuous stirred tank reactors
DGGE	denaturing gradient gel electrophoresis
DMPK	drug metabolism and pharmacokinetics
DOPA	3,4-dihydroxy-L-phenylalanine
E17bG	estradiol-17b-D-glucuronide
ECM	extracellular matrix
ECs	endothelial cells

ECS	extra capillary space
ELAD	extracorporeal liver assist device
EPA	eicosapentaenoic acid
ESCs	embryonic stem cells
FBF	fluidized bed fermentation
FCS	fetal calf serum
GHG	greenhouse gas
H_2	hydrogen
HFBs	hollow fiber-based bioreactors
hGH	human growth hormone
hiPSCs	human induced pluripotent stem cells
HKC	human kidney proximal tubular cell line
hPTCs	human PTCs
iPSCs	induced pluripotent stem cells
MBs	membrane bioreactors
MBs	membrane bioreactors
MDR1	multidrug resistance protein
MELS	modular extracorporeal liver support system
MLSS	mixed liquor suspended solids
MRP2	multidrug resistance-associated protein
OAT	organic anion transporter
OATP1B1	organic anion transporting polypeptide 1B1
PBMEC/C1-2	porcine brain microvascular endothelial cells
PBRs	packed bed bioreactors
PCE	photon conversion efficiency
PCL	polycaprolactone
pCO_2	produced carbon dioxide
PCR	polymerase chain reaction
PDMS	polydimethylsiloxane
Pe	Peclet number
PET	polyethylene terephthalate-degrading
PGF2a	prostaglandin F2a
P-gp	P-glycoprotein
PHA	polyhydroxyalkanoate
PTCs	proximal tubule cells
RAD	renal assist device
RDBs	rotating drum bioreactors
Re	Reynolds number
SmFs	submerged fermentations
SSFs	solid-state fermentations
STR	stirred tank reactors
T-RFLP	terminal restriction fragment length polymorphism

9.1 Introduction

The limited availability and fast depletion of fossil fuels in order to fulfill the increasing energy needs all over the world has cast a negative impact on the environment, leading to the search for sustainable and alternative sources of energy (Taghizadeh-Alisaraei et al., 2019). Biomass is a form of renewable energy that can be used for biofuel production. It has several benefits over the traditional fossil fuels as it reduces the greenhouse gas (GHG) (Rajarathinam et al., 2019). In ancient times, the people of Egypt, Mesopotamia, Rome, and Greece used ceramic containers that can be considered as bioreactors for the fermentation of food like wine, bread, meat, and beer without having a clue of microbes and fermentation technology. Louis Pasteur in the 19th century discovered microbes and brought to light the fermentation process by producing butanol using a mixed culture of bacteria with at least one Clostridial strain. Chaim Weizmann also used a *Clostridium* strain for acetone production while dealing with artillery projectiles during the First World War. The designing of bioreactors increased after the discovery of penicillin by Alexander Fleming, which boosted the antibiotics market, resulting in the production of 7 trillion penicillin units in 1945. By the middle of the 20th century, great advancements occurred in the bioreactor design process, comprising methods of sterilization, agitation, and the aeration systems, and the involvement of multivessel systems so as to produce a wide variety of products at a time, followed by the production of equipment with high holding capacity. By the 1980s, bioreactors were utilized not only for the cultivation of microbes but also for animal and plant cells; therefore, new types of bioreactors were built to meet the demands of the growing industries (Spier et al., 2011). With the advancement of technologies like three-dimensional (3D) printing, artificial intelligence, and automation in the 21st century, bioreactor design has also advanced a lot, resulting in the cost-efficient manufacture of a wide variety of biotechnological products. The bioprocessing industry solely depends upon the bioreactors, and each type of bioreactor has its own pros and cons. About a 1200% increase has been noticed in the designing of bioreactors between the years 2003 and 2023, especially in the industrial sectors dealing with fermentation processes. In the present-day scenario, bioreactors are key components of laboratory studies dealing with small-scale bioprocess experimental trials and large-scale industrial bioprocesses (Palladino et al., 2024).

Bioreactors are devices used for the cultivation of microbes, plant cells, or animal cells on either a small or a large scale. They are equipped with unique systems to control aeration, agitation, pH, temperature, nutrient intake, and the removal of the final product. Bioreactors provide an optimal environment for cells, thereby resulting in the production of the desired product by the smooth functioning of bioprocesses in a cost-effective manner and reduced time (Aghajani & Wang, 2022). To carry out a bioprocess in a cost-effective manner, the selection of an appropriate bioreactor is crucial, as these are immensely important biotechnological tools participating in the production of a wide variety of highly important food as well as valued products (Palladino et al., 2024). Bioreactors can be described as vessels that are utilized to carry out the activity of a biological agent in order to attain a desired product by chemical conversion. They usually provide a biochemical as well as a biomechanical platform, controlling nutrient exchange as well as transfer of oxygen to the cells followed by desired products from cells (Sharma, 2012). Bioreactors can also be defined as devices intended for the ideal growth followed by metabolic activity of microbes via biocatalysts or enzymes or other microbes or plant or animal cells. The crude material for bioreactors can be an organic material, an inorganic chemical product, or a complex product. The conversion product may comprise a single-cell protein, baker's yeast, animal feed, microbial cell culture, or other substrates like vitamins, amino acids, polysaccharides, organic acids, or ethanol, as well as antibiotics. Bioreactors can also be used for the biotransformation or bioconversion of products like steroid biotransformation, L-sorbitol, or enzymes like cellulase, amylase, or lipase, as well as recombinant products like vaccines or growth hormones. Bioreactor design also focuses on the type of substrates as well as biocatalysts involved in a biochemical process (El Haj et al., 2005). Bioreactors are different from classic chemical reactors as they utilize and govern biological agents. Since the microbes involved are much more susceptible and less durable than chemicals, bioreactor design should be strong enough in order to control the growth environment and minimize contamination levels (Williams, 2002). The conditions inside the bioreactor must favor microbial growth, leading to the production of the desired products, which demands the insertion of special features for effective biocatalytic conversion (Gudin & Chaumont, 1991). Maintenance of the required biological activity and keeping the unwanted activities at a reduced level are some of the challenges of bioreactors, as microbes are prone to mutation, and any alteration can affect the nature as well as properties of the desired product produced by the microbes used (Williams, 2002). A bioreactor is sometimes compared with the fermenter, where living organisms carry out a biochemical reaction. Fermentation can be defined as a process in which microorganisms grow on food either aerobically or anaerobically. Commercial fermenters are composed of glass, exotic alloys, plastic tanks, glass-lined steel, and stainless steel assembled with gauges. They are utilized for the growth of pure bacterial, fungal, or yeast cultures for the production of a wide variety of commercially available fermented food items. The size of a bioreactor can vary from a microbial cell ranging a few mm^3 to a 100–1000 mL shake flask, to a 1–50 L laboratory scale fermenter, or 0.3–10 m^3 on a pilot level, or 2–500 m^3 on a plant scale for industrial production of products. Special consideration is paid to a number of biotechnological processes while assembling a bioreactor. For example, maintenance of the environmental conditions is essential for microbial cell growth as well as for the reaction rate. Flow of gases like carbon dioxide, nitrogen, and oxygen, as well as air, followed by pH, dissolved oxygen levels, temperature, circulation rate, and agitation speed, are all components that need to be closely checked and monitored (Chen & Hu, 2006; Figure 9.1).

Bioreactors play a crucial role in drug delivery as well as tissue engineering, as they can be used for the generation of tissue constructs for clinical trials, followed by investigation of knowledge of tissue integrity and disease onset, as well as increased cell production for consistent high-quality cell output (Zhang et al., 2021). Tissue engineering is very much dependent on bioreactors as they promote monitoring as well as manipulation of different environmental parameters in a controlled fashion, and by utilizing them,

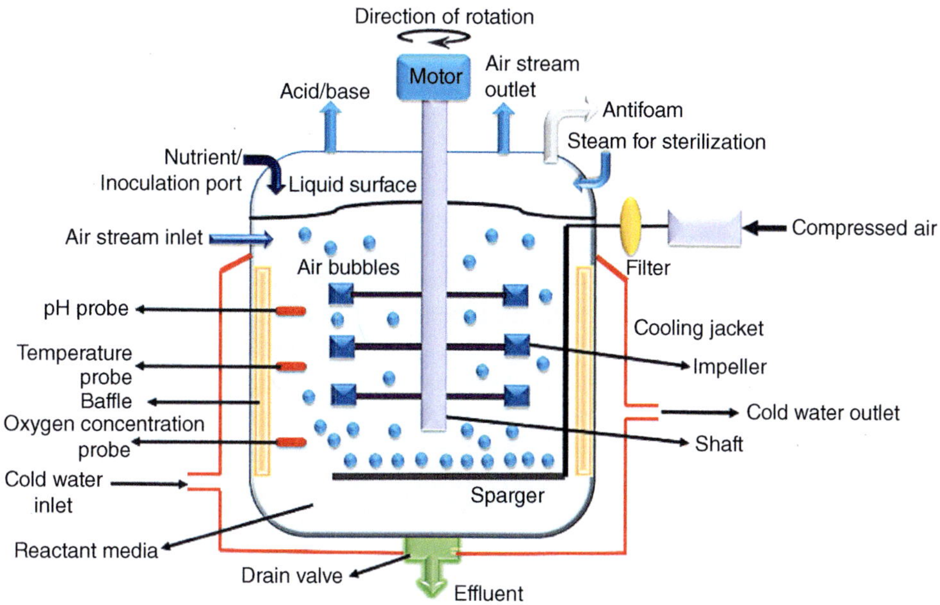

Figure 9.1 Bioreactor general setup (*Source:* Adapted from Tiwari et al. (2023) / with permission of Taylor & Francis).

several experiments can be conducted (Ma et al., 2023). Bioreactors can be classified into batch, fed-batch, or continuous bioreactors depending upon their method of operation. Also, they can be divided on the basis of the culture conditions as dynamic, perfusion, rotating wall vessel, and static systems (Iravani & Varma, 2021). Every type of bioreactor has its own advantages as well as disadvantages pertaining to the desired results as well as the applications. For instance, dynamic bioreactors that are composed of spinner flasks as well as the bioreactors with rotating wall vessel offer a fantastic environment mimicking *in vivo* conditions by promoting cell growth and differentiation, as well as the formation of tissues. An ideal bioreactor construction must ensure a perfect environment for cell growth and formation of a tissue-specific extracellular matrix (ECM), resulting in effective tissue functioning (Naghdi et al., 2022). Bioreactors can also be employed for the introduction of biophysical stimuli for promoting cellular response and development of tissue (Wei et al., 2017). With the recent advances in biotechnology, different types of bioreactors have been developed for specific tissues. For example, the usage of bioreactors in the cartilage tissue engineering perfectly mimics the *in vivo* environment (Nikolova & Chavali, 2019). Perfusion bioreactors are being used in bone tissue engineering for the maintenance of nutrient flow and cell growth promotion, followed by engineered bone tissue mineralization. Stem cell-based therapies are also utilizing the bioreactor system, as they provide an environment for stem cell production on a mass scale by keeping their self-renewal property intact (Rosser & Thomas, 2018). Bioreactors provide future hope for drug delivery research, as they can be utilized to study the complex relation of release kinetics with the biological response of drugs in an *in vitro* environment (Reyhanoglu & Gokturk, 2019).

9.2 Bioreactor Design

Designing a bioreactor and its operation is an ambidextrous area where both biology and engineering, along with technology, meet to run a biological process for a variety of functions. Bioreactors are those tools where, under optimal growth conditions, living cells like yeast, bacteria, or even mammalian cells can be cultured to yield highly valuable materials, for instance, biofuels, pharmaceuticals, chemicals, enzymes, and other desired materials. The major success behind the culture of microbes and the yielding of desired products relies upon different factors like temperature, nutrient supply, pH, agitation, and oxygen levels. With the advancement of biotechnology, bioreactors have been key elements in various sectors like agriculture, environmental remediation, and healthcare (Mandenius, 2016). Microorganisms growing inside the bioreactor can be genetically modified for the production of desired products; for instance, vaccines and monoclonal antibodies on a commercial scale are in general produced inside a bioreactor with the utilization of genetically altered mammalian cells.

Bioreactor design is a complex approach where one design may not be suitable for another, as pH, temperature, nutrient supply, and oxygen levels need to be controlled for better cell growth that results in efficient production of the desired products. The overall success of bioreactors relies on these key components; for example, bacterial cells require different levels of oxygen in comparison to those of mammalian cells; therefore, differently designed reactors are needed for their cultures. Scalability is one of the key components required in the design of bioreactors. Laboratory-scale bioreactors are used for the development and augmentation of cultures on a small scale, while for mass production, larger vessels are required. Setting up bioreactors of gigantic sizes for industrial production is a challenge in itself that not only requires space but also requires the optimization as well as maintenance of the controlled conditions for microbial growth. In the field of bioprocess engineering, the scalability of bioreactor design is key, considering features (Townsend, 2018).

The functioning of the bioreactor is as crucial as its design, and actual control over the different parameters leads to the successful production of the desired products. Control over the temperature inside the bioreactor is crucial, as it ensures that organisms live in their set range of growth. Monitoring pH levels is important as it can affect the growth as well as the desired product formation. Oxygen levels, especially for aerobic microbes, must be maintained at optimum levels, as it can also affect growth as well as product formation. Nutrient supply is also one of the most important aspects to be monitored, as living organisms have specific nutrient requirements like nitrogen, minerals, and carbon sources for desired product formation. In general, continuous feeding is required so that the cells can grow continuously, as any hindrance in the supply of the required nutrients can alter the whole process of product generation. The management of the waste as well as the by-products is also monitored in the bioreactors, as microbes, despite producing the desired valued products, also generate waste products which, if not monitored properly, can inhibit the whole process; for example, organic acid production by microbes can result in a pH drop, resulting in poor growth and low production. The whole process of bioreactor design as well as function is linked with the methods of control and monitoring. Measurement of the key components like product concentration, metabolite levels, and cell density helps in the adjustment of the operational parameters, and precise adjustment of these factors can minimize human error (Eibl et al., 2008).

Heat transfer, proper mass transfer (oxygen), and efficient substrate feeding with restrictions of overfeeding as well as flow conditions must be monitored appropriately within the bioreactor. An adequate supply of sufficient substrate, nutrient salts, suspension of solids, and vitamins must be maintained with the availability of oxygen and water in case of aerobic processes. Removal of the by-product and the management of the generated gases must be controlled. Monitoring of the sterilization, process control devices, construction and measuring, scale-up, compatibility with upstream and downstream processes, regulating techniques, antifoaming measures, and flexibility in operations are key features that must be kept in mind while designing the bioreactors (Sharma, 2012; Table 9.1). The key features of the bioreactors comprise different parameters as described in detail below (Alaghlavi, 2013):

a) **Headspace volume**
The effective space of the bioreactor is the fragment of its total space occupied by the microbes and medium, as well as the gas bubbles, followed by the remaining space defined as the headspace. The working volume, in general, occupies about 70–80% of the total volume of the reactor, which depends upon the rate of foam formation in the reactor (Van't Reit & Tramper, 1991).

b) **Agitator system**
It is composed of an external power drive along with an impeller as well as the baffles, leading to the profound mixing and enhanced mass transfer via the bulk liquid and the layers of bubble boundary. This provides efficient conditions needed for breaking the bubbles. A Rushton turbine-type impeller is most commonly used in the majority of microbial fermentations.

c) **Air delivery system**
It is made of a compressor, a sterilization system, inlet air, an air sparger, and an exit air sterilization system to avoid contamination.

d) **Foam control system**
It is a key component of a bioreactor, as too much foam formation can block the air exit filters and generate high pressure inside the bioreactor, leading to imbalanced growth conditions. This system keeps control over the foam formation inside the bioreactor.

e) **Temperature control system**
It is composed of temperature probes and a heat transfer system (jacket, coil). Electric heaters, as well as the steam produced by the boilers, result in heating, while cooling towers or refrigerants like ammonia provide cooling water.

f) **pH control system**
It uses neutralizing agents for pH control, and these agents must be non-toxic to the growing cells, as well as non-corrosive when mixed with the growth medium. Sodium carbonate is the most commonly used agent.

g) **Sampling ports**
These are used for the addition of the nutrients and salts, as well as water, and also for sample collection.

h) **Cleaning and sterilization system**
This system is essential to keep track of contamination. In the large-scale equipment, thermal sterilization is generally preferred. In the heat-sensitive equipment, sterilization

Table 9.1 Scalability of Bioreactors.

Reactor Type	Mass Transfer	Heat Transfer	Shear Stress	Medium Homogeneity	Scalability	Oxygen Transfer
Stirred tank reactor (STR)	🟢 High, mechanical agitation facilitates efficient transfer	🟢 High, controlled with external heat jackets	🔴 High, due to mechanical agitation	🟢 High, mechanical mixing ensures uniform conditions	🟢 Excellent, most scalable type of bioreactor	🟢 High, efficient oxygen transfer with proper agitation
Modified stirred tank reactor (STR)	🟢 High, mechanical stirring provides excellent mass transfer	🟢 High, effective due to stirring and jacketed vessels	🟡 Moderate, controlled with proper agitation	🟢 High, stirring ensures even distribution	🟢 Good, highly scalable, common for large-scale production	🟢 High, impeller design can enhance oxygen transfer
Airlift bioreactor	🟢 High, due to circulation and organized gas–liquid flow	🟢 High, due to efficient fluid circulation	🟡 Low to moderate, more controlled shear	🟢 High, consistent circulation ensures good mixing	🟢 Good, maintains performance at larger scales	🟢 High, good oxygen transfer via gas-lift mechanism
Bubble column bioreactor	🟡 Moderate, depends on gas flow and bubble size	🟡 Moderate, relies on gas–liquid interactions	🟢 Low, ideal for shear-sensitive organisms	🟡 Moderate, can be less uniform in larger volumes	🟡 Easy, but efficiency decreases with larger volumes	🟡 Moderate, depends on bubble size and gas flow rate
Wave bioreactor	🟡 Moderate, depends on wave motion for oxygen transfer	🟡 Moderate, good for small volumes	🟢 Low, gentle wave motion	🟢 High, wave motion provides good fluid mixing	🟢 Easy to scale, widely used for cell culture applications	🟡 Moderate, limited by wave amplitude and frequency
Disposable bioreactor	🟡 Moderate, depends on bag design and aeration	🔴 Limited, generally relies on external temperature control	🟢 Low, typically used for shear-sensitive cultures	🟡 Moderate to high, depending on design and agitation	🟢 Easy, designed for modular scale-up	🟡 Moderate, typically relies on sparging or external aeration

Membrane bioreactor (MB)	🟢 High, depends on membrane type and configuration	🟢 High, membranes can provide efficient heat exchange	🟢 Low, no agitation involved	🟢 High, consistent flow across the membrane	🟡 Moderate, membrane surface area limits scalability	🔴 Low, oxygen transfer is limited by membrane diffusion
Solid-state fermentation	🔴 Limited, relies on diffusion in solid medium	🔴 Challenging due to solid substrates	🟢 Low, typically no agitation	🔴 Low, difficult to maintain uniform conditions	🔴 Low, scalability is complex due to heat and mass limitations	🔴 Low, oxygen transfer is poor in solid media
3D-printed bioreactor	🟡 Moderate, highly customizable mass transfer	🟡 Moderate, depending on design	🟢 Low, can be optimized for shear-sensitive processes	🟢 High, design flexibility allows for uniformity	🟡 Good, but depends on printing technology and design	🟡 Moderate, can vary depending on the design and material
3D-printed microbioreac-tor	🟢 Moderate, highly customizable with good surface-to-volume ratio	🟢 Good, depends on material properties and design	🟢 Low, can be tailored for shear- sensitive cultures	🟢 High, uniform conditions due to small size	🟡 Moderate, scalability is design-dependent	🟡 Moderate, can vary depending on the design and material

🟢: High efficiency; 🟡: medium efficiency; 🔴: low efficiency.

Source: Adapted from Palladino et al. (2024) / MDPI / CC BY 4.0.

by chemicals is done. UV or X-rays are also used for sterilization. Membrane filters with a uniform pore size are also used sometimes.

i) **Charging and emptying**
These are required for the input of the materials as well as for the removal of the desired products from the bioreactor (Singh et al., 2014).

A comprehensive description of the relevant aspects of the bioreactor design, as well as the scale-up aspects, is discussed below.

9.2.1 Scale Up

For the freely suspended batch or fed-batch stirred tank animal cell culture reactors, the operational or production volumes have been reported to range from 8 to 10 m^3 (Nelson, 1988a, 1988b) and 15 m^3 (Werner, 1994). For the scaling of a reactor to meet the requirement, the type of reactor used must be kept in mind, followed by the selection of the cell line as well as the quality of the desired product. Monoclonal antibodies used for diagnostic purposes are at the lower end of the scale; however, some proteins used for therapeutic purposes must be produced in bulk quantities, reaching hundreds of kg $year^{-1}$ to >10 000 L. For the successful and cost-effective scale-up production, it is essential to keep a track of the market need, as well as the cost of the new product, which is minimized. The success of scale-up operations is dependent upon the quality of the product and yield at different scales, where the productivity of the bioreactor reaches mg product L^{-1} day^{-1}. To reach the maximum production by the scaling operation, it is mandatory to have (i) adequate and anticipated fusion of solid suspension and gas–liquid dispersion, followed by satisfactory homogenization of materials for efficient physiological and biochemical homogeneity (Votruba & Sobotka, 1992); (ii) adequate and foreseeable transfer of gases as well as nutrients at mass scale; and (iii) minimization of cell damage as well as cell physiology alterations. A successful scale-up must utilize a shorter time period with cost-effective production and higher benefits (Reisman, 1993).

9.2.2 Mixing

Mixing can be assessed as well as amplified by different means, as mixing times offer potential information regarding the time for reaching the limit of homogeneity. Mixing time measurements are generally achieved by measuring the time during the addition of a salt, a base, or an acid in a pulse or step, and changes in the conductivity, concentration, and pH with time are recorded. The 95% mixing time is calculated when the concentration or conductivity reaches 95% of its final value. Investigation of the response curves to the addition of tracers gives information about the circulation time as well as about the dispersion of the tracer element during the circulation process. Currently, numerous methods can be employed for analyzing the tracer response curves (Lu et al., 1994). The Bodenstein number (Bo) is often used for the comparison of mixing in different areas of the reactor and is also a measure of longitudinal mixing. The Bo number has no dimension (Lv/D_1, where L = characteristic length [recirculation length], v = average velocity, and D_1 = axial diffusivity). When the Bo number reaches 0, the reactor behaves like that of a stirred tank reactor, while when the Bo number reaches infinity, the reactor is similar to that of an ideal plug-flow reactor. The Peclet number (Pe) is also used for measuring the mixing time, where Pe = vD/D_1 (D = tank diameter). Both Bo and Pe numbers are used not only in the stirred tanks or air lift reactors but also in other types of reactors. To boost up a scale-up process, uniform mass distribution as well as energy transfer are important and are quite difficult to ensure. Mixing is crucial for animal cell culture, as they are more sensitive to environmental changes. Earlier, it was believed that lower mixing and circulation times while growing microbial cells were less critical as they have low substrate consumption with lower changes of pH and temperature when grown freely. However, with the advancement of the technology, it has become evident that dispersion of nutrients, pH gradients, gas dispersion, and elimination of waste products are of utmost importance (Wayte et al., 1997). Dispersion of gas is another important factor, as removal of carbon dioxide (CO_2) is crucial; if not removed, it can cause adverse effects. Tramper (1995) has calculated that animal cells grown in stirred vessels, airlift reactors, and bubble columns have oxygen gradients of 10 m^3. The mixing process is better understood in the case of continuous systems than in batch systems.

9.2.3 Mass Transfer

Mass transfer, followed by gas distribution, toxic products, and nutrients, is essential in animal cell culture. Oxygen and carbon dioxide transfer is important, especially CO_2, which is required for pH control. Oxygen requirements are dependent on the cell lines, ranging from 0.05–0.5 mmol oxygen L^{-1} h^{-1} for 10^6 cells mL^{-1} for human cells (Fleischaker & Sinskey, 1981) to 0.1–2.5 mg·10^8 $cell^{-1}$ h for industrially relevant cell lines (Aunins & Henzler, 1993). The gas transfer rate is:

$$\text{Gas transfer rate} = k_L a\,(C^* - C) \tag{9.1}$$

where k_La is the overall mass transfer coefficient, C is the dissolved gas concentration, and C^* is the saturation concentration of the dissolved gas. For the microbial systems, a wide diversity of empirical correlations relating k_La or k_L to power input are available, and the general form is:

$$k_L a = c\left(\frac{P}{v}\right)^m U_s^n \tag{9.2}$$

where P/V is power input per unit volume of dispersion, U_s is the superficial gas velocity, and c, m, and n are empirical constants. In general, most correlations for oxygen k_La are obtained in microbial fermentations at high agitation speeds, which is not suitable for animal cell cultures. The overall mass transfer coefficient can be defined as a function of the physical properties of the liquid phase, where k_La can vary by 50% based on the composition of the medium (Dorresteijn et al., 1994). There are several methods of k_La measurement, and new methods are continuously being developed for managing CO_2 and oxygen mass transfer in bioreactors. Smith et al. (1990) studied control of dissolved gas concentrations using *Catharanthus roseus* cells and found out that injection of gas in the fermenters can affect the mass transfer rate, thereby influencing the interfacial area used for mass transfer. Oxygen transfer via liquid surface is also important, especially for small-scale cell cultures. Hu et al. (1986) found out that surface aerators can boost oxygen transfer without cell alteration, but only in small-scale reactors. Gas can be introduced in the system by direct sparging with air or oxygen and gas-permeable membranes. Other methods recommended at a small scale but that can be used a large scale include (i) *in situ* oxygen generation by utilizing, for example, hydrogen peroxide and the enzyme peroxidase, or (ii) pure oxygen sparging in a vibrating wire gauze oxygenator. Oxygen transfer can be enhanced by the addition of oxygen-complexing agents in the medium, such as fluorocarbon oils or modified hemoglobin, but it can affect downstream processing (Katinger et al., 1979). Moreira et al. (1995) compared oxygen transfer using different aeration methods like sparger, surface, and membrane aeration. Correlations for power as a function of Reynolds number (Re), followed by bubble size for each sparger design and k_La, were reported for each method. The findings suggested that surface aeration was suitable only for small volumes, and for sparged aeration, higher k_{La} values were found. Addition of polymers like Pluronic F-68 reduces k_La. Chisti (1993) described a marine impeller-fitted 300 L fermenter holding the growth of 20 L hybridoma cells by considering two sparger designs: one with porous metal spargers and other with larger multi-hole spargers. The findings suggested that larger multi-hole spargers were better, as foaming issues were generated in the porous metal spargers. Designing gas injection techniques has also been a focus in order to achieve perfect dissolution of bubbles before they reach the surface and cause damage to cell cultures, as bubbles keep on collapsing at the liquid interface. Vibrating silicone tubes and gas-permeable membranes were developed for gas–liquid transfer without bubble formation (Monahan & Holtzapple, 1993). There are a few drawbacks of membrane diffusion, like difficulty in maintaining, especially at a larger scale, followed by reduced interfacial area (a), which in turn affects k_La. Zhang et al. (1992) used three ranges of bubbles (5 mm, 1 mm, and 100 μm in diameter) to measure the oxygen transfer rate in an aerated animal cell bioreactor and found that oxygen transfer rate increases as the bubble size decreases. Newborn calf serum, as well as Pluronic F-68, added to the medium also affect the size of the bubbles. Bubble mass is crucial, as it governs the size of the bubble, followed by the interfacial area and k_La; however, direct evidence is not available yet (Reuss, 1993). The rate of gas flow increases the bubble formation by separate bubble formation, chain bubbling, jet regime, and turbulent regime. Jetting regime can be achieved when the Reynolds number (Re) >2000, where gas is flowing through the orifice. However, reaching this regime is quite unlikely in case of animal cultures. So, separate bubble formation followed by chain bubbling regimes is preferable. For separate bubble formation:

$$dv_s = 1.7\left\{\frac{\sigma d_o}{\Delta\rho g}\right\}^{1/3} \tag{9.3}$$

where dv_s is the Sauter mean diameter, σ is the surface tension, d_o is the orifice diameter, $\Delta\rho$ is the density difference, and g is the gravitational acceleration.

For chain bubbling regime:

$$dv_s = 1 \cdot 17 v_o^{0.4} d_o^{0.8} g^{-0.2} \tag{9.4}$$

where v_o is the gas velocity at the sparger. The above-mentioned equations were made for bubble size in turbulent flow, coalescing, and non-coalescing solutions (Machon et al., 1997). Generally, in the animal cell cultures, lower gas flow as well as lower agitation rates are used, and the size of the bubble is likely to be similar to the one at the sparger, falling in the range of 4–6 mm. In animal cultures, the partial pressure of metabolically produced carbon dioxide (pCO_2) and, therefore, carbon dioxide mass transfer from the medium to the gas bubbles is crucial. A drop in pH is observed when pCO_2 rises, as it causes an increase in the levels of H_2CO_3, so this parameter must be monitored carefully. Higher levels of dissolved CO_2 can alter the pH of the growth medium and, therefore, can affect cell growth.

9.2.4 Shear

Numerous studies have been carried out focusing on the shear-sensitive nature of plant, animal, and microbial cells, orienting the design and operation of the reactors (Papoutsakis, 1991). Several factors can cause damage to cells in reactors; for instance, pressure differences between the front and back sides of the impellers, direct collisions of cells with the stirrers or baffles, micro-eddies with a mean diameter smaller than the cellular diameter, and turbulence mediated by differences in the vector length of the vector field of non-laminar flow (Leist et al., 1990). However, there are two prime factors that govern cell damage: first, when Kolmogorov eddy size is similar to or smaller than the size of the cell (Kunas & Papoutsakis, 1990). Kioukia et al. (1992) tested batch cultures in baffled and unbaffled bioreactors and concluded that the bubble disengagement region is the prime area where the majority of the cell damage is caused; higher the passage of cells through this region, the greater the damage. The second factor to be proposed for cell damage is cell–bubble interaction, which occurs in three areas: (i) bubble generation area, (ii) bubble rising area, and (iii) bubble disengagement area. Current research suggests that rising bubbles have minimal effect on cell damage, but bubble bursting and bubble generation can cause major damage to the cells (Michaels et al., 1994). Kunas and Papoutsakis (1990) suggested that at a speed of >700 rpm, cell damage occurs in a specific hybridoma cell line due to micro-scale turbulent eddies; however, at lower speeds, cell damage was caused by breakup and gas entrainment. The exact reason or conditions where real cell damage occurs depend on several factors, like the geometry of the reactor as well as the resistance of selected cell lines to shearing. Cherry (1993) found that no damage occurs to hybridoma cell lines with an agitation speed of 700 rpm in the absence of air. Chalmers (1994) found that hybridoma cell lines can tolerate speeds between 100 and 450 rpm when grown in 1 L bioreactors, and cell damage was observed only when air entered the system; however, with the removal of the air–liquid interface from the vessel's top, an agitation speed of 700 rpm resulted in no cell damage. Leist et al. (1990) reported that human melanoma cells grown in 5% fetal calf serum (FCS) at speeds >1.5 ms^{-1} stirrer tip speed exhibited stress effects, while Chinese hamster ovary (CHO) cells when grown in 10% FCS tolerated speeds <1 ms^{-1}. Nelson (1988a) suggested that shear damage is simply exaggerated, as no altered effects were observed on the hybridoma cell line when grown in a 13,000 L reactor with an agitator speed of 500 rpm (6 ms^{-1} tip speed). However, several studies have indicated that sparging and tip speed can cause cell death or cell damage, and different cell lines have different shear sensitivities (Gardner et al., 1990).

Bliem and Katinger (1988) reported that many cell lines used on a large scale are rather shear-insensitive; for example, the Namalwa, BHK 21, CHO, and Vero hybridoma cell lines.

Advanced designs of impellers and aeration methods can reduce cell damage. Bubble-free aeration system is feasible at a small scale, but as the scale progresses, gas introduction into the reactor becomes mandatory. Impellers are used to overcome this issue by reducing mixing times, lowering agitation rate, and then reducing shear stress (Moreira et al., 1995). In the bioreactor and viscometric studies, it has been proposed that:

(i) the inoculum state greatly affects the ability of the cells to the shear force
(ii) cells are more prone to shear in the stationary and lag phases of growth; however, their robustness increases in the exponential phase of growth
(iii) cells cultivated in systems with uniform agitation are less prone to shearing than the ones grown in stationary or slow agitated ones
(iv) an inhibitory environment also increases the sensitivity of the cells to shear
(v) lower pH also makes the cells shear sensitive (Reuveny & Lazar, 1989).

Several studies have also indicated that in the agitation tanks, increased agitation can alter sensitivity of the cells to shear. Alternative factors that can also cause cell damage include: (i) cell concentration, (ii) temperature of the fluid in the system, (iii) nature and extent of container surface, and (iv) deformation and stress history of the material (Cherry & Papoutsakis, 1990). Different equations have been proposed in order to detect cell death due to cell contact with bubbles containing surfactants:

$$k_d = \frac{4Fx'}{\pi D^2 H} = \frac{F}{v}x' \tag{9.5}$$

where k_d is the death rate, F is the gas flow rate, V is the reactor volume, X' is a hypothetical killing volume, and H is the reactor height (Tramper et al., 1987). It is evident from the equation that cell death increases if F/V increases, so it is crucial to boost mass transfer from bubble to the liquid by minimizing the gas flow rate per unit volume. Jobses et al. (1991) suggested the following equation:

$$k_d = \frac{Q_G}{T^2 H} \tag{9.6}$$

where QG is the volumetric gas flow rate, T is the impeller diameter, and H is the reactor height. Kioukia et al. (1992) further developed this equation by taking into account the diameter of the bubble:

$$k_d \alpha = \frac{Q_G^{4/3}}{T^{1/3} H d_b^3} \tag{9.7}$$

where db is the bubble diameter.

This equation suggests that with the increase in the aeration rate, the death rate also increases, and increasing the bubble height and the diameter should also decrease cell death; however, smaller bubbles often lead to a higher fraction of cell death. Al-Rubeai et al. (1995) found that apoptosis and necrosis are basically dependent on the energy dissipation rates of the system. Born et al. (1992) suggested that micromanipulation of animal cells using laminar shear stresses can indicate the perceptivity of cells to damage in bioreactors. Shear not only results in cell death but also affects other components, like cell suspension, fluid mixing, mass transfer, cell growth, cell-to-cell or cell-to-substrate adhesion, and product formation (Bliem & Katinger, 1988). Tramper et al. (1993) suggested that it is almost impossible to figure out the methods for the calculation of the shear stress in an air-lift reactor. Pluronic surfactants (polymers of poly(oxypropylene) and poly(oxyethylene)) are widely utilized as protective agents in serum-free media for reducing foaming and providing additional protection (Cherry & Papoutsakis, 1990). Leist et al. (1990) studied the effect of serum along with polymers in order to find out protection against cell damage caused by agitation or sparging and found that serum provides protection against sparging and enhances foaming. Cherry and Papoutsakis (1990) found out that use of 10% fetal bovine serum can decrease cell damage and promote cell growth with higher agitation rates and surface aeration in bioreactors. Zhang and Thomas (1993) suggested that medium additives can also play a role in preventing cell damage by avoiding bubble bursting by foam formation. Kioukia et al. (1992) suggested that Pluronics can enhance culture activity as they directly interact with the cell membranes and, thus, can directly affect cell growth; however, the exact mechanism behind this is unknown but additives protect cells from bubble bursting. The true effect of these protective agents is dependent upon the cell line, culture conditions, as well as medium composition (Papoutsakis, 1991). Handa-Corrigan (1990) emphasized that cell damage and oxygenation can be minimized by (i) designing stirred tank reactors (STRs) with independent mixing and oxygen transfer (improved impeller design and bubble disruption, as well as dispersion, can be minimized), (ii) minimizing the size of the bubble so that stable foam can be produced, and (iii) utilizing surface-active components in media formulation in order to produce stable foams. Bliem and Katinger (1988) emphasize understanding micro-turbulence is crucial to understanding the relationship between cell damage and shear.

9.2.5 Pressure

Pressure affects the solubility of gases, namely oxygen and CO_2, gas bubble size, as well as cell viability followed by cell physiology. Scant literature is available on the effect of pressure on cell culture. Takagi et al. (1995) reported that hydrostatic head (0.1–0.9 MPa) had very little effect on the growth or glucose consumption of a hybridoma cell metabolism. However, a 20% increase in the production of monoclonal antibody was noticed when pressure was increased, as it affected the glucose-to-lactate and glutamine consumption rates.

9.2.6 Scale-Up Criteria

Two typical criteria for initiating the scale-up process are known: (i) scaling up is required in an already existing reactor and (ii) designing of a new fermenter. The below-mentioned parameters are generally kept constant, as suggested by Ju and Chase (1992):

- reactor dimensions,
- $k_L a$,
- shear rate (maximum value),
- P/V,
- volumetric gas flow rate per unit volume of liquid,
- superficial gas velocity,
- mixing time, and
- impeller tip speed (for stirred tanks).

However, all these parameters cannot be set constant at a given time point; for instance, if constant P/V and geometric similarity are utilized, shear rate, mixing time, as well as impeller tip speed cannot be kept constant for scale-up. Superficial gas velocity cannot be maintained constant with gas volumetric flow rate. Geometry is a major scaling criterion. For microbial cell systems, the third criterion is constant power per unit volume. Constant mixing time is generally not considered as a scaling criterion.

9.2.7 Scale Down

Scale down is an important process generally noticed on a large scale, offering possibilities for improvements. Sola and Godia (1995) explained the regime analysis of this phenomenon. This comprises sole understanding of the rate control mechanism to be considered as a scale-down variable, and to do so, mass transfer and hydrodynamics must be taken into consideration; for instance, in the stirred tanks, both mixed as well as non-mixed zones can be easily figured out. Kossen (1995) pointed out that in the scale-down experiments, geometrical similarity plays an important role in the maintenance of the scale-up process.

9.3 Types of Bioreactors

In recent years, the use of bioreactors for solid-state fermentations (SSFs) and submerged fermentations (SmFs) has been emphasized due to the advancement of industrial biotechnological processes for the production of vaccines, biopharmaceuticals, as well as value-added products. Biotechnological processes can be established in a solid medium having low water activity (SSF), with or without agitation and aeration (Kaur & Sharma, 2020). Biochemical engineering can be set up in liquid medium (SmF) along with microbes, and three types of bioreactors are used for this purpose, namely the STRs, the airlifts, and the bubble column bioreactors (Zhong, 2011). The size of the bioreactor market in 2022 was 9.3 billion dollars, and it is supposed to grow further in the next ten years, reaching up to 30.42 billion dollars by the year 2032. The most widely used bioreactor is the aerated STR, as it offers multiple advantages, like simple infrastructure, versatile nature, high scalability, optimal control of cultures, and minimal risk of contamination. This bioreactor is widely used in the culturing of bacterial or yeast cells due to their lower sensitivity to shear stress. Mechanical agitation system is one of the key components of the STRs, using a wide variety of blades, turbines, or propellers for the proper mixing of the culture medium. Agitation is mandatory, as it ensures efficient nutrient homogenization, proper oxygen distribution, and maintenance of ideal conditions within the reactor volume (Spier et al., 2011). Rushton turbines and the pitched-blade impellers are commonly used for agitation. Rushton turbines are generally utilized for cultures of bacterial and yeast cells, as they are shear resistant, while pitched blades are used for low-shear cultures, reducing the risk of cell damage. Pitched blades are perfect for the culture of insect and mammalian cells thriving on microcarriers or in suspension, or for the growth of filamentous fungi. However, combination of different impellers for increased mixing and reduced shear stress is also possible. With the advancement of SmF for animal cells utilized in vaccine or biopharmaceutical production, modified STRs have come into light (Kaiser et al., 2018). Animal cells are more prone to shear stress; therefore, modifications of STRs for biopharmaceutical production focus on (Karnieli, 2016)

i) height-to-diameter ratio <2,
ii) perforated sprinklers are used for the crops that need more aeration. These sprinklers control the speed of gas spray as well as bubble size for lower shear stress therefore, provide homogenous mixing with zero foam formation,
iii) usage of Marine-type impeller.

Apart from STR bioreactors, bubble column and airlift or pneumatic bioreactors have also been used in recent years. Although they are used on a smaller scale in comparison to STRs, they are also important; however, rigorous agitation is hard to maintain as microbial cells are sensitive to shear stress. Airlift bioreactors follow a draft-tube principle and are composed of a cylindrical vessel fitted with an aeration system. This cylindrical vessel from inside is composed of two parts: an ascending tube, or riser, where gas is pushed in or allowed to go out, and a descending tube, or degassed chamber (de Assis & Belini, 2023). Air movement and aeration of the system are well maintained in airlift bioreactors. Bubble column bioreactors, on the other hand, are also composed of a cylindrical vessel similar to airlift bioreactors, where the aeration system is installed in the lower part of the reactor. These pneumatic bioreactors have several advantages, as they offer (i) efficient mass transfer, (ii) efficient mixing with reduced shear stress, (iii) lower energy utilization, (iv) smooth operation in a sterile environment, and (v) uncomplicated design. However, these bioreactors have the disadvantage of difficult scale-up for larger volumes (Behin & Amiri, 2023). Wave bioreactors have been designed as an alternative to STRs for submerged plant and animal cell cultivation, as this device was developed in the 1990s for the crops that have low demand of oxygen and do not require submerged gasification and for cells that have anchorage-dependent growth (Zhang et al., 2019). They have a simple setup resembling the Cellbag, which contains a hydrophobic filter enabling gaseous exchange for the oscillatory movements along with temperature control. Although Cellbag is disposable and sterile, it has limitations like a restricted scale expansion to 500 L. This type of reactor is used in vaccine production, as it is suitable for animal cell cultivation and some bioconjugates (Decarli et al., 2018). Disposable bioreactors are composed of high-quality plastics, eliminating the need for cleaning and sterilization between different batches of cultivation, thereby reducing the risk of cross-contamination as well as simplifying the regulatory compliance and validation processes. They are efficient and flexible compared to stainless-steel or glass bioreactors and can vary in sizes reaching a volume of up to 2000 L (Markarian, 2019). Membrane bioreactors (MBs) have also come into light in recent years, as they are widely used for the effluent treatment on the industrial scale and are composed of semi-permeable membranes allowing selective passage of materials of different sizes while avoiding the passage of cells from the reactor. These membranes are ultrafiltering, microfiltering, or nanofiltering depending on the size of particle to be filtered from the culture medium. The major profits of MBs include (i) effective separation, (ii) continuous operation, (iii) contamination control, and (iv) saving of the space. However, they have certain disadvantages, like (i) costs, (ii) maintenance complexity, and (iii) flow limitations (Xiao et al., 2019). Apart from being used in the treatment of effluents, MBs can also be used for the production of valued products by taking into account upstream and downstream steps simultaneously, where the membrane system purifies the molecules of interest (Akkoyunlu et al., 2021). Static or agitated bioreactors can be utilized in SSF, and the common types utilized are (i) trays, (ii) fixed beds, and (iii) stirred drums. Although they are widely used on an industrial scale, there are still some challenges to be considered (Ge et al., 2017):

- The majority of SSF uses filamentous fungi for fermentation, and their hyphae are susceptible to damage by rigorous agitation.
- Agglomeration often happens in the solid medium, resulting in the uneven growth of the microbial biomass due to heat transfer and aeration.
- Large-scale inoculation, control, and media sterilization are quite difficult.

Current research has been focused on the innovative designs of bioreactors with automated control for enhanced production of desired products (Manan & Webb, 2017). The use of 3D printing technology has gained interest in the past few years, as this technique offers the design of both micro- and macrobioreactors for enzyme and bench-scale cultivation. This technique can also be used for manufacturing matrices and micromixers for cell immobilization as well as enzymes (Gkantzou et al., 2023). 3D printing also provides a solution for both SSF and SmF processes by aiding in bioreactor design, enabling easy operation. Micro-bioreactors are generally used for cellular response optimization, drug testing, and disease modeling in the pharmaceutical processes, while macro-bioreactors are used for the cultivation of functional tissues for implantation in tissue engineering studies (Altmann et al., 2020). Operational safety focuses on the prevention of contamination and waste management by PDMS (polydimethylsiloxane), which is of utmost importance. PDMS is a well-known silicon-based organic polymer with a strong flexibility, gas permeability, and transparency, and these properties ensure its usage in microdevice preparation by ensuring control and integrity of bioprocesses (Prado & Borges, 2018). Currently, the advancement of artificial intelligence (AI) and machine learning algorithms that improve efficiency and reduce waste have resulted in proper energy consumption as well as process productivity. These algorithms can define energy requirements and help in the development of control strategies that are useful in cultivation processes (Bradford et al., 2018). Industrial-scale bioreactors face challenges pertaining to safety, energy efficiency, and engineering. Culture homogeneity maintenance and operational safety are crucial factors. Selection of an appropriate impeller combined with AI results in increased efficiency of bioreactor operation. Frontistisf et al. (2023) used an MB control system coupled with neural networks for wastewater treatment. They integrated neural networks with algorithms implemented with Internet of Things (IoT) devices and new-generation sensors and found that such complex system can be used for wastewater treatment, leading to the development of self-adaptive and intelligent systems. Table 9.2 shows the comparison between different types of bioreactors along with STRs and modern bioreactors, focusing on critical features (Palladino et al., 2024):

- **Mass transfer:** The potential of the bioreactor to facilitate the transfer of gases, nutrients, and substrates between different stages, which is crucial for cell viability and yield of the final product.

Table 9.2 Advantages and Disadvantages of Different Types of Bioreactors.

Type of Bioreactor	Advantages	Disadvantages
Stirred tank reactor (STR)	• Simple design • Versatility and scalability • Greater control of cultivation conditions • Reduced risk of contamination	• High shear stress, not suitable for shear-sensitive cells • Complex mechanical agitation system
Modified STR	• Suitable for animal cells—lower height/diameter ratio • Reduced shear stress • Homogeneous mixture	• Specific design modifications required • Potentially higher costs
Airlift bioreactor	• High mass transfer • Good mixing with low-shear stress • Low energy consumption • Easy operation in sterile conditions	• Difficult lo scale up • Limited use compared to STR in industrial settings
Bubble column bioreactor	• High mass transfer—low shear stress • Low energy consumption • Simple design	• Difficult to scale up • Limited use compared to STR
Wave bioreactor	• No need for sterilization (disposable) • Suitable for low oxygen demand cells • Simple configuration	• Limited scale expansion (up to 500 L) • Not suitable for high oxygen-demanding crops
Disposable bioreactor	• Eliminates the need for cleaning and sterilization • Reduces contamination risk • Flexible and efficient • Suitable for various scales	• Environmental concerns (plastic waste) • Potentially higher long-term costs
Membrane bioreactor (MB)	• Efficient separation • Continuous operation • Contamination control • Space saving	• High costs • Complex maintenance • Flow limitations
Solid-state fermentation	• Suitable for low water activity media • Can use various substrates—less energy intensive	• Damage to filamentous fungi • Uneven biomass growth • Issues with aeration/heat transfer—inoculation and sterilization challenges
3D-printed bioreactor	• Customizable design • Efficient process optimization • Application in various bioprocesses	• High initial setup cost, complexity in design, and manufacturing

Source: Adapted from Palladino et al. (2024) / MDPI / CC BY 4.0.

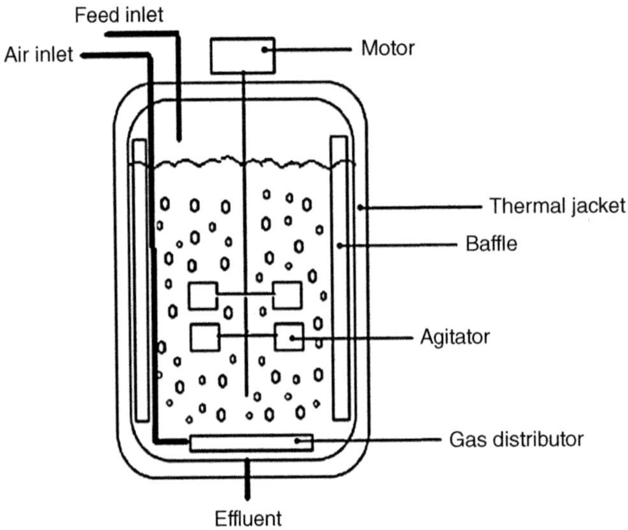

Figure 9.2 General setup of a stirred tank reactor (STR) (*Source:* Adapted from Bakratsas et al. (2021)) / with permission of Elsevier.

- **Heat transfer:** The maintenance of optimal temperature, which affects the reaction rate as well as microbe performance.
- **Shear stress:** Mechanical forces that can damage the cell integrity.
- **Medium homogeneity:** Uniform nutrient, gas, and other substrate distribution throughout the culture medium for efficient cell growth.
- **Scalability:** Increased reactor size for the large-scale production of value-added products without altering the bioreactor performance.
- **Oxygen transfer:** Efficient oxygen exchange in the medium for the efficient cell production.

9.3.1 Stirred Tank Reactors (STRs)

STRs are widely used on the industrial scale due to their flexible and versatile nature (Figure 9.2). Najafpour et al. (2009) summarized three types of fermenters that are used for the industrial applications:

1) Non-mixed + non-aerated systems: ~70%
2) Non-mixed + aerated systems: ~10%
3) Mixed + aerated systems: ~20%

Non-aerated + non-mixed tanks are generally utilized for the production of traditional items like cheese, wine, and beer. However, mixed + aerated tanks are used for the production of newer products by the culturing of microbes. Bioreactors in general provide a controlled environment for the growth of cells, leading to the generation of the desired products. STRs are used at an industrial scale due to their low costs of operation, and their size ranges from dm^3 to few hundred m^3. The tanks used at the laboratory scale are composed of glass with a capacity of about 20 L, but they are composed of steel for higher production; however, different combinations of metal and other materials are also used depending upon the purpose. The ratio between the height and diameter lies between 2:1 and 6:1, which is dependent upon heat removal (Najafpour et al., 2009). A 1:1 ratio is the best, as it offers a low superficial area with less demand for material, but when aeration is needed, it must be maintained high in order to promote better contact between the liquid and the air bubbles, and also high hydrostatic pressure maintenance is needed. Bubble dispersion as well as homogeneity in the bioreactor is obtained by agitation, which has a higher energy intake. A wide variety of impellers are now available for discrete flow patterns in the tank. About 70–80% of the overall volume of the STRs is liquid-filled; therefore, it has a headspace for the exhaustion of gas as well as foam formation, which is under the control of a foam breaker. STRs are largely implied for free and stationary cells (Doran, 1995).

9.3.1.1 Continuous Stirred Tank Reactors (CSTRs)

CSTRs are STRs that operate continuously with a regular supply of substrate and continuous removal of energy and mass. These devices are constructed to provide efficient mixing without considering temperature, fluid properties, concentration, or reaction rate hindrances, indicating that both the inlet + outlet streams have the same properties as the medium in the bioreactor. Batch reactors and CSTRs share the same properties, with the main differences being in instrument design and control system. Basically two operation strategies are known for CSTRs. First is the chemostat, which is used for the cultivation of cells in which all the nutrients are replenished in abundance, and the volume of the liquid is maintained constant by equalizing the inlet and outlet flow rates. The second is the turbidostat, in which the concentration of the cells is kept constant by tracking the optical density of the culture, and the liquid volume is kept constant by equalizing the outlet flow rate with the inlet flow rate. CSTRs can be connected to more than one reactor in a series by maintaining different conditions in each reactor (Laska & Cooney, 1999; Figure 9.3).

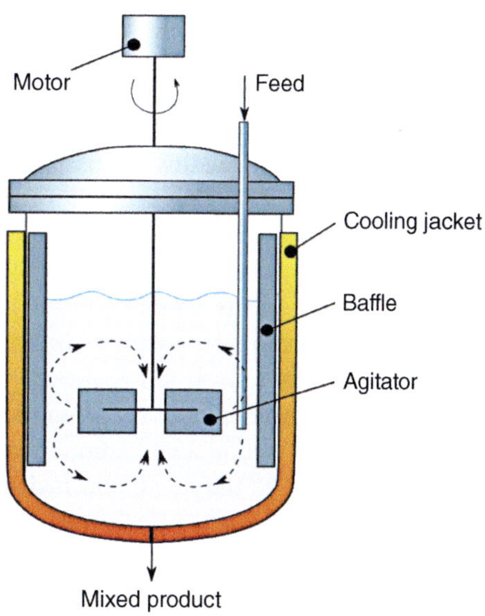

Figure 9.3 General setup of a continuous stirred tank reactor (CSTR) (*Source:* adapted from Wikimedia Commons / CC BY-SA 3.0).

9.3.1.2 Features of the STRs

a) **Impellers**
Impellers are usually placed at the center and located in the superior part of the reactor for performing mechanical mixing. Many types of impellers are now available for different processes and can be divided into two categories: the turbine and the paddle impellers. The selection of the impeller depends on the dimension of the bioreactor, type of medium, total volume, and viscosity and density of the fluid used. Impeller type is defined by the viscosity scale with the range within 1–104 cP. During a perfect reaction, a strong interaction occurs between the substrate and the microbes + enzymes, resulting in efficient mixing, leading to ideal heat and mass transfer. The mixing intensity is dependent on the process type. Excessive shear stress applied to the cells can result in damage or death of the cells resulting in no production. Therefore, appropriate impeller must be used. Plant and mammalian cells are rather sensitive; therefore, selection of impeller must be done cautiously. Impeller designs also contribute to the cell growth and have three profiles, namely axial, tangential, and radial (Mirro & Voll, 2009).

b) **Magnetic mixing**
Magnetic impellers are also used apart from the mechanical impellers at the industrial scale. Magnetic agitators are usually attached to the tank's bottom therefore, can result in efficient mixing. This feature removes the demand of sealing, leading to contamination and fluid loss. The magnetic agitators need small engines in comparison to mechanical impellers and promote bioreactor cleaning (Mirro & Voll, 2009).

c) **Baffles**
Baffles are vertical metal strips attached to the tank wall and are used for vortex and swirl reduction of the fluid. Baffle attachment to the tank is accompanied by brackets that are fixed by welding. In general, 4 baffles are used in a tank. The ideal baffle width ranges from 1/10 to 1/12 of the diameter of the tank. Baffles are attached perpendicularly to the tank wall in case of low viscous fluids. They can also be set at a required angle to the tank. These flexible attachments of the baffles initiate efficient mixing of the cell cultures (Doran, 1995).

d) **Mass and heat transfer in STRs**
Power-law model explains the pseudoplastic, non-Newtonian rheological behavior of the mycelial cells growing in the fermentation broths, as this behavior affects the overall performance of the bioreactor by influencing the mixing efficiency, power demand, and with heat and mass transfer (Gavrilescu et al., 1993). Increasing the broth apparent viscosity (μ_{ap}) in an aerobic fermentation can be influenced by an increase in the conditions (N and Q) to maintain uniform k_La values. However, higher impeller speed (N) forms higher shear zones next to the impellers, leading to consequent physical cell damage and therefore reducing the production rate (Smith et al., 1990). Because STRs are employed in a large number of industrial processes, k_La prediction is almost impossible due to the gas–liquid hydrodynamic complexity. Several researchers have related the k_La values to the power density (Pg/V) along with superficial gas velocity (vs) over similar-sized vessels, and two types of correlations are used for the k_La prediction. The first one used a nondimensional criterion, where k_La is directly related

to the gassed power consumption per unit volume of the broth (Pg/V) along with the superficial gas velocity (vs), as suggested by Cooper et al. (1944):

$$K_L a \propto \left(\frac{P_g}{v}\right)^{a_1} \cdot (v_s)^{b_1}$$

where the constants a_1 and b_1 can have variable values that depend upon the geometry of the system, vessel characteristics, composition of the media, and microbe morphology and concentration (Badino et al., 2001).

e) **Scale-up of STRs**

Scale-up can be defined as the method of producing results in a plant-scale device from an already established laboratory or pilot device (Hubbard, 1987). This process directly affects the effectiveness and production potential of bioprocesses by maintaining some parameters constant, such as volumetric oxygen transfer coefficient $k_L a$, reactor geometry, maximum shear, volumetric gas flow rate per unit volume of liquid (Q/V or VVM), superficial gas velocity (vs), power input per unit volume of liquid (Pg/V), impeller Reynolds number (Re), momentum factor, and mixing time (Ju & Chase, 1992). The degree of freedom limits the number of realizable factors, and the maximum number of criteria to be used in a scale-up is three.

9.3.2 Bubble Column Bioreactors (BCRs)

Bubble column reactors (BCRs), also known as the pneumatically mixed reactors, were designed for the cultivation of susceptible microbial cells like mammalian or plant cells, as well as the filamentous fungi. They are basically composed of a cylindrical vessel fitted with a gas supplier at the bottom. The gas enters the reactor as bubbles and directly mixes with the liquid or the liquid–solid suspension. BCRs are often designated as slurry bubble column reactors when a solid phase is present (Figure 9.4).

Bubble columns are widely used for multiphase reactors by various biochemical, chemical, metallurgical, and petrochemical industries (Kantarci et al., 2004). Both multiphase contactors as well as the reactors in BCRs use bubbles with a ratio of high:diameter of 2:1, 3:1, or 6:1, in some cases, which differs from the STRs, providing better air contact with the liquid. The larger diameter of the BCR cylinder promotes bubble liberation and foam break. Aeration of the bubble column is done via compressed air with the help of spargers connected at the tank's bottom, which are not the internal parts. The gas sparger affects the bubble properties, therefore affecting the gas holdup and other bubble-column parameters. Sparger also affects the bubble size. Smaller opening plates result in the formation of smaller bubbles. Gas holdup determines the transport processes of the BCRs and is defined as the volume portion of the gas taken up by the bubbles (Luo et al., 1999). Bubble columns are widely used for the industrial production of proteins, enzymes, and antibiotics. They are easy to design, and their scale-up requires a better understanding of the fluid dynamics as well as its effects. The major key parameters for BCRs are bubble ascending velocity, hold up, residence time, mass transfer, and interfacial area. Hold up can be defined as the portion of the liquid taken up by the bubble or gas in relation to liquid (Kantarci et al., 2004).

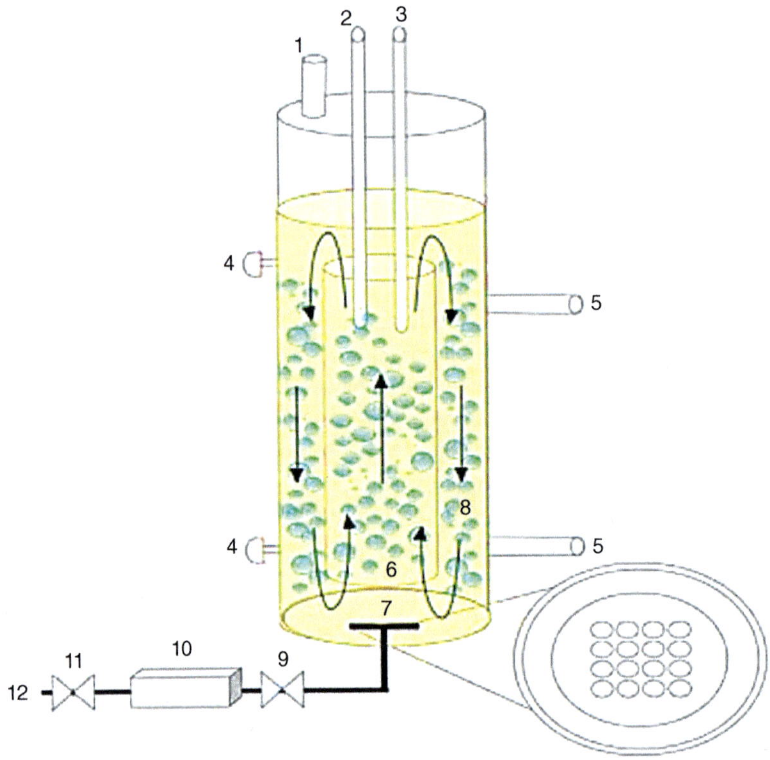

Figure 9.4 Bubble column bioreactor. 1. Gas outlet, 2. pH electrode with temperature meter, 3. Oxygen probe, 4. Sample collection, 5. Translators' pressure, 6. Sparger, 7. Gas filter, 8. Rotameter, 9. Pressure regulator, and 10. Gas inlet (*Source:* Adapted from Pino et al. (2018) / with permission of Elsevier).

9.3.3 Airlift Bioreactors (ARLs)

Airlift bioreactors (ARLs) are simply modified BCRs, with a difference being the presence of a central tube or other channels that aid in efficient component mixing as well as fluid recirculation, and therefore, they prevent bubble coalescence and minimize shear stress (Figure 9.5).

The term airlift is used because pneumatic contact of gas–liquid–solid or gas–liquid interface of the fluids results in a cyclic circulation pattern. Two basic configurations of the ALRs include external- and internal-loop reactors. In the former, fluid circulation occurs through distinctive channels, while in the latter, a barrier is present inside a unique vessel that forms channels for circulation. The structures of both external and internal loop reactors vary from each other, and with the advancement of technology, new possibilities for better gas exhaustion from the liquid phase can be developed. Scale-up studies of ARLs resemble the ones carried out for BCRs (Flickinger & Drew, 1999).

9.3.4 Packed Bed Bioreactors (PBRs)

PBRs are basically composed of a packed bed that adheres the cells on or inside the carriers with a reservoir that recirculates the oxygenated nutrient medium via the bed. Two

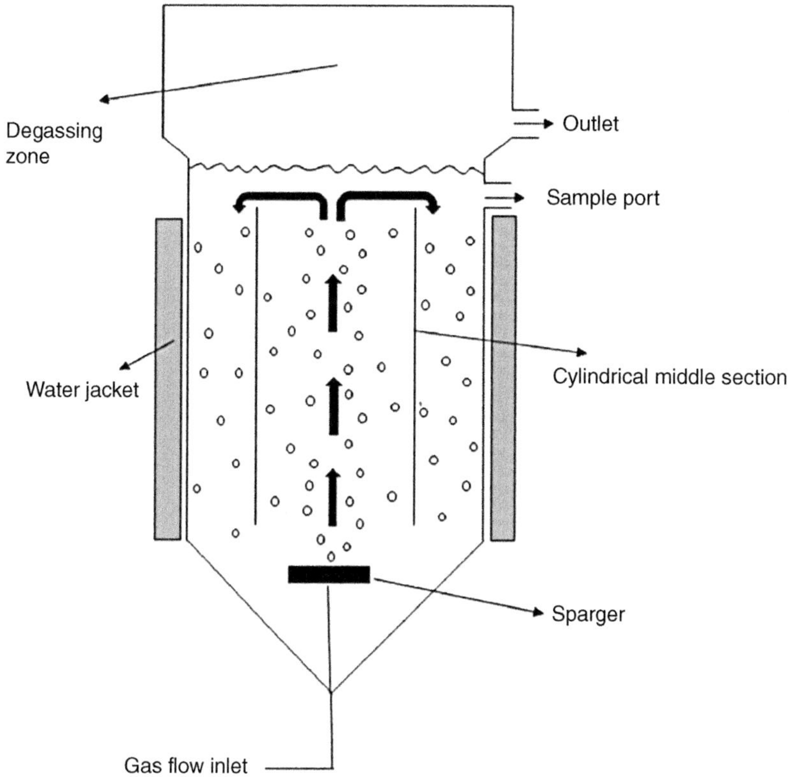

Figure 9.5 Setup of airlift bioreactors (*Source:* Adapted from Bakratsas et al. (2021) / With permission of Elsevier).

configurations are available for PBRs: one in which the packed-bed section is located either outside or within the reservoir (Wang, Zhang, Freedman, et al., 1992; Wang, Zhang, Jacklin, et al., 1992; Figure 9.6). Usually small-scale models of PBRs are developed first in order to figure out the ideal packing matrix of the selected cell line. The optimal matrix provided an adequate combination of cell anchorage, followed by proliferation, as well as productivity. This matrix plays an essential role in setting up PBR trials at a laboratory scale. Several studies have been conducted in order to find out materials that can support cell growth. Utilization of packing materials like disks, glass fibers, ceramic spheres, polyvinyl foams, or polyurethane foams resulted in high porosity with a size ranging from 0.80 to 0.95 (Meuwly et al., 2007). They have been widely used for culturing immobilized mammal cells and are often considered as "artificial organs," with one such application being the bioartificial liver (BAL) device in the biomedical field (Allen & Bhatia, 2002).

9.3.5 Fluidized Bed (FB) Reactors

FBs are typically composed of tall columns, where the height: diameter ratio (aspect ratio) is >10:1. Currently, fermenters with the aspect ratio of 20:1 or 40 :1 have already been designed and tested. The diameter of the column must be approximately 50 times of the diameter of the particle, but on a laboratory scale, this ratio is compromised to a smaller value (Figure 9.7).

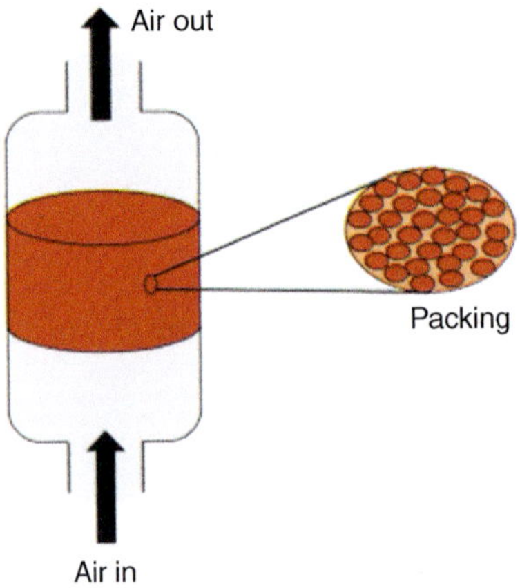

Figure 9.6 Packed bed bioreactor (PBR) (*Source:* Adapted from Krishania et al. (2018) / with permission of Elsevier).

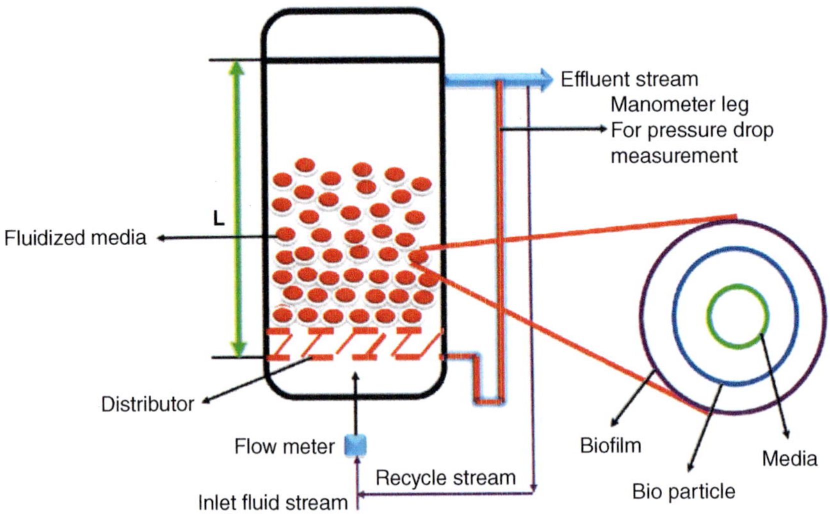

Figure 9.7 Fluidized bed (FB) reactors (*Source:* Adapted from Tiwari et al. (2023) / with permission of Taylor & Francis).

On a laboratory scale, the aspect ratio of the fermenters generally lies between 25:1 and 40:1. The fluid medium can be liquid, gas, or a combination of the two, with the flow oriented upwards for particles that are heavier than the fluid or downwards for the particles that are less dense than the fluid. Fluidized bed fermentation (FBF) usually uses broth as the fluid medium, while for the biological fluidized bed (BFB), wastewater usually flows through the bed. FBs offer a great deal in wastewater management, microbial metabolites, or biomass production by pure culture fermentation (Dempsey, 1994).

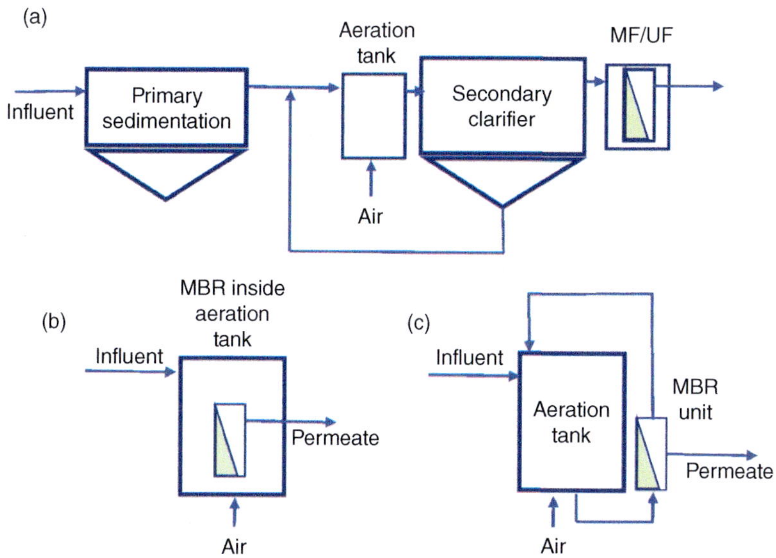

Figure 9.8 Operation of the MBRs in wastewater treatment (*Source:* Adapted from Rahman et al. (2023) / MDPI / CC BY 4.0).

9.3.6 Membrane Bioreactors (MBs)

MBRs are constructed in such a way that they have an external configuration, where the membrane modules are fitted outside the reactor and recirculation of the biomass occurs *via* a filtration loop. MBR technology is widely advancing around the globe, and it is widely utilized on the commercial scale with the development of several generations of MBRs. Till today, MBRs play a pivotal role in wastewater management, including industrial, municipal, and domestic wastewaters, where little footprint, water reuse, and rigorous discharge standards are needed (Meng et al., 2009). In wastewater treatment, MBRs follow a classic activated sludge method, where biomass separation occurs via membrane with a pore size of <0.1 μm. They make the biomass concentrated and therefore do not require a large tank and also enhance the biotreatment efficiency. They furnish great concentrations of mixed liquor suspended solids (MLSS) with lower sludge production and increased biological oxygen demand (BOD) as well as chemical oxygen demand (COD) followed by water reclamation. Membrane deterioration is one of the major drawbacks of MBRs (Santos et al., 2011; Figure 9.8).

9.3.7 Perforated Trays

These, in general, are simple systems where the substrate is spread on the trays that are perforated for promoting air movement. The substrate to be used is usually placed in thin layers on the trays, and the trays are then stacked upon each other in a room or a chamber under controlled humidity and temperature environments. Scaling up of these bioreactors is simple, as it increases with increasing the tray number, but it demands a larger operational

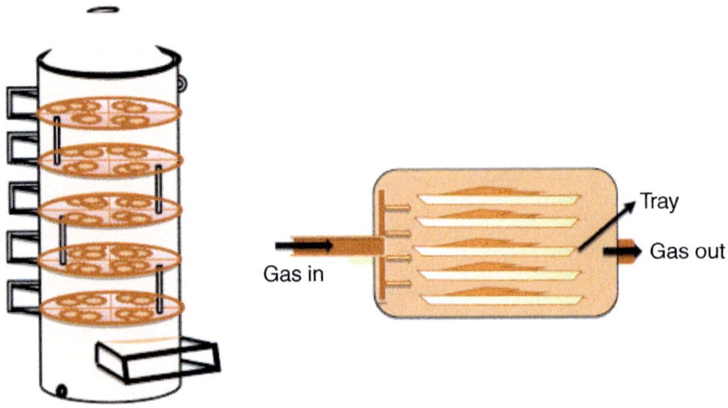

Figure 9.9 Perforated tray type bioreactor (*Source:* Adapted from Krishania et al. (2018) / with permission of Elsevier).

area, is labor-intensive, and is difficult to maintain a sterile environment. Oxygen transfer can be a hindrance in the operation, as the diffusion rate of oxygen transport depends upon the properties of the tray. In the beginning of the process after substrate inoculation, oxygen level is uniform in all substrate layers. However, with the passage of time as the substrate layer shrinks due to mycelium formation, porosity changes resulting in effective diffusion. CO_2 release and heat restrict oxygen movement to the bottom of the reactor. In the deeper layers, oxygen levels reach zero, and therefore, reaction stops due to lack of oxygen. This indicates that oxygen must be present at all times for an effective reaction, leading to effective production (Spier et al., 2011; Figure 9.9).

9.3.8 Fixed Bed Bioreactors

They differ from tray-type bioreactors as they have a closed system with forced aeration. They are useful due to their process control and heat removal efficiency. Although the process control element contributes to several advantages, forceful aeration can result in evaporation of water, causing the substrate to dry. This can result in unwanted low water activity followed by humidity imbalance, leading to poor microbial growth. Column reactors, however, are widely studied at the laboratory scale, as they reduce the issue of temperature gradients: the air directly enters the reactor, and CO_2 produced as a result of the microbial activity is removed, promoting air replacement. Temperature is controlled in these reactors by putting them either in bath or by covering them with a circulating coolant. Porosity reduction with the increase in the fermentation process is a major drawback of these reactors. Column reactors have lower growth of bacteria due to their closed nature, but they can be used for both fermentation as well as for final product extraction (Pandey et al., 2008; Figure 9.10).

9.3.9 Drum Bioreactor

Agitated drum reactors use agitation while operating and are classically designed in a drum shape with or without baffles. Such types of static reactors have also been designed where

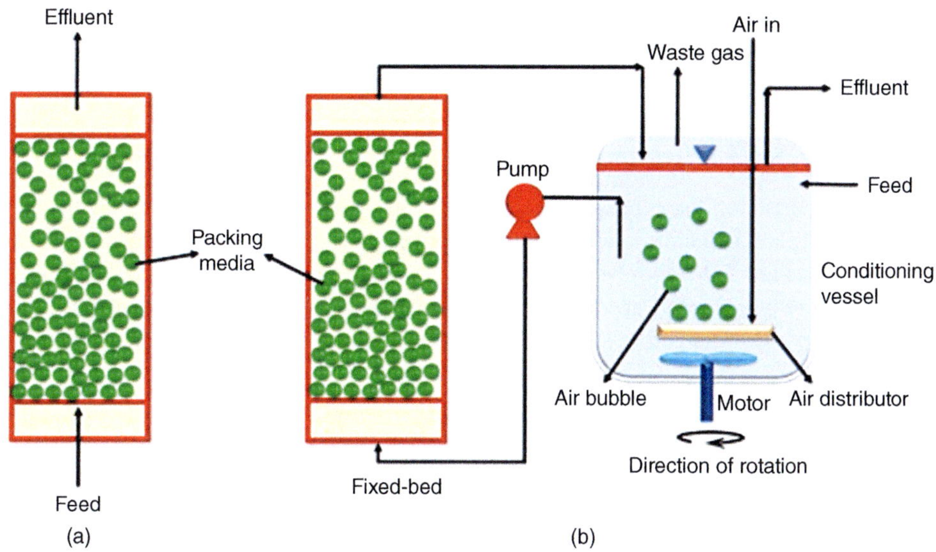

Figure 9.10 Fixed bed reactors (*Source:* Adapted from Tiwari et al. (2023) / with permission of Taylor & Francis).

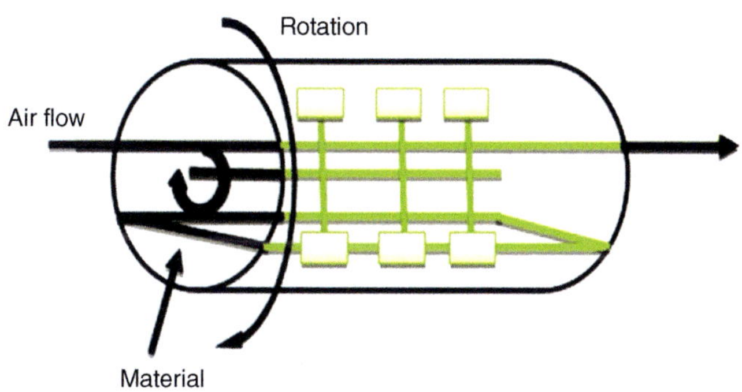

Figure 9.11 Rotating drum bioreactor (*Source:* Adapted from Krishania et al. (2018) / with permission of Elsevier).

horizontal cylindrical vessels ensure mixing of the media and the substrate, and agitation in such reactors can be either continuous or sporadic, which can damage the mycelial structure. Rotating drums are widely used in the SSF for the industrial production of value-added products. Rotating drum bioreactors (RDBs) offer affable and uniform mixing due to better baffle construction, as an agitator is absent from the substrate bed. Recently, RBDs are used for biofuel production (Wang et al., 2010; Figure 9.11).

9.3.10 Wave Bioreactors

Wave bioreactors have been largely used for tobacco, apple, and grape suspension cells. Eibl and Eibl (2006) obtained higher biomass production of grape cells with no damage to the cell morphology by using wave bioreactors in comparison to STRs. These reactors save a lot of time, with reduced foam formation, and are easy to operate with a much lower contamination risk.

9.3.11 Disposable Bioreactors

Disposable bioreactors have passed GMP regulations, as they can be used for industrial cultivation of plant cell culture with reduced production cost. They are usually composed of plastics that are FDA approved, for example, polystyrene, polyethylene, polytetrafluoroethylene, ethylene vinyl acetate, and polypropylene. They have certain advantages, like no need of cleaning and re-sterilization with minimized culture contamination risks. (Xu et al., 2011).

9.3.12 Photobioreactors

Photobioreactors are widely used in processes involving photosynthesis, biomass growth, or growth of microalgae under controlled environments. One of the major drawbacks of these reactors in the algal mass cultivation is mass transfer, which is important for effective working of the algal cells. Algal biomass is usually used in water treatment, in aquaculture, for human and animal feed supplement production, and also in heavy metal remediation and CO_2 fixation. These reactors can be designed from a small laboratory scale to large industrial scales, and based upon their design, they can be classified as closed, open ponds, flat plate, horizontal/serpentine tubular, airlift, or inclined tubular photobioreactors (Ugwu et al., 2008; Figure 9.12). The cultivation of microalgae in these bioreactors has resulted in higher productive ratio with minimal risk of contamination. These bioreactors can be set-up outdoors. Laboratory-scale photobioreactors are usually illuminated by an artificial light source (Tredici, 2004). The common types of photobioreactors are described below.

a) **Open ponds**
   They are easy to build and use but have certain drawbacks, like poor utilization of light by the cells, loss by evaporation, CO_2 diffusion into the environment, demand for large areas of land, predator contamination, and fast heterotrophic growth. Lack of proper agitation system can result in poor mass transfer rates, and therefore, the productivity is also very low. Putt et al. (2011) used a system for proper carbonation and achieved high productivity of biomass while culturing algal cells by overcoming the CO_2 diffusion issue of the open pond photobioreactors. They suggested that these devices can be coupled with a gas stream to maintain high CO_2 concentrations that promote algal growth. Park et al. (2011) obtained high algal biomass for biofuel conversion by growing *Chlorella* sp. and *Spirulina* sp. in wastewater using the open pond photobioreactors.
b) **Flat-plate bioreactors**
   This photobioreactor focuses on the cultivation of photosynthetic microorganisms in a largely illuminated surface. These photobioreactors are composed of transparent materials for the utilization of maximum solar energy. In these devices, the dissolved oxygen concentration is relatively low in comparison to horizontal tubular photobioreactors, and with them, photosynthetic efficiency can reach its peak, and they are

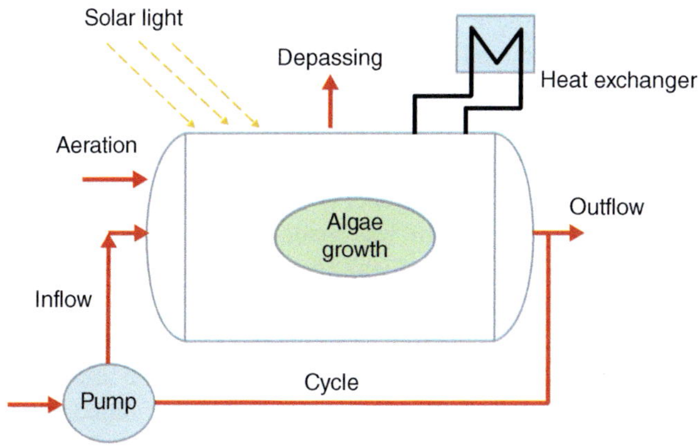

Figure 9.12 Photobioreactor for algal growth (*Source:* Adapted from Delavar and Wang (2022) / with permission of Elsevier).

excellent for algal cultivation (Richmond & Cheng-Wu, 2001). Tamburic et al. (2011) selected the geometry of the flat-plate reactor for the production of H_2 because its great surface-to-volume ratio led to increased photochemical efficiency. This type of photobioreactor is composed of two chambers: the first chamber consists of algal culture for the control measurements, and the second chamber holds turbulent culture mixing by a circulating gas-lift system fitted with a liquid diaphragm pump. They achieved a high yield of H_2 with very high photochemical conversion efficiency. Cheng-Wu et al. (2001) used a flat-plate photobioreactor for *Nannochloropsis* sp., a marine unicellular photoautotrophic algal species known for the production of eicosapentaenoic acid (EPA). The photobioreactor was made of many thick glass plates separated by dividers, and the device required a small ground area and could use the light energy to the maximum with a very high surface-to-volume ratio. The yield was relatively high with the lowest investment cost per liter of the culture.

c) **Tubular photobioreactors**

These are used for the outdoor culture of the biomass, with a higher requirement of surface area for illumination, and are made up of either plastic or glass tubes. They can be either conical, inclined, near horizontal, or horizontally serpentine photobioreactors. Improvement of the mixing system inside the tubes can increase the light distribution to the cell culture. The cultures undergo recirculation either by an airlift system or by a pump. Bergeijk et al. (2010) used a tubular photobioreactor (outdoor) for the biomass culture of *Isochrysis* sp., microalgae that has higher levels of docosahexaenoic fatty acid. The reactor had a horizontal serpentine-like design with acrylic tubes connected with the fiberglass gas exchange tank that contained the culture medium. The tubes were dipped in a water bath to cope with temperature fluctuations. This study suggested that tubular photobioreactors are prone to fluctuations in environmental conditions, and therefore, further research is needed for more stable design development.

d) **Vertical-column photobioreactors**

These photobioreactors are complex, low cost, easy to handle, and very promising for mass-scale algal biomass cultivation. It has been reported that both airlift and bubble-column photobioreactors can achieve a biomass concentration along with a growth rate that is comparable to that obtained from the narrow tubular photobioreactors (Sánchez-Mirón et al., 2002). Sometimes, draft tubes or split cylinders are installed in bubble column photobioreactors. In draft tube photobioreactors, mixing occurs between the upper and the lower zones via the draft tube walls. Rasoul-Amini et al. (2011) cultivated *Chlorella* sp. by using a bubble-column photobioreactor for the production of biodiesel. Their photobioreactor was made up of a glass cylinder with four plastic tubes that served as a gas inlet as well as a gas outlet, followed by a feeding tube and a sampling tube. Bubbling of the air was done by a porous bulk stone fitted with air pump, a plastic cover at the top, and four ports for plastic tubes. The biomass yield of *Chlorella* sp. doubled in the exponential phase as a good interface between the light–dark cycles, and a low surface-to-volume ratio was maintained via this photobioreactor. Ranjbar et al. (2008) used an airlift photobioreactor for culturing *Haematococcus pluvialis* cells for the production of astaxanthin. Their reactor was made up of Pyrex glass with a draft tube internal loop providing a high working volume with exposure to certain light–dark cycles and mixing inside the reactor. They achieved a high yield of cell numbers, indicating that proper light–dark cycles coupled with laminar flow positively affected the accumulation of astaxanthin. Increasing the intensity of light during astaxanthin accumulation led to higher yields, many times higher than the ones reported earlier. Sánchez-Mirón et al. (2002) compared airlift and outdoor bubble column bioreactors with the monospecific fed-batch culture of the microalga *Phaeodactylum tricornutum* and found out that all the three types of reactors yielded a similar amount of biomass. Light-limited culture leads to the production of pigment-rich biomass. Carotenoid content of *P. tricornutum* are susceptible to light in light-limited culture compared to chlorophyll contents. Airlift and bubble column photobioreactors can generate final biomass levels that are more or less similar to those generated in narrow tubes of classic horizontal tube reactors. The excellent performance of vertical reactors depends on the lack of photo-oxidation as well as photoinhibition of the biomass and oxygen inhibition of photosynthesis. Illumination of some photobioreactors can be achieved using artificial light sources coupled with impellers used for agitation. Cultures are also supplied with air as well as CO_2, and such a photobioreactor can also use solar energy as the light source.

9.4 Recent Advances in Bioreactor Applications

9.4.1 Production of Biopharmaceuticals

The presence of real-time direct measurements required for bioreactor control is one of the greatest challenges in the field of bioreactors due to the complexity and tough variability of biological processes, and single-use technology has come into light to overcome this challenge. Single-use reactors can be used on a pilot scale and have resulted in high yield with a significant reduction of costs (Khanal & Lenhoff, 2021). Computational Fluid Dynamics (CFD) can be used in the design and operation of the bioreactors, as the requirement for empirical tests can be curtailed and costs can be reduced. The risk of contamination is minimized by using single-use sensors, as they are sterilized by gamma radiation and are ready to use (Babi et al., 2022). Corbin et al. (2020) tested a techno-economic model for SuperPro Designer and found that this model resulted in significant cost reduction in comparison to the traditional batch experiment of butyrylcholinesterase (BChE) large-scale production by a two-stage semi-continuous bioreactor, indicating that a computational system can be coupled with biotechnology. Ruiz-Molina et al. (2022) tested the production of complement regulator MFHR1 by *Physcomitrella* moss in photobioreactors and found that the production of the recombinant protein was doubled via batch or fed-batch in comparison to the semi-continuous operation and the productivity increased significantly. They also tested an unstructured kinetic model for the prediction of nitrate uptake and protein production, as well as biomass growth by optimization and process control, and found that auxin addition can result in a good yield of MFHR1 in the shaking flasks, but higher yields were obtained when grown in bioreactors, suggesting the great efficiency of *Physcomitrella* moss in the field of biopharmaceuticals. Bioreactors offer controlled conditions of temperature, pH, shear stress, and nutrient supply for cell culture, but mass oxygen transfer is a problem that still causes issues with culturing. For instance, cardiac muscle tissue is packed with a tough network of cells and has a high demand for oxygen, which can be fulfilled *in vivo* by a meshwork of capillaries *via* the tissue. To understand the true nature of the cardiac tissue and its effect on the cardiac construct synchronization, a bioreactor-based model was designed (Radisic et al., 2005). In this device, rodent cardiomyocytes and fibroblasts were cultured on scaffolds having a parallel array of perfused channels to imitate a capillary network; blood was imitated by the medium, and the model was used for the assessment of different parameters like the effect of perfusion rates of oxygen on cell density, viability, and conditions required for the culture of cardiac constructs, mimicking relevant thickness (Wendt et al., 2012). Hollow fiber as well as perfusion systems are widely used for the designing of *in vitro* models, and advancements of these systems have led to the designing of complex bioreactors that can mimic the physiological environment, and such bioreactors are composed of three sets of closely packed hollow fibers, two used for medium perfusion and one for gas supply. Primary human liver cells are generally cultured in the extra capillary space (ECS) and have shown the exhibition of cytochrome P450 (CYP) enzyme activity when grown for three weeks, opening the way for future pharmacological studies (Zeilinger et al., 2011). Some of the recent discoveries in the biopharmaceutical field by using bioreactors are detailed below.

9.4.1.1 Bioartificial Liver Devices

The liver is one of the essential excretory organs of the body, and apart from the removal of waste products via bile, it also participates in a wide variety of specialized functions such as intermediate metabolism, oxidative detoxification, nutrient supply, protein and macromolecule synthesis, as well as modulation of immune and hormonal systems (Gerlach et al., 2008). Failure to carry out any of these functions can cause acute or chronic liver failure, resulting in death; therefore, extensive research has been done worldwide for the development of *in vitro* models to overcome such problems, and the BAL device is one such model that has been used in liver failure treatment (Park & Lee, 2005). The flat plate, as well as monolayer bioreactors that are used, have a uniform cell distribution along with unique microenvironments with a low surface-to-volume ratio. Perfused beds or scaffolds mimic the complex infrastructure of the liver. The cells have shown considerably good metabolic and synthetic activities, but the risk of shear forces leading to cell damage demands a better-designed bioreactor. Traditional hollow fiber cartridges, however, provide a vast surface area favorable for the immunoisolation as well as attachment of the cells when coupled in endothelial cells (ECs), and damage from the shear forces is also minimized. But non-uniform distribution of the cells in the fibers can create problems, as membranes can restrict the movement of nutrients (Ni et al., 2011).

Hepatocytes are the prime choice for the BAL, as they are closely related to the liver and are able to conduct *in vivo* metabolic activities like amino acid metabolism, albumin synthesis and excretion, production of urea, and processing, as well as drug and toxin elimination (Gómez-Lechón et al., 2007). Phase I and II drug-metabolizing enzymes play an important role in drug processing in the liver, so these enzymes must be preserved inside the model to mimic the *in vivo* cellular environment. Instead of primary hepatocytes, different cell lines are preferred these days due to their prolonged survival and easy maintenance. HepG2/C3A cell line, obtained from hepatomas, is one such frequently used cell line in pharmacological studies and is used in the microfluidic bioreactors for the examination of susceptibility of these models against xenotoxins (Snouber et al., 2013). The functionality of these cell lines often gets lost or altered in comparison to the primary hepatocytes. For example, HepG2 cells express phase I enzymes at relatively low levels when compared with the primary human hepatocytes, suggesting that primary hepatocytes still remain a choice for *in vitro* studies (Wilkening et al., 2003). Recently, it has been suggested that merging the HepaRG human hepatoma cell line with the 3D BAL device results in preservation of the metabolic and transporter abilities of the cells in suspension for up to one week, suggesting their use in *in vitro* pharmacological studies (Darnell et al., 2012). Primary porcine hepatocytes have been widely coupled in BAL devices, as they are easily accessible and can maintain their differentiable metabolic activities under culture conditions with high cell yield and resemble human hepatocytes. However, factors affecting the sustainability of this differentiable hepatic function *in vitro* have not been identified yet (Allen & Bhatia, 2002). The utilization of the primary human adult hepatocytes is highly preferable for studying BAL system functionality and biocompatibility, but they suffer from loss of function over time *in vitro* studies, resulting in reduced cytochrome P450 and other essential enzyme activities, followed by the downregulation of CCAAT/enhancer-binding protein alpha (transcription factors) (Vinken et al., 2012). Usage of chemically complex media has promoted the growth of hepatocytes in the culture media, leading to their morphological stability and improved liver-related functions. Co-culturing of the hepatocytes with the non-parenchymal cells helped in the preservation of liver-specific functions, leading to physiological cell reorganization and a canalicular transporter distribution pattern similar to those *in vivo* (Zeilinger et al., 2011).

Inside the liver, the hepatocytes form a 3D lobular network of hexagonal nature, which in turn is surrounded by a central blood vessel. Individual hepatocytes maintain the polarity between the apical domain facing the bile canaliculi and the basal domain facing the sinusoid for specific activities. Within a culture, hepatocytes form flat layers while attaching to the surface; however, they can form cell–cell junctions, but the disruption of the cytoskeleton occurs that can result in cell polarity loss as well as loss of liver-specific functions (Gerlach et al., 2008). The extracorporeal liver assist device (ELAD) is composed of a dual pump dialysis unit attached to hollow fiber cartridges filled with C3A cells in the ECS, where blood is filtered through the pump attached to the fibers along with the plasma that passes through the ECS. This type of direct interaction results in the separation of cellular debris and cells from going back to the bloodstream. The liver has its own oxygen-rich environment governed by the hepatic arteries as well as the portal vein. C3A cells grown in hollow fiber-based bioreactors (HFBs) resulted in improved functionality when the culture medium was impregnated with red blood cells, indicating improved oxygen supply for *in vitro* cell integrity maintenance (Ding & Shi, 2011). The Excorp medical BAL support system (BLSS) has a heat exchanger and an oxygenator attached with a blood pump fitted to an HFB containing primary porcine hepatocytes in the ECS, where the blood passes through the fibers and protects the cells from the unwanted immunological responses of the host (Mazariegos et al., 2002). Modular extracorporeal liver support system (MELS) is another example of a BAL that focuses on the biliary system of the liver responsible for the excretion of the metabolites and conjugated bilirubin. Human hepatocytes collected from the discarded donor livers are also seeded with the hollow fiber bundles for the removal of albumin-bound toxins (Sauer et al., 2002).

Bioreactors have recently been used in the drug metabolism and pharmacokinetics (DMPK) studies to investigate the functionality of the cells in relation to drug transporters, along with metabolic enzyme expression + function. Darnell et al. (2012). while studying the diclofenac metabolism, found that human hepatocytes grown in multicompartmental bioreactors resulted in lower production of the hydroxylated product of diclofenac (produced by the cytochrome P450 enzyme CYP2C9) in comparison to the cells grown in 2D conditions (Zeilinger et al., 2011). Diclofenac metabolite, diclofenac acyl glucuronide (intermediary product), was absent in the human hepatocyte suspensions but was present in hepatocytes cultured in a bioreactor that mimicked the *in vivo* environment, promoting the efflux of metabolites far from the cells and therefore stopping further activity. However, the levels of the metabolites generated in the suspensions were similar to the ones from the 3D cultures. Transporter expression of the hepatocytes is another aspect that makes them reliable to be used in the DMPK studies. Inside the bioreactor, the expression of both bile canalicular efflux transporters (multidrug resistance-associated protein [MRP2] and multidrug resistance protein [MDR1]) was studied for two weeks and was found to be successful (Zeilinger et al., 2011). Organic anion transporting polypeptide 1B1 (OATP1B1) expression was also studied in a 3D system, and the combination of the substrates estradiol-17b-D-glucuronide (E17bG) and atorvastatin acid (ATA), either in the presence or absence of the inhibitor estrone-3-sulphate (E3S), resulted in the functional activity of the protein that was preserved for 9 days, compared to the OATP1B1/1B3 in plated primary human hepatocytes, where a decrease in the functional activity was observed after just two hours (Ulvestad et al., 2011).

9.4.1.2 Blood-Brain Barrier (BBB) Models

The BBB forms one of the most crucial lines of protection against brain infections by serving as a bridge between the central nervous system (CNS) and systemic circulation, therefore maintaining brain homeostasis. The capillary network of the brain differs from the network of the peripheral tissues due to tight junction proteins present between the ECs; therefore, it restricts the movement of molecules and contact of ECs with astrocytes for support. This network also protects other neurons from glycemia and hypoxia

(Abbott et al., 2006). Multidrug transporters like P-glycoprotein (P-gp, also known as MDR1) can restrict transport of bigger drug molecules into the CNS, suggesting a demand for the development of *in vitro* models for ADMET (absorption, distribution, metabolism, excretion, and toxicology) studies. The models that are available are not emphasized well like that of the liver devices due to the absence of an extracorporeal device, but just like the BAL devices, HFBs have been developed to be used as BBB models. Like the liver models as well as the bioartificial devices, none of the cell lines can be used to mimic the BBB, as most of the models known to date are based on two types of cells, namely glial cells (astrocytes) and the ECs. Models of the past used either EC cells alone or ECs combined with glial cells seeded on the transwell plates, and this approach tended to be useful due to its simple nature, but ECs *in vitro* lose their BBB properties due to the lack of the stimuli that are regularly present *in vivo* (Cucullo et al., 2005). P-glycoprotein (P-gp) transporters are downregulated without astrocyte-derived soluble agents, and therefore, abnormal permeability occurs close to the monolayer, but such a process can be initiated by the addition of serum on both sides of the EC monolayer (Hori et al., 2004). Models incorporated with glial cells mimic the *in vivo* environment more precisely, as the expression of transporters, brain endothelial marker enzymes, and transendothelial electrical resistance (TEER) was observed across the monolayer. For BBB models, factors like shear stress promote inhibition of growth and EC differentiation, thereby enabling the trafficking of metabolites reaching the brain to first be diffused into animal models before being used for human models (Cucullo et al., 2002).

Dynamic (flow-based) models use HFBs loaded with ECs that are coated with fiber layers and glial cells. Stanness et al. (1997) developed a model where ECs and glial cells originating from humans, rat, and calves were cultured by using an HFB layered with ProNectin. Their results indicated that growth patterns and cell morphology were similar for all cell types used, but culturing different types of ECs and glial cells, for example, when human umbilical cord ECs were cultured with C6 cell line glial cells, BBB properties failed to develop. In the absence of glial cells from the HFBs, poor EC growth was observed due to low transendothelial resistance. Neuhaus et al. (2006) used a commercial bioreactor seeded with porcine brain microvascular ECs (PBMEC/C1-2) (immortalized) merged with the glial cell line (C6 cells) in the ECS, where medium was introduced in the system via shear stress in order to mimic the brain capillary network. Cell growth and benzodiazepine transport results were compared with the flow-based and models. The results indicated that PBMEC/C1-2 cells coated on the transwell inserts showed detachment from the walls 5 days after the culturing, suggesting that for long-term studies the use of transwell systems is not suitable. However, the morphology and growth of ECs in the HFB were influenced by flow exposure, increasing the survival ratio of the cells. Benzodiazepine permeability coefficient was also low in the HFB model due to tight cell junctions that strengthened the barrier function in this device.

9.4.1.3 Bioartificial Kidney (BAK) Devices

Kidneys participate in the excretion of waste products like ammonia and urea in the form of urine from the body. They also have secondary roles like homeostasis maintenance. Cases of chronic kidney disease (CKD) are increasing worldwide, demanding early disease diagnosis, cost-effective and efficient renal transplant therapies, and effective drug treatment, orienting the focus of researchers for the development of ideal kidney models for pharmacological studies. Already available bioartificial kidney (BAK) devices were developed for renal replacement therapy and use both biological as well as mechanical processes for the blood filtration by the device, renal assist device (RAD). RAD houses renal cells filled in the hollow fibers of the hemofiltration cartridge. *In vitro* studies conducted by using the BAK device showed that differential active transport occurred similar to the ones occurring during dialysis (Humes & Szczypka, 2004). Humes et al. (1999) developed a model for treating patients with acute renal failure and multiple organ failure and got some success, while Saito et al. (2006) concentrated on the treatment of long-term complications in the patients undergoing dialysis by a model developed by them. Both these models used hollow fiber modules coated with proximal tubule cells (PTCs) that allowed passage of ultrafiltered blood before going back into the patient's system. Humes et al. (2004) also used human PTCs (hPTCs) seeded on the commercially available hemofiltration cartridges. For the development of a reliable BAK device, renal PTCs are incorporated in these devices, as these PTCs play numerous roles *in vivo* systems (Zhang et al., 2009). Renal PTCs are equipped with different transporters like organic anion and cation ones aiding in basolateral drug uptake, and ATP binding cassette (ABC) ones for luminal excretion into the urine. Some of these transporters can bind with molecules of different sizes, therefore showing substrate overlapping characteristics (Masereeuw & Russel, 2010). Clinical BAK devices use primary hPTCs but their supply is limited as the culture of vast amount of cells is quite difficult therefore, alternative cell lines are required. Organic anion transporter (OAT) 1 and OAT3 are the two transporters that were thought to be important for the basolateral uptake of different organic ions *in vivo* from blood. OAT3 was found to be responsible for rosuvastatin uptake in the presence or absence of probenecid (a uricosuric drug) *in vitro* (Verhulst et al., 2008). Probenecid inhibits OAT2 and OAT4 and it has been found that in the presence of this drug, prostaglandin F2a (PGF2a) uptake by OAT2 as well as estrone-3-sulfate uptake by OAT4 was stopped. OCT2 located on the basolateral membrane of the hPTCs was believed to be responsible for organic cation uptake. It has been found that in the presence or absence of quinine, the ASP+ (a fluorescent dye) uptake was reduced in hPTCs (Pietig et al., 2001). P-gp also aids in the xenobiotic transport in the hPTCs, like methotrexate, similar to the liver (Leslie et al., 2005). It has also been documented that the expression of both phase I and II enzymes has been retained in the hPTC cultures. A confluent monolayer is a mandate in the BAK devices, and when hPTCs are grown on them, it has been found that both phase I and II enzyme expression was uniform, although they were lower than that of the native tissue cells (Lash et al., 2008). Human tubular kidney cell culture grown on a permeable filter was shown to express drug transporters, resulting in the culture of distal and proximal tubular cells. These findings strengthened the fact that hPTCs can be efficient tools for the clinical BAK devices (Masereeuw & Russel, 2010). Utilization of immortalized cell lines can help to overcome the problem of the demand for a higher number of cells. Techniques like cloning, formation of cell lines from cancer cells, or viral transfection can generate genetically stable but fully functional cell lines. Human cells are hard to immortalize, as they fail to express lots of required features when grown *in vitro*; therefore, HK2 cell lines have been produced from the infected human PTCs combined with an immortalizing construct derived from the human papilloma virus. Human kidney proximal tubular cell line (HKC) is the most recently developed cell line generated from the Adeno 12:SV40 hybrid virus exposure to the renal epithelial cells and has been tested in the clinical BAK devices (Racusen et al., 1997). Other available animal cell lines include LLC-PK1, MDCKII, and LLC-RK1, which have also been tested.

Porcine PTCs resemble the hPTCs in their anatomical and physiological features and have been proven to be reliable for kidney research. Porcine PTCs, when used in a BAK device during the preclinical trials, have shown greater tolerance toward the uremic environment, opening the gate for clinical trials (Humes et al., 1997). Embryonic stem cells (ESCs) can serve as potential sources of PTCs in the BAK devices due to their multipotent and self-renewal ability when they come in contact with correct environmental cues (Humes & Szczypka, 2004). Utilization of human ESCs for renal cell differentiation has not been tested on a great scale, but primary results have shown that undifferentiated and mesoderm-derived ESCs have the potential to form nephrons and are able to express numerous renal markers by exhibiting the upregulation of kidney precursor markers (LIM1, CD24, and EYA1) (Batchelder et al., 2009). When the commercially used membranes, like PSF or PVP and polyethersulfone (PES), regenerated cellulose (RC), or polyvinylpyrrolidone (PVP) layered with ECM, were tested with primary hPTCs, it was found that the cell performance did not changed at all, suggesting that these materials are not good to be used in BAK device construction. Coating of fibers with the modified ECM has been tested. Different combinations of single and double coatings have been tried with different materials like collagen IV and laminin blends, along with growth factors (Zhang et al., 2009). hPTC performance showed improved performance when grown in single-coated fibers, while hPTCs grown in an HFB with a double coating of 3,4-dihydroxy-L-phenylalanine (DOPA) and collagen IV showed functional epithelia formation. In order to mimic the kidney, the devices require a tube composed of a hollow fiber having an extracapillary space along with shear rates to generate a native environment; this can be achieved in the ECS more easily than in the fibers. Researchers at the Wyss Institute have developed "organs on a chip," where hPTCs are seeded inside a microfluidic channel joined with a polyester membrane. This channel remains submerged in the medium, and shear stress is applied to the medium via the channel. Tight junction proteins, cytoskeletal proteins, and ion transporters have shown improved results in a fluidic model (Oo et al., 2011).

9.4.2 Production of Biofuels

Production of biofuel has proved to be an alternative to replacing fossil fuels. Biodiesel production from microalgae has been investigated globally as the third-generation biofuels, as their cultivation in open ponds is cheaper and easy to handle, but the risk of contamination cannot be minimized as algal cells can be easily influenced by environmental fluctuations. Photobioreactors can be used as an alternative for microalgae cultivation but have operational costs; therefore, many alternatives with low cost demands have been developed, like microalgal culture in aquaculture water in open pond systems, and production of materials of high value (antioxidants, astaxanthin, pigments, and hormones) in the photobioreactors (Rafa et al., 2021). Production of biojet fuels has also been done from genetically altered microbes. For the carbon source, lignocellulosic biomass is used for carrying out the fermentation that requires pretreatment + saccharification in sequence, making the whole process expensive due to scale-up maintenance. Confined bioprocessing seems to be a promising alternative where cellulase production and saccharification, followed by fermentation, can occur together and has many ecological profits for value-added product generation (José et al., 2022). Selection of the production of the right choice is very important, as it affects the energy balance throughout the process. In downstream processes, harvesting of gaseous products is easy and cheap, but the harvest of intracellular products is quite complex and expensive. In biodiesel production, separation of lipid from the residual component takes place via mechanical as well as liquid solvent extraction techniques, but the chemical properties of the produced oil are different from the crude oil; therefore, chemical purification is executed to achieve the refined final product. Polysaccharide and protein recovery can also be done from the residual biomass. Biogas production can be carried out by chemical processes like coal gasification or biologically by anaerobic fermentation, where bacteria degrade the biomass into methane and CO_2. Hydrogen production has also been investigated from some microalgal strains when cultivated in anaerobic conditions, and it is a two-step process, and H_2 (hydrogen) can be produced via photosynthesis in a photobioreactor. Polysaccharides can also be another alternative for biofuel production. Red microalgae are capable of producing sulfated extracellular polysaccharides that have high industrial demand, as the final product is excreted from the algal cells and its harvesting doesn't require cell alteration; therefore, the process can be carried out regularly (Gasljevic et al., 2009).

Photoautotrophic organisms require regular light as a source of energy as well as CO_2 as a source of carbon for maintaining their activities, and both these demands are challenging while designing a photobioreactor. In general, only a scant amount of energy is enough for biomass formation as well as production of desired products, which can be calculated by photon conversion efficiency (PCE). PCE can be defined as the fraction of solar energy required by the organism for biomass formation and is limited by many factors, like the physical nature of light itself, that can affect the rate of photosynthesis. Loss of energy during transfer from the atmosphere cannot be minimized and the other fraction of light holds light of different spectra from infrared to UV, where photosynthesis can only occur in the wavelength range of 400–700 nm (photosynthetically active radiation). PCE is affected by the loss of energy via reflection and loss of pigments as a result of fluorescence, followed by heat loss. Respiration cannot be ignored, as oxygen is evolved as a result of photosynthesis. Cellular compound formation from the products of photosynthesis also leads to lower production, and efficient production of biofuel is dependent upon the energy required for the production of the desired product inside the microbial cells. PCE is a crucial factor in designing bioreactors, as it determines the potential of the bioreactor for ideal algal growth as well as sets the limit of energy harvest (Zhu et al., 2008). Different types of airlift bioreactors have been used for large-scale outdoor purposes with a large volume. Gas spargers cause upward stirring in the parallel chambers while baffles induce vortexing, resulting in the perfect mixing of the light with the liquid medium. The system has several advantages, like perfect mixing, low cost, and use of the flashing light effect without any exposed areas, with reduced energy demand (Degen et al., 2001).

Scarabaeid larvae (order Coleoptera) are either saprophagous or herbivorous; however, many species thrive in soil and eat plant roots or organic matter. The intestinal tract of soil-thriving (humivorous) scarabaeid larvae has several interesting physicochemical properties, like a very long midgut almost the length of the body cavity, an expanded hindgut (fermentation chamber), an alkaline midgut loaded with alkali-stable hydrolases, and specialized microflora. These larvae play an essential role in the decomposition of dung and wood decomposition and therefore can play a role in the utilization of biowastes as sources of energy (Zhang & Jackson, 2008). Huang et al. (2010) studied the digestive enzymes and gut bacteria of scarab larvae to understand the digestion of (hemi) celluloses that can help in bioreactor design for biofuel generation. The midgut of the scarabaeid larva is highly alkaline and is different from other coleopterans. For example, the midgut as well as the hindgut pH of the *Costelytra zealandica* larvae was >8 and varied with the gut location. The pH of the pericecal region was >8, which increased to >10 in the anterior part of the midgut, increasing further in the middle midgut (>10) but dropping in the posterior region of the midgut (9), while the pH of the hindgut was also >8 (Biggs & McGregor, 1996). Similar findings were observed in the *Pachnoda ephippiata* larva, where a gut pH >8 was found. In the case of *P. ephippiata* larva, the anterior midgut had a pH between 10.1 and 10.7 that dropped to 8.4 while reaching the third row of caeca. The midgut pH of *Melolontha melolontha* larva was rather constant (7.9–8.2) but increased to 8.6 in the hindgut paunch but dropped to neutral in the colon + rectal regions. Similar findings were noticed in the *Sericesthis geminata* larvae as well (Egert et al., 2005). This range of pH helps in the digestion mechanism, thereby aiding in nutrient absorption.

Physiological conditions of the gut, like pH range, O_2 concentration, and redox potential, determine the microbial activity; for instance, the growth and development of strictly anaerobic methanogenic bacteria residing in the proctodeum of *Oryctes* spp. require reducing conditions. The electric potential of the midgut of the *Oryctes nasicornis* larvae lies between +50 and −10 mV, while in the fermentation chamber (i.e., proctodeum) it lies between −40 and −100 mV as these redox potentials participate in the generation of the reducing conditions inside the intestine (Bayon & Etievant, 1980). The midgut of the *P. ephippiata* (first instar larvae) has oxidizing conditions; however, the redox potential turns to a reducing one in the second and third instars, indicating that redox potential changes with the development of the midgut, but the proctodeum maintains the reducing conditions in all instars. Both redox potential and pH affect the oxidative state of phenolics, tannins, and other components that enter the gut, thereby governing the activity of the digestive enzymes. It has been established that decomposition of organic matter is linked to the redox potential; for example, a redox potential of < + 300 mV creates reducing conditions, therefore initiating organic matter decomposition leading to the accumulation of soluble products. When negative redox potential is reached (below −100 mV), strict anaerobic conditions are created, resulting in complete carbohydrates degradation leading to the production of methane gas. This type of environment is naturally present in the hindgut of scarab larva, therefore, has the potential for methane generation (Egert et al., 2005). The hindgut of the scarab larvae has the highest concentration of microbes and can be compared with the rumen of higher mammals, where plant organic matter is fermented by microbes. Studies have revealed that the *C. zealandica* hindgut is a hub of flagellate protozoa along with bacteria ($2–5 \times 10^{10}$ bacteria/g wet weight of the gut) that can digest xylan and pectin (Bauchop & Clarke, 1975). Zhang and Jackson (2008) investigated the diversity of the gut bacteria of *C. zealandica* larvae by plate counting and polymerase chain reaction (PCR)-denaturing gradient gel electrophoresis (DGGE) methods and found that the diversity of the bacteria was higher in the hindgut in comparison to the midgut. The midgut bacterial population changed with collection site and diet but the hindgut population remained unchanged by the external factors. PCR-DGGE confirmed the abundance of bacteria belonging to the *Clostridium* genus, followed by bacteria of δ-proteobacteria, bacteroidetes, and β-Proteobacteria. Egert et al. (2005) used the terminal restriction fragment length polymorphism (T-RFLP) technique to find bacterial 16S rRNA from the gut of *M. melolontha* larvae and found that the midgut has a simple but variable non-specific population; however, the hindgut had a diverse population of gut-specific bacteria that aid in digestion and most of these bacteria belonged to *Clostridiales*; however, populations of Bacillales, Actinobacteria, γ-proteobacteria, and Lactobacillales were also found. Higher acetate levels were present in both midgut and hindgut, suggesting fermentation processes by microbes in both these chambers. Egert et al. (2003) tested the gut of *P. ephippiata* larvae and found that four fermentative metabolism bacterial groups, namely Clostridiales, Bacillales, Cytophaga, Flavobacterium-Bacteroides (CFB) phylum, and Lactobacillales, were abundant. The 16S rRNA technique indicates that Actinobacteria are abundant in the midgut, while the CFB phylum dominates the hindgut. Lemke et al. (2003) found the abundance of Clostridiales, members of the CFB phylum, Lactobacillales, and clones connected to *Turicibacter sanguinis* in the gut of *Pachnoda* spp. larvae, responsible for maintaining higher concentrations of lactate and acetate in the hindgut. Similarly, high levels of acetate were found in the hindgut of *M. melolontha* larvae, suggesting that methanogenesis occurs in this chamber by microbial fermentation (Egert et al., 2005). The abundance of Clostridial bacteria in the gut of the scarab larvae has developed a symbiotic contact with the larvae (Zhang & Jackson, 2008). Bayon and Mathelin (1980) investigated cellulose degradation in the *O. nasicornis* midgut and proctodeum by injecting 14C-labeled cellulose *in vitro* and found that cellulose was completely broken down. Hemicellulose digestion usually occurs in the hindgut paunch by the microbes, as these bacteria are not present in the midgut (Douglas, 2009). In the midguts of *O. nasicornis* and *P. marginata* larvae, lower β-glucosidase, carboxymethyl cellulase (CMC-ase), and xylanase activities have been found. The activity of these enzymes in the alkaline conditions of the midgut is not known; therefore, it is believed that midguts assist in the predigestion of lignocellulose and further degradation occurs in the anaerobic hindgut (Zverlov et al., 2003). The dominance of the genus *Clostridium* in the scarab hindguts suggests that they help in root and organic matter decomposition. Cellulose is broken down by the bacteria and forms volatile fatty acids (VFAs) that can serve as a substrate for the methanogenic bacteria, assisting in methane production in the hindgut. Hemicellulolytic activity by the facultative and strictly anaerobic bacteria has been identified in the *P. marginata* larvae's hindgut, and the dominant species carrying out this task was *Promicromonospora pachnodae*, which produces endoglucanase and xylanase under aerobic as well as anaerobic conditions by acting on polymers of plant origin. *Cellulomonas pachnodae* was also isolated from the *P. marginata* larvae's hindgut as well as two new genes, *xyn11A* as well as *xyn10B*, expressed in *Escherichia coli* participating in xylanase formation, were identified as well (Cazemier et al., 2003). Geissinger et al. (2009) could isolate *Elusimicrobium minutum*, a bacterium belonging to the termite group 1 phylum, from a humivorous scarab larva and found that it grows on sugar heterotrophically and can ferment a wide variety of sugars as well as *N*-acetyl-D-glucosamine to value-added products. Cellulase absence in the fluid of the hindgut confirms that cellulose digestion is governed by the gut bacteria rather than the gut enzymes. Acetic acid is produced as the end product of the cellulose breakdown that passes through the hindgut to be used by the insect (Martin, 1983). This efficiency of these larvae can open the doors for biofuel production in the future.

9.4.3 Bioremediation

Bioreactors can serve as a tool for bioremediation, where pollutants can be broken down under controlled conditions by the microbes as they degrade the pollutants into compounds that can be reused. Remediation by microbes can be either *ex situ* or *in situ*. The *in situ* process is slow and requires complete control, and bioreactors can be used for such purposes (Srivastava et al., 2021). Fluidized-bed bioreactors, stirred-tank bioreactors, fixed-bed bioreactors, packed-bed bioreactors, and airlift bioreactors are the common types of bioreactors implied in the bioremediation process. Bioreactors have the potential of maintaining microbial activity while the whole bioprocess is ongoing, and this involves designing of appropriate bioreactor that is resistant to corrosion, especially while wastewater treatment is done. Based on the current research, bioreactors are advantageous for the bioremediation process as they provide a controlled environment for efficient microbial activity. They provide an appropriate environment that offers suitable conditions of mass transport and conditions for microbes to thrive easily. Agitation, pH, temperature, aeration, and moisture, as well as pollutant and nutrient concentrations, must be controlled within the bioreactor, and laboratory testing can help in the standardization of these parameters. Recently, many bioremediation applications are being tested using bioreactors (Sonwani et al., 2022).

Marine environments are heavily contaminated with hydrocarbons, resulting in oil spills that can enhance the growth of microbes that can thrive in such polluted environments, and the isolation of these microbes has oriented the use of these microbes for pollutant degradation. A number of laboratory experiments have been conducted using different setups of bioreactors with the suggestion of restoration of oil-infested areas (Narciso-Ortiz et al., 2020). Airlift and bubble column bioreactors have been tested as they have lower consumption of energy and provide perfect mixing and simple construction, aiding in the generation of perfectly homogeneous shear stress. Tiwari et al. (2023) tested airlift bioreactors against marine water polluted with hydrocarbons (gasoline, pyrene, phenanthrene, diesel, and hexadecane) by using a bacterial conglomerate and found 90% degradation of the hydrocarbons used. They also tested a bubble column bioreactor for Mayan crude oil degradation and got total degradation in a period of 12 days. Similar results have also been obtained with the use of immobilized cell systems, slurry bioreactors, and multistage-designed bioreactors (Antwi-Akomea et al., 2018). Rodríguez-Calvo et al. (2020) designed a bioreactor sorbent with a combination of granulated hydrophobic cork and melt blown polypropylene as a filling material, where industrial wastewater polluted with hydrocarbon was circulated for 7 days along with native microbes, and found that 97.5% of the petroleum was degraded. Antwi-Akomea et al. (2018) tested a multistage bioreactor with three media filters with mixed microbial conglomerate for salt water contaminated with hydrocarbons and came to the conclusion that the petroleum removal rate was 99%. Germano-de Almeida et al. (2017) used a stirred bioreactor to study the degradation of marine fuel by a seawater-isolated microbial conglomerate and found a 93.5% success rate.

Degradation of plastic depends upon polymer properties, salinity, biological activity, and environmental conditions. The degradation process involves biodeterioration and biofilm formation on the plastic, followed by metabolite production (Jeyavani et al., 2021). *Ideonella sakaiensis* was reported to degrade polyethylene terephthalate-degrading (PET) by the production of MHETase and PETase enzymes (Yoshida et al., 2016). Carniel et al. (2020) used stirred reactors for PET hydrolysis by cutinase produced by *Humicola insolens* and obtained an increased hydrolysis rate. Lopardo and Urakawa (2019) investigated biodegradation of a bacterial-based polymer (polyhydroxyalkanoate [PHA]) and a petroleum-based polymer (polycaprolactone [PCL]) with natural microflora by recirculating in bioreactors containing marine aquaculture wastewater and found increased C/N ratio as well as increased nutrient removal potential. They also obtained higher degradation efficiency for both PHA and PCL during the removal of nitrate-nitrite from the medium. Briassoulis et al. (2020) studied biodegradation of non-floating plastics in the coastal marine systems using bioreactors with native bacterial populations in stirring environments and obtained 80% degradation of polyethylene.

Heavy metals enter the environment from a wide variety of sources and can be toxic if their concentration increases, affecting plant, animal, as well as human health. Lots of efforts have been made to remove them from water, and bioreactors appear to be such an alternative. Several microbes have the potential for heavy metal removal by accumulating them within the cellular components. Almost all types of bioreactors known to date have been tested for Cr(VI) biosorption (Vendruscolo et al., 2017). Corral-Bobadilla et al. (2019) used the spent mushroom substrate of *Agaricus bisporus* in the bioreactor for heavy metal removal from industrial wastewater and obtained a 50% removal rate, where pH became the limiting parameter. Ibrahim et al. (2020) used bacterial isolates of a polluted lake for heavy metal ion removal from polluted wastewaters via fixed bed column bioreactor fitted with a solid support and got successful results. Fulekar et al. (2012) tested copper, cadmium, and iron bioremediation with the microbial culture obtained from a heavy metal polluted area using a salty medium in a laboratory stirred bioreactor and obtained a very high success rate. De et al. (2006) found that *Alcaligenes faecalis* and *Pseudomonas aeruginosa* are mercury-resistant marine bacterial strains and can remove about 70% of cadmium from the seawater broth. Tec-Caamal et al. (2021) used archaea *Acidianus brierleyi* for removing arsenic by an airlift bioreactor and obtained successful results. MBs have also been proven their worth in bioremediation. Dash and Osborne (2020) found that *Kosakinia oryzae* is capable of producing organophosphate-degrading enzymes when cultivated with *Agaricus bisporus* in a vertical-flow packed bed biofilm bioreactor. Their results showed a higher organophosphorus biodegradation ratio of 90%. Ravi et al. (2021) studied biodegradation of chlorpyrifos in a two-stage bioreactor with bacterial conglomerate obtained from cow dung slurry as the liquid medium and hexadecane as the organic component. Their results indicated a higher rate of pesticide degradation.

9.4.4 Cell Culture and Tissue Engineering

Bioreactors have been proven important tools in tissue engineering, as functional tissues can be generated *in vitro* under controlled environments. Mammalian cells require strict temperature, pH, and oxygen-controlled conditions for constant cell viability as well as functionality (Chico et al., 2008). Tissue growth in bioreactors is promising, as every tissue needs a specific environment for growth, and such requirements can be maintained within them. Eukaryotic bioreactors resemble a laboratory in themselves, as a controlled and ideal environment can be maintained for the cultivation of both plant and animal cells. They have been used in scientific as well as industrial sectors for manufacturing medicines as well as value-added products. Generally, bioreactors used for eukaryotic cell culture are used in the production of monoclonal antibodies, therapeutic proteins, and cell biology research. With the development of tissue engineering, the challenge of cultivation of cells in 3D as well as generation of whole functional organs has increased (Kobayashi, 2017). Cell therapy has paved the way for the utilization of bioreactor lymphocytes and macrophages obtained from human-induced pluripotent stem cells (iPSCs) for studying many diseases, but higher numbers of cells are needed for transplantation. Ackermann et al. (2022) devised a method for bulk production of macrophages of human induced pluripotent stem cell (iPSC-Mac) origin in scalable suspension culture using a stirred-tank bioreactor. Abdin et al. (2023) tried to scale up human macrophage production by human induced pluripotent stem cells (hiPSCs) combined with chimeric antigen receptors (CARs) known as CAR-iMacs. However, in 2D systems, CAR macrophage production has already been known, but their upscaled production in a 3D system is a new advancement. A specialized bioreactor has been designed for the growth of the cells in a controlled environment where a perfusion system is attached to the cell culture that separates and keeps the cells within the bioreactor even after the final product harvest (Matanguihan & Wu, 2022). This setup differs from the conventional bioreactors as continuous and fresh nutrients keep entering the reactor and dead cells are removed continuously during the operation. For the cultivation of plant cells, different types of bioreactors are used, and in designing such models, computational dynamics plays an essential role (Verdú-Navarro et al., 2023). Recently, the use of small well-mixed bioreactors has been emphasized for cell multiplication where the cells are cultivated on microcarriers. Such systems are used for culturing encapsulated cells and neural stem cells in single-cell suspension culture (Nagel-Heyer et al., 2005). In the fluidized and fixed bed bioreactors, immobilization of the cells occurs in the macroporous carriers, where these carriers are aligned in a column, either tightly packed or floating. The column of the bioreactor is perfused with a monitored medium connected with a reservoir in a circulating loop manner. Such a type of reactor is good for the long-term mammalian cell cultivation for biopharmaceutical production (Nehring et al., 2004). In the HFBs, cells are grown in a compartment with tissue-like densities where different types of membranes are used for oxygen and nutrient supply as well as for the toxic metabolite removal. HFBs have been used for the production of monoclonal antibodies, and many examples of modified membrane reactors for 3D cell culture are known to date. Many tissue-specific bioreactors and culture systems have been known, of which the rotating-wall vessel is the most commonly used type, where a construct is maintained in a free-fall state in the medium with minimal shear stress with a high rate of mass transfer. This system has been used for a wide range of applications (Martin et al., 2004). A new type of perfused flow-chamber bioreactor has been developed recently where tissue-specific constructs of different types of tissues were incorporated (Nagel-Heyer et al., 2005).

9.4.5 Food Production

Food production by filamentous fungi by SSFs using oriental cultures is widely known. *In vitro* myogenesis has been used for artificial meat production, and meats grown in the laboratory have certain pros, like a reduction of greenhouse gases in the atmosphere and a decrease in the usage of animals for meat (Kulus et al., 2023). Studies have been conducted to produce meat on a pilot as well as bench scale by using STRs of fed-batch, batch, and continuous types. Lindskog (2018) figured out that continuous mode STRs are more profitable than the batch STRs due to better cell growth, as constant levels of nutrients can be maintained. Perfusion reactors, a type of continuous batch reactors, allow culture medium change without causing damage to the cells, therefore providing an opportunity for medium recycling; however, they demand cell retention devices for continuous cell culture. HFBs and airlift reactors have also been tested for cell cultivation as they offer low shear stress, easy nutrients, and gaseous exchange between cells and media, but have limited scalability and high costs (Baba & Sankai, 2017).

Eukaryotic microalgae can be cultured in bioreactors. Genetic engineering has had major success in the last 10 years with many diatom and green algal species. Microalgae have been implicated in the production of insecticidal proteins, antibodies, as well as vaccines. Nuclear transformation of microalgae is one of the focuses of the biotechnology industries, and this has led to the advanced designing of bioreactors, but only three species have been routinely transformed, namely, *Chlamydomonas reinhardtii*, *Volvox carteri*, and *Phaeodactylum tricornutum*. Light is a crucial factor for the growth of algal cells in open pond systems, as it cannot penetrate deeper if the culture is too dense; for instance, total solar energy conversion was found to be only 3–4% in the outdoor cultivation systems as well as in the laboratory conditions, as observed by Benemann and Oswald (1996). To overcome this issue, mutagenesis is being tested to increase the rate of photosynthesis and productivity of the cultured *Dunaliella* and *Chlamydomonas* under controlled conditions. Transformation from photoautotrophic microalgae to heterotrophic one can be a solution for minimizing the effect of light when grown in a photobioreactor. The conversion of photoautotrophic diatom *P. tricornutum* to heterotrophic ones has solved the light issue as dry biomass production increases 10–50 times in comparison to the cells grown in light-oriented culture setup; however, this system is profitable only if a cheaper carbon source is used (Zaslavskaia et al., 2001). Chloroplast genetic engineering is gaining more popularity these days; for example, expression of human somatotropin was successfully carried out in the total soluble protein (TSP) in chloroplasts, which was 300 times higher than the gene expression done by nuclear transformation (Staub et al., 2000). De Cosa et al. (2001) could produce higher concentrations of Btcry2Aa2 proteins in genetically engineered chloroplasts. Tetanus toxin was also found to be accumulated in the TSP in chloroplasts of transgenic tobacco, and the recombinant protein could produce higher antibody levels in mice (Tregoning et al., 2003). Chloroplast transformation technique has several advantages, like higher expression levels in comparison to the nucleus, the ability to use genes like operons, and unique and target-oriented DNA recombination by homologous recombination for desired gene expression. Researchers have transformed *C. reinhardtii* chloroplast for the expression of large single-chain antibody against herpes simplex virus glycoprotein D and the results indicated that a fully active recombinant antibody was found to accumulate in the transgenic chloroplasts and this was the first report of microalgal use as a bioreactor system (Mayfield et al., 2003). Studies are also being conducted for the antibody expression in the *C. reinhardtii* nuclear genome. Sun et al. (2003) investigated the foot and mouth disease virus VP1 as well as cholera toxin B subunit fusion protein expression in the chloroplast of *C. reinhardtii;* a good success rate was obtained. It has been found that the feeding of transgenic *Chlorella* expressing trypsin-modulating isostatic factor to the mosquito larvae can generate larvicide that can be used for mosquito control (Borovsky, 2003). Use of *C. sorokiniana* and *C. vulgaris* as bioreactors has been tested for human growth hormone (hGH) expression (Hawkins & Nakamura, 1999). Mature rabbit neutrophil peptide-1 (NP-1) gene has been successfully coded in *C. ellipsoidea* cells, and higher expression has been noticed (Chen et al., 2001).

9.5 Conclusion and Future Perspectives of Bioreactors

Bioreactor designing and their utilization in pilot- and large-scale experiments are in progress, and under controlled conditions, great yields can be obtained. Therefore, there is a need for improvements in designs as well as better performance of the bioreactors for suitable and high productivity. Apart from the costs of designing and low energy loss, high yields of value-added products can be achieved via bioreactors. With the use of advanced models, bioreactors can be a boon for the industrial as well as societal sectors with the use of complex sensors, and control of different parameters that control cell cultivation in them. The advanced designs and control can result in the production of products not only of pharmaceutical interest but also of food and biofuel production. Development of the disposable reactors can also minimize the risk of contamination and cleaning, and will orient the direction toward the usage of more and more recyclable materials.

Acknowledgments

I thank Gábor Draskovits, Laboratory Researcher, Dr. József Marek Animal Health Laboratory, Prophyl Kft., Dózsa György út 18, Mohács-7700, Hungary, for motivating me in the completion of this chapter. I also thank Dr. Pradeep Kumar Shukla, Associate Professor, Department of Biological Sciences, Faculty of Science, Sam Higginbottom University of Agriculture, Technology and Sciences, Prayagraj-211007, UP, India, for guidance. Last but not least, Prof. (Dr.) Pramod W. Ramteke (now retired), former Dean of PG Studies and Head, Department of Biological Sciences, Faculty of Science, Sam Higginbottom University of Agriculture, Technology and Sciences, Prayagraj-211007, UP, India, for always being on my side.

References

Abbott, N. J., Rönnbäck, L., & Hansson, E. (2006). Astrocyte–endothelial interactions at the blood–brain barrier. *Nature Reviews. Neuroscience*, *7*, 41–53. https://doi.org/10.1038/nrn1824.

Abdin, S. M., Paasch, D., Kloos, A., Oliveira, M. C., Jang, M.-S., Ackermann, M., Stamopoulou, A., Mroch, P. J., Falk, C. S., von Kaisenberg, C. S., Schambach, A., Heuser, M., Moritz, T., Hansen, G., Morgane, M., & lachmann, N. (2023). Scalable generation of functional human iPSC-derived CAR-macrophages that efficiently eradicate CD19-positive leukemia. *Journal for Immunotherapy of Cancer*, *11*, e007705. https://doi.org/10.1136/jitc-2023-007705.

Ackermann, M., Rafiei Hashtchin, A., Manstein, F., Carvalho Oliveira, M., Kempf, H., Zweigerdt, R., & Lachmann, N. (2022). Continuous human iPSC-macrophage mass production by suspension culture in stirred tank bioreactors. *Nature Protocols*, *17*, 513–539. https://doi.org/10.1038/s41596-021-00654-7.

Aghajani, D. M., & Wang, J. (2022). Modeling microbial growth of dynamic membrane in a biohydrogen production bioreactor. *International Journal of Hydrogen Energy*, *47*, 7666–7681. https://doi.org/10.1016/j.ijhydene.2021.12.090.

Akkoyunlu, B., Daly, S., & Casey, E. (2021). Membrane bioreactors for the production of value-added products: Recent developments, challenges and perspectives. *Bioresource Technology*, *341*, 125793. https://doi.org/10.1016/j.biortech.2021.125793.

Alaghlavi. (2013). Design of fermenter and kinetics. *Bioprocess Engineering*, 23–30.

Allen, J. W., & Bhatia, S. N. (2002). Improving the next generation of bioartificial liver devices. *Seminars in Cell & Developmental Biology*, *13*, 447–454. https://doi.org/10.1016/S1084952102001337.

Al-Rubeai, M., Singh, M. H., Goldman, M. H., & Emery, A. N. (1995). Death mechanisms of animal cells in conditions of intensive agitation. *Biotechnology and Bioengineering*, *45*, 463–472. https://doi.org/10.1002/bit.260450602.

Altmann, B., Grun, C., Nies, C., & Gottwald, E. (2020). Advanced 3D cell culture techniques in micro-bioreactors, part II: Systems and applications. *Processes*, *9*, 21. https://doi.org/10.3390/pr9010021.

Antwi-Akomea, S., Fei-Baffoe, B., Belford, E. J. D., & Borigu, M. (2018). Hydrocarbon contaminated water remediation using a locally constructed multi-stage bioreactor incorporated with media filtration. *Global Journal of Environmental Science and Management*, *4*(4), 413–426. https://doi.org/10.22034/gjesm.2018.04.00.

Aunins, J. G., & Henzler, H. J. (1993). Aeration in cell culture bioreactors. In G. Stephanopoulos (Ed.), *Biotechnology V 3: Bioprocessing* (pp. 219–281). VCH. (Chapter 11).

Baba, K., & Sankai, Y. (2017). Development of biomimetic system for scale up of cell spheroids-building blocks for cell transplantation. *Proceedings of the 2017, 39th annual international conference of the IEEE Engineering in Medicine and Biology Society (EMBC), Jeju, Republic of Korea*, pp. 1611–1616.

Babi, D. K., Griesbach, J., Hunt, S., Insaidoo, F., Roush, D., Todd, R., Staby, A., Welsh, J., & Wittkopp, F. (2022). Opportunities and challenges for model utilization in the biopharmaceutical industry: Current versus future state. *Current Opinion in Chemical Engineering*, *36*, 100813. https://doi.org/10.1016/j.coche.2022.100813.

Badino, A. C., Jr., Facciotti, M. C. R., & Schmidell, W. (2001). Volumetric oxygen transfer coefficients (kLa) in batch cultivations involving non-Newtonian broths. *Biochemical Engineering Journal*, *8*, 111–119. https://doi.org/10.1016/S1369-703X(01)00092-4.

Bakratsas, G., Polydera, A., Katapodis, P., & Stamatis, H. (2021). Recent trends in submerged cultivation of mushrooms and their application as a source of nutraceuticals and food additives. *Future Foods*, *4*, 100086. https://doi.org/10.1016/j.fufo.2021.100086.

Batchelder, C. A., Lee, C. C. I., Matsell, D. G., Yoder, M. C., & Tarantal, A. F. (2009). Renal ontogeny in the rhesus monkey (*Macaca mulatta*) and directed differentiation of human embryonic stem cells towards kidney precursors. *Differentiation*, *78*, 45–56. https://doi.org/10.1016/j.diff.2009.05.001.

Bauchop, T. R., & Clarke, T. J. (1975). Gut microbiology and carbohydrate digestion in the larva of Costelytva zealandica (Coleoptera: Scarabaeidae). *New Zealand Journal of Zoology*, *2*, 237–243. https://doi.org/10.1080/03014223.1975.9517875.

Bayon, C., & Etievant, P. (1980). Methanic fermentation in the digestive tract of a xylophagous insect: *Oryctes nasicornis* L. larva (Coleoptera; Scarabaeidae). *Cellular and Molecular Life Sciences*, *36*, 154–155. https://doi.org/10.1007/BF01953702.

Bayon, C., & Mathelin, J. (1980). Carbohydrate fermentation and by-product absorption studied with labelled cellulose in Oryctes nasicornis larvae (Coleoptera: Scarabaeidae). *Journal of Insect Physiology*, *26*, 833–840. https://doi.org/10.1016/0022-1910(80)90100-6.

Behin, J., & Amiri, P. (2023). A review of recent advances in airlift reactors technology with emphasis on environmental remediation. *Journal of Environmental Management*, *335*, 117560. https://doi.org/10.1016/j.jenvman.2023.117560.

Benemann, J. R., & Oswald, W. J. (1996). *Systems and economic analysis of microalgae ponds for conversion of* CO_2 *to biomass*. Final report to the US Department of Energy. Pittsburgh Energy Technology Center.

Bergeijk, S. A., Salas-Leiton, E., & Cañavate, J. P. (2010). Low and variable productivity and low efficiency of mass cultures of the haptophyte *Isochrysis* aff. *galbana* (T-iso) in outdoor tubular photo bioreactors. *Aquacultural Engineering*, *43*(1), 14–23. https://doi.org/10.1016/j.aquaeng.2010.03.001.

Biggs, D. R., & McGregor, P. G. (1996). Gut pH and amylase and protease activity in larvae of the New Zealand grass grub (*Costelytra zealandica*; Coleoptera: Scarabaeidae) as a basis for selecting inhibitors. *Insect Biochemistry and Molecular Biology*, *26*, 69–75. https://doi.org/10.1016/0965-1748(95)00064-X.

Bliem, R., & Katinger, H. (1988). Scale up engineering in animal cell technology: Part II. *Trends in Biotechnology*, *6*, 224–230. https://doi.org/10.1016/0167-7799(88)90045-5.

Born, C., Zhang, Z., Al-Rubeai, M., & Thomas, C. R. (1992). Estimation of disruption of animal cells by laminar shear stress. *Biotechnology and Bioengineering*, *40*, 1004–1010. https://doi.org/10.1002/bit.260400903.

Borovsky, D. (2003). Trypsin-modulating oostatic factor: A potential new larvicide for mosquito control. *Journal of Experimental Biology*, *206*, 3869–3875. https://doi.org/10.1242/jeb.00602.

Bradford, E., Schweidtmann, A. M., Zhang, D., Jing, K., & del Rio-Chanona, E. A. (2018). Dynamic modeling and optimization of sustainable algal production with uncertainty using multivariate Gaussian processes. *Computers & Chemical Engineering*, *118*, 143–158. https://doi.org/10.1016/j.compchemeng.2018.07.015.

Briassoulis, D., Pikasi, A., Papardaki, N. G., & Mistriotis, A. (2020). Aerobic biodegradation of bio-based plastics in the seawater/sediment interface (sublittoral) marine environment of the coastal zone–test method under controlled laboratory conditions. *Science of the Total Environment*, *722*, 137748. https://doi.org/10.1016/j.scitotenv.2020.137748.

Carniel, A., Gomes, A. D. C., Coelho, M. A. Z., & de Castro, A. M. (2020). Process strategies to improve biocatalytic depolymerization of post-consumer PET packages in bioreactors, and

investigation on consumables cost reduction. *Bioprocess and Biosystems Engineering*, *44*(3), 507–516. https://doi.org/10.1007/s00449-020-02461-y.
Cazemier, A. E., Verdoes, J. C., Reubsaet, F. A. G., Hackstein, J. H. P., van der Drift, C., & Op den Camp, H. J. (2003). *Promicromonospora pachnodae* sp. nov., a member of the (hemi)cellulolytic hindgut flora of larvae of the scarab beetle *Pachnoda marginata*. *Antonie van Leeuwenhoek*, *83*, 135–148. https://doi.org/10.1023/A:1023325817663.
Chalmers, J. J. (1994). Cells and bubbles in sparged bioreactors. *Cytotechnology*, *15*, 311–320. https://doi.org/10.1007/BF00762406.
Chen, H.-C., & Hu, Y.-C. (2006). Bioreactors for tissue engineering. *Biotechnology Letters*, *28*(18), 1415–1423. https://doi.org/10.1007/s10529-006-9111-x.
Chen, Y., Wang, Y., Sun, Y., Zhang, L., & Li, W. (2001). Highly efficient expression of rabbit neutrophil peptide-1 gene in *Chlorella ellipsoidea* cells. *Current Genetics*, *39*, 365–370. https://doi.org/10.1007/s002940100205.
Cheng-Wu, Z., Zmora, O., Kopel, R., & Richmond, A. (2001). An industrial-size flat plate glass reactor for mass production of *Nannochloropsis* sp. Eustigmatophyceae. *Aquaculture*, *195*, 35–49. https://doi.org/10.1016/S0044-8486(00)00533-0.
Cherry, R. S. (1993). Animal cells in turbulent fluids—Details of the physical stimulus and the biological response. *Biotechnology Advances*, *11*(2), 279–299. https://doi.org/10.1016/0734-9750(93)90043-M.
Cherry, R. S., & Papoutsakis, E. T. (1990). Fluid mechanical injury of cells in bioreactors. In R. E. Spier & J. B. Griffiths (Eds.), *Animal cell biotechnology* (Vol. *4*, pp. 71–121). (Chapter 3).
Chico, E., Rodriguez, G., Figueiredo, A. (2008). Biorreatores para Células Animais. In Roca Barcelona, *Tecnologia do Cultivo de Células Animais: De Biofármacos a Terapia Gênica*; 1st edition Spain: pp. 216–251.
Chisti, Y. (1993). Animal cell culture in stirred bioreactors: Observations on scale up. *Bioprocess Engineering*, *9*, 191–196. https://doi.org/10.1007/BF00369402.
Cooper, C. M., Fernstrom, G. A., & Miller, S. A. (1944). Performance of agitated gas-liquid contactors. *Industrial and Engineering Chemistry*, *36*, 504–509. https://doi.org/10.1021/ie50414a005.
Corbin, J. M., McNulty, M. J., Macharoen, K., McDonald, K. A., & Nandi, S. (2020). Technoeconomic analysis of semicontinuous bioreactor production of biopharmaceuticals in transgenic rice cell suspension cultures. *Biotechnology and Bioengineering*, *117*, 3053–3065. https://doi.org/10.1002/bit.27475.
Corral-Bobadilla, M., González-Marcos, A., Vergara-González, E. P., & Alba-Elías, F. (2019). Bioremediation of waste water to remove heavy metals using the spent mushroom substrate of *Agaricus bisporus*. *Water*, *11*(3), 454. https://doi.org/10.3390/w11030454.
Cucullo, L., Aumayr, B., Rapp, E., & Janigro, D. (2005). Drug delivery and in vitro models of the blood-brain barrier. *Current Opinion in Drug Discovery & Development*, *8*, 89–99.
Cucullo, L., McAllister, M. S., Kight, K., Krizanac-Bengez, L., Marroni, M., Mayberg, M. R., Stanness, K. A., & Janigro, D. (2002). A new dynamic in vitro model for the multidimensional study of astrocyte–endothelial cell interactions at the blood–brain barrier. *Brain Research*, *951*, 243–254. https://doi.org/10.1016/S0006-8993(02)03167-0.
Darnell, M., Ulvestad, M., Ellis, E., Weidolf, L., & Andersson, T. B. (2012). *In vitro* evaluation of major in vivo drug metabolic pathways using primary human hepatocytes and HepaRG cells in suspension and a dynamic three-dimensional bioreactor system. *The Journal of*

Pharmacology and Experimental Therapeutics, *343*, 134–144. https://doi.org/10.1124/jpet.112.195834.

Dash, D. M., & Osborne, W. J. (2020). Rapid biodegradation and biofilm-mediated bioremoval of organophosphorus pesticides using an indigenous *Kosakonia oryzae* strain-VITPSCQ3 in a vertical-flow packed bed biofilm bioreactor. *Ecotoxicology and Environmental Safety, 192*, 110290. doi:10.1016/j.ecoenv.2020.110290.

de Assis, G. M., & Belini, G. B. (2023). Biorreator airlift: Estudo sobre suas aplicac.es. *Environmental Science & Technology Innovation, ISSN 2965-1158 2*, 27–34.

De Cosa, B., Moar, W., Lee, S. B., Miller, M., & Daniell, H. (2001). Over expression of the *Bt* cry2Aa2 operon in chloroplasts leads to formation of insecticidal crystals. *Nature Biotechnology*, *19*, 71–74. https://doi.org/10.1038/83559.

De, J., Sarkar, A., & Ramaiah, N. (2006). Bioremediation of toxic substances by mercury resistant marine bacteria. *Ecotoxicology*, *15*(4), 385–389. https://doi.org/10.1007/s10646-006-0066-4.

Decarli, M. C., dos Santos, D. P., Astray, R. M., Ventini-Monteiro, D. C., Jorge, S. A. C., Correia, D. M., de Sá da Silva, J., Rocca, M. P., Langoni, H., Menozzi, B. D., Pereira, C. A., & Suazo, C. A. T. (2018). Drosophila S2 cell culture in a WAVE Bioreactor: Potential for scaling up the production of the recombinant rabies virus glycoprotein. *Applied Microbiology and Biotechnology*, *102*, 4773–4783. https://doi.org/10.1007/s00253-018-8962-0.

Degen, J., Uebele, A., Retze, A., Schmid-Staiger, U., & Trosch, W. (2001). A novel airlift photobioreactor with baffles for improved light utilization through the flashing light effect. *Journal of Biotechnology*, *92*(2), 89–94. https://doi.org/10.1016/S0168-1656(01)00350-9.

Delavar, M. A., & Wang, J. (2022). Bioreactor concepts, types, and modeling. In *Advanced methods and mathematical modeling of biofilms* (pp. 195–245). Academic Press.

Dempsey, M. J. (1994). Biofilms and fluidized bed fermentation. *International Biodeterioration & Biodegradation*, *34*(3–4), 237–244. https://doi.org/10.1016/0964-8305(94)90085-X.

Ding, Y.-T., & Shi, X.-L. (2011). Bioartificial liver devices: Perspectives on the state of the art. *Frontiers of Medicine*, *5*, 15–19. https://doi.org/10.1007/s11684-010-0110-x

Doran, P. M. (1995). *Bioprocess engineering principles*. Elsevier Science & Technology Books.

Dorresteijn, R. C., de Gooijer, C. D., Tramper, J., & Beuvery, E. C. (1994). A simple dynamic method for on-line and off-line determination of k_La during cultivation of animal cells. *Biotechnology Techniques*, *8*(9), 675–680. https://doi.org/10.1007/BF00241697.

Douglas, A. E. (2009). The microbial dimension in insect nutritional ecology. *Functional Ecology*, *23*, 38–47. https://doi.org/10.1111/j.1365-2435.2008.01442.x.

Egert, M., Stingl, U., Bruun, D. L., Wagner, B., Brune, A., & Friedrich, M. W. (2005). Structure and topology of microbial communities in the major gut compartments of *Melolontha melolontha* larvae (Coleoptera: Scarabaeidae). *Applied and Environmental Microbiology*, *71*, 4556–4566. https://doi.org/10.1128/AEM.71.8.4556-4566.2005.

Egert, M., Wagner, B., Lemke, T., Brune, A., & Friedrich, M. W. (2003). Microbial community structure in midgut and hindgut of the humus-feeding larva of *Pachnoda ephippiata* (Coleoptera: Scarabaeidae). *Applied and Environmental Microbiology*, *69*, 6659–6668. https://doi.org/10.1128/AEM.69.11.6659-6668.2003.

Eibl, R., & Eibl, D. (2006). Design and use of the wave bioreactor for plant cell culture. *Plant Tissue Culture Engineering*, *6*, 203–227. https://doi.org/10.1007/978-1-4020-3694-1_12.

Eibl, R., Eibl, D., Portner, R., Catapano, G., & Czermak, P. (2008). *Cell and tissue reaction engineering*. Springer.

El Haj, A. J., Wood, M. A., Thomas, P., & Yang, Y. (2005). Controlling cell biomechanics in orthopaedic tissue engineering and repair. *Pathologie et Biologie*, *53*(10), 581–589. https://doi.org/10.1016/j.patbio.2004.12.002.

Fleischaker, R. J., & Sinskey, A. (1981). Oxygen demand and supply in cell culture. *European Journal of Applied Microbiology and Biotechnology*, *12*, 193–197. https://doi.org/10.1007/BF00499486.

Flickinger, M. C., & Drew, S. W. (1999). *Encyclopedia of bioprocess technology: Fermentation, biocatalysis, and bioseparation* (1st ed.). Wiley.

Frontistisf, Z., Lykogiannis, G., & Sarmpanis, A. (2023). Artificial neural networks in membrane bioreactors: A comprehensive review—Overcoming challenges and future perspectives. *Science*, *5*, 31. https://doi.org/10.3390/sci5030031.

Fulekar, M. H., Sharma, J., & Tendulkar, A. (2012). Bioremediation of heavy metals using biostimulation in laboratory bioreactor. *Environmental Monitoring and Assessment*, *184*(12), 7299–7307. https://doi.org/10.1007/s10661-011-2499-3.

Gardner, A. R., Gainer, J. L., & Kirwan, D. J. (1990). Effects of stirring and sparging on cultured hybridoma cells. *Biotechnology and Bioengineering*, *35*(9), 940–947. https://doi.org/10.1002/bit.260350912.

Gasljevic, K., Hall, K. A., Oakes, S., Chapman, D. J., & Matthys, E. F. (2009). Increased production of extracellular polysaccharide by *Porphyridium cruentum* immobilized in foam sheets. *Engineering in Life Sciences*, *9*(6), 479–489. https://doi.org/10.1002/elsc.200800096.

Gavrilescu, M., Roman, R. V., & Efimov, V. (1993). The volumetric oxygen mass transfer coefficient in antibiotic biosynthesis liquids. *Acta Biotechnology*, *13*, 59. https://doi.org/10.1002/abio.370130114.

Ge, X., Vasco-Correa, J., & Li, Y. (2017). Solid-state fermentation bioreactors and fundamentals. In C. Larroche, M. A. Sanromán, & A. Pandey (Eds.), *Current developments in biotechnology and bioengineering* (pp. 381–402). Elsevier. https://doi.org/10.1016/B978-0-444-63663-8.00013-6.

Geissinger, O., Herlemann, D. P. R., Morschel, E., Maier, U. G., & and Brune, A. (2009). The ultramicrobacterium "*Elusimicrobium minutum*" gen. nov., sp. nov., the first cultivated representative of the termite group 1 phylum. *Applied and Environmental Microbiology*, *75*, 2831–2840. https://doi.org/10.1128/AEM.02697-08.

Gerlach, J. C., Zeilinger, K., & Patzer, J. F., II. (2008). Bioartificial liver systems: Why, what, whither? *Regenerative Medicine*, *3*, 575–595. https://doi.org/10.2217/17460751.3.4.575.

Germano-de Almeida, G., da Silva, M.D.G.C., Barbosa, R., de Silva, D. D. S. P., O Silva, R., Lima, G. M. S., de Gusmao, N. B, de Sousa, M. D. F. V. D Q. (2017). Biodegradation of marine fuel MF-380 by microbial consortium isolated from seawater near the petrochemical Suape port, Brazil. *International Biodeterioration & Biodegradation 116*: 73–82. doi: https://doi.org/10.1016/j.ibiod.2016.09.028.

Gkantzou, E., Weinhart, M., & Kara, S. (2023). 3D printing for flow biocatalysis. *RSC Sustainability*, *1*, 1672–1685. https://doi.org/10.1039/D3SU00155E.

Gómez-Lechón, M. J., Castell, J. V., & Donato, M. T. (2007). Hepatocytes: The choice to investigate drug metabolism and toxicity in man: In vitro variability as a reflection of in vivo. *Chemico-Biological Interactions*, *168*(1), 30–50. https://doi.org/10.1016/j.cbi.2006.10.013.

Gudin, C., & Chaumont, D. (1991). Cell fragility: The key problem of microalgae mass production in closed photobioreactors. *Bioresource Technology*, *38*(2–3), 145–151. https://doi.org/10.1016/0960-8524(91)90146-B.

Handa-Corrigan, A. (1990). Oxygenating animal cell cultures: The remaining problems. In R. E. Spier & J. B. Griffiths (Eds.), *Animal cell biotechnology* (Vol. *4*, pp. 123–132). Academic Press (Chapter 4).

Hawkins, R. L., & Nakamura, M. (1999). Expression of human growth hormone by the eukaryotic alga, Chlorella. *Current Microbiology*, *38*, 335–341. https://doi.org/10.1007/PL00006813.

Hori, S., Ohtsuki, S., Tachikawa, M., Kimura, N., Kondo, T., Watanabe, M., Nakashima, E., & Terasaki, T. (2004). Functional expression of rat ABCG2 on the luminal side of brain capillaries and its enhancement by astrocyte-derived soluble factor(s). *Journal of Neurochemistry*, *90*, 526–536. https://doi.org/10.1111/j.1471-4159.2004.02537.x.

Hu, W. S., Meier, J., & Wang, D. I. C. (1986). Use of surface aerator to improve oxygen transfer in cell culture. *Biotechnology and Bioengineering*, *28*, 122–125. https://doi.org/10.1002/bit.260280120.

Huang, S.-W., Zhang, H.-Y., Marshall, S., & Jackson, T. A. (2010). The scarab gut: A potential bioreactor for bio-fuel production. *Insect Science*, *17*, 175–183. https://doi.org/10.1111/j.1744-7917.2010.01320.x.

Hubbard, D. W. (1987). Biotechnology processes. In *Scale-up and mixing* (pp. 168–184). AIChE.

Humes, H. D., MacKay, S. M., Funke, A. J., & Buffington, D. A. (1997). The bioartificial renal tubule assist device to enhance CRRT in acute renal failure. *American Journal of Kidney Diseases*, *30*, S28–S31. https://doi.org/10.1016/S0272-6386(97)90539-4.

Humes, H. D., Mackay, S. M., Funke, A. J., & Buffington, D. A. (1999). Tissue engineering of a bioartificial renal tubule assist device: In vitro transport and metabolic characteristics. *Kidney International*, *55*, 2502–2514. https://doi.org/10.1046/j.1523-1755.1999.00486.x.

Humes, H. D., & Szczypka, M. S. (2004). Advances in cell therapy for renal failure. *Transplantation Immunology*, *12*, 219–227. https://doi.org/10.1016/j.trim.2003.12.015.

Humes, H. D., Weitzel, W. F., Bartlett, R. H., Swaniker, F. C., Paganini, E. P., Luderer, J. R., & Sobota, J. (2004). Initial clinical results of the bioartificial kidney containing human cells in ICU patients with acute renal failure. *Kidney International*, *66*, 1578–1588. https://doi.org/10.1111/j.1523-1755.2004.00923.x.

Ibrahim, S., Yu, G., & Li, Y. (2020). Efficient treatment of recalcitrant textile wastewater using two-phase mesophilic anaerobic process: Bio-hythane production and decolorization improvements. *Journal of Material Cycles and Waste Management*, *22*, 515–523. https://doi.org/10.1007/s10163-019-00944-z.

Iravani, S., & Varma, R. S. (2021). MXenes and MXene-based materials for tissue engineering and regenerative medicine: Recent advances. *Materials Advances*, *2*(9), 2906–2917. https://doi.org/10.1039/D1MA00189B.

Jeyavani, J., Sibiya, A., Shanthini, S., Ravi, C., Vijayakumar, S., Rajan, D. K., & Vaseeharan, B. (2021). A review on aquatic impacts of microplastics and its bioremediation aspects. *Current Pollution Reports*, *7*(3), 286–299. https://doi.org/10.1007/s40726-021-00188-2.

Jobses, I., Martens, D., & Tramper, J. (1991). Lethal events during gas sparging in animal cell culture. *Biotechnology and Bioengineering*, *37*(5), 484–490. https://doi.org/10.1002/bit.260370510.

José, Á. H. M., Moura, E. A. B., Rodrigues, D., Jr., Kleingesinds, E. K., & Rodrigues, R. C. L. B. (2022). A residue-free and effective corncob extrusion pretreatment for the enhancement of high solids loading enzymatic hydrolysis to produce sugars. *Industrial Crops and Products, 188*, 115655. https://doi.org/10.1016/j.indcrop.2022.115655.

Ju, L.-K., & Chase, G. G. (1992). Improved scale up strategies of bioreactors. *Bioprocess Engineering, 8*, 49–53. https://doi.org/10.1007/BF00369263.

Kaiser, S. C., Werner, S., Jossen, V., Blaschczok, K., & Eibl, D. (2018). Power input measurements in stirred bioreactors at laboratory scale. *Journal of Visualized Experiments, 135*, 56078. https://doi.org/10.3791/56078.

Kantarci, N., Borak, F., & Ulge, K. O. (2004). Bubble column reactors. *Process Biochemistry, 40*, 2263–2283. https://doi.org/10.1016/j.procbio.2004.10.004.

Karnieli, O. (2016). Bioreactors and downstream processing for stem cell manufacturing. In J. M. S. Cabral, C. S. de Silva, & M. M. Diogo (Eds.), *Stem cell manufacturing* (pp. 141–160). Elsevier. https://doi.org/10.1016/B978-0-444-63265-4.00006-6.

Katinger, H. W. D., Scheirer, W., & Kromer, E. (1979). Bubble column reactor for mass propagation of animal cells in suspension culture. *German Chemical Engineering, 2*, 31–38.

Kaur, I., & Sharma, A. (2020). Bioreactor: Design, functions and fermentation innovations. *Research & Reviews in Biotechnology & Biosciences, 8*, 116–125. https://doi.org/10.5281/zenodo.5775455.

Khanal, O., & Lenhoff, A. M. (2021). Developments and opportunities in continuous biopharmaceutical manufacturing. *MAbs, 13*, 1903664. https://doi.org/10.1080/19420862.2021.1903664.

Kioukia, N., Nienow, A. W., Emery, A. N., & Al-Rubeai, M. (1992). The impact of fluid dynamics on the biological performance of free suspension animal cell culture: Further studies. *Food and Bioproducts Processing: Transactions of the Institution of Chemical Engineers, Part C, 70*, 143–148. https://doi.org/10.1080/19420862.2021.1903664.

Kobayashi, E. (2017). Challenges for production of human transplantable organ grafts. *Cell Medicine, 9*, 9–14. https://doi.org/10.3727/215517916X69311.

Kossen, N. W. F. (1995). Scale-up. In E. Galindo & O. T. Ramirez (Eds.), *Advances in bioprocess engineering* (pp. 53–65). Kluwer Academic Publishers.

Krishania, M., Sindhu, R., Binod, P., Ahluwalia, V., Kumar, V., Sangwan, R. S., & Pandey, A. (2018). Design of bioreactors in solid-state fermentation. In A. Pandey, C. Larroche, & R. R. Singhania (Eds.), *Current developments in biotechnology and bioengineering* (pp. 83–96). Elsevier. https://doi.org/10.1016/B978-0-443-27624-8.00004-X.

Kulus, M., Jankowski, M., Kranc, W., Golkar, N. A., Farzaneh, M., Dzięgiel, P., Zabel, M., Antosik, P., Bukowska, D., Mozdziak, P., & Kempisty, B. (2023). Bioreactors, scaffolds and microcarriers and in vitro meat production—Current obstacles and potential solutions. *Frontiers in Nutrition, 10*, 1225233. https://doi.org/10.3389/fnut.2023.1225233.

Kunas, K. T., & Papoutsakis, E. T. (1990). Damage mechanisms of suspended animal cells in agitated bioreactors with and without bubble entrainment. *Biotechnology and Bioengineering, 36*(5), 476–483. https://doi.org/10.1002/bit.260360507.

Lash, L. H., Putt, D. A., & Cai, H. (2008). Drug metabolism enzyme expression and activity in primary cultures of human proximal tubular cells. *Toxicology, 244*, 56–65. https://doi.org/10.1016/j.tox.2007.10.022.

Laska, M. E., & Cooney, C. L. (1999). *Encyclopedia of bioprocess technology: Fermentation, biocatalysis, and separation* (1st ed.). Wiley.

Leist, C. H., Meyer, H. P., & Fiechter, A. (1990). Potential problems of animal cells in suspension culture. *Journal of Biotechnology*, *15*, 1–46. https://doi.org/10.1016/0168-1656(90)90049-H.

Lemke, T., Stingl, U., Egert, M., Friedrich, M. W., & Brune, A. (2003). Physicochemical conditions and microbial activities in the highly alkaline gut of the humus-feeding larva of *Pachnoda ephippiata* (Coleoptera: Scarabaeidae). *Applied and Environmental Microbiology*, *69*, 6650–6658. https://doi.org/10.1128/AEM.69.11.6650-6658.2003.

Leslie, E. M., Deeley, R. G., & Cole, S. P. C. (2005). Multidrug resistance proteins: Role of P-glycoprotein, MRP1, MRP2, and BCRP (ABCG2) in tissue defense. *Toxicology and Applied Pharmacology*, *204*, 216–237. doi:10.1016/j.taap.2004.10.012.

Lindskog, E. K. (2018). The upstream process: Principal modes of operation. In G. Jagschies, E. Lindskog, & P. Galliher (Eds.), *Biopharmaceutical processing* (pp. 625–635). Elsevier. https://doi.org/10.1016/B978-0-08-100623-8.00031-1.

Lopardo, C. R., & Urakawa, H. (2019). Performance and microbial diversity of bioreactors using polycaprolactone and polyhydroxyalkanoate as carbon source and biofilm carrier in a closed recirculating aquaculture system. *Aquaculture International*, *27*(5), 1251–1268. https://doi.org/10.1007/s10499-019-00383-5.

Lu, W. J., Hwang, S. J., & Chang, C. M. (1994). Liquid mixing in internal loop airlift reactors. *Industrial & Engineering Chemistry Research*, *33*, 2180–2186. https://doi.org/10.1021/ie00033a023.

Luo, X., Lee, D. J., Lau, R., Yang, G., & Fan, L. (1999). Maximum stable bubble size and gas holdup in high-pressure slurry bubble columns. *AIChE Journal*, *45*, 665–685. https://doi.org/10.1002/aic.690450402.

Ma, W., Hua, D., Xiong, R., & Huang, C. (2023). Bio-based stimuli-responsive materials for biomedical applications. *Materials Advances*, *4*(2), 458–475. https://doi.org/10.1039/D2MA01011A.

Machon, V., Pacek, A. W., & Nienow, A. W. (1997). Bubble sizes in electrolyte and alcohol solutions in a turbulent stirred vessel. *Chemical Engineering Research and Design*, *75*(3), 339–348. https://doi.org/10.1205/026387697523651.

Manan, M. A., & Webb, C. W. (2017). Design aspects of solid state fermentation as applied to microbial bioprocessing. *Journal of Applied Biotechnology & Bioengineering*, *4*, 511–532. https://doi.org/10.15406/jabb.2017.04.00094.

Mandenius, C. F. (2016). Challenges for bioreactor design and operation. In C.-F. Mandenius (Ed.), *Bioreactors: Design, operation and novel applications* (pp. 1–34). Wiley.

Markarian, J. (2019). Sustainability in disposal of single-use systems. *BioPharm International*, *32*(2), 19.

Martin, I., Wendt, D., & Heberer, M. (2004). The role of bioreactors in tissue engineering. *Trends in Biotechnology*, *22*, 80–86. https://doi.org/10.1016/j.tibtech.2003.12.001.

Martin, M. M. (1983). Cellulose digestion in insects. *Comparative Biochemistry and Physiology*, *75*(3), 313–324. https://doi.org/10.1016/0300-9629(83)90088-9.

Masereeuw, R., & Russel, F. G. M. (2010). Therapeutic implications of renal anionic drug transporters. *Pharmacology & Therapeutics*, *126*, 200–216. https://doi.org/10.1016/j.pharmthera.2010.02.007.

Matanguihan, C., & Wu, P. (2022). Upstream continuous processing: Recent advances in production of biopharmaceuticals and challenges in manufacturing. *Current Opinion in Biotechnology*, *78*, 102828. https://doi.org/10.1016/j.copbio.2022.102828.

Mayfield, S. P., Franklin, S. E., & Lerner, R. A. (2003). Expression and assembly of a fully active antibody in algae. *Proceedings of the National Academy of Sciences of the United States of America*, *100*, 438–442.

Mazariegos, G. V., Patzer, J. F., II, Lopez, R. C., Giraldo, M., DeVera, M. E., Grogan, T. A., Zhu, Y., Fulmer, M. L., Amiot, B. P., & Kramer, D. J. (2002). First clinical use of a novel bioartificial liver support system (BLSS). *American Journal of Transplantation*, *2*(3), 260–266. https://doi.org/10.1034/j.1600-6143.2002.20311.x.

Meng, F., Chae, S.-R., Drews, A., Kraume, M., Hang-Sik Shin, H.-S., & Yang, F. (2009). Recent advances in membrane bioreactors (MBRs): Membrane fouling and membrane material. *Water Research*, *43*, 1489–1512. https://doi.org/10.1016/j.watres.2008.12.044.

Meuwly, F., Ruffieux, P.-A., Kadouri, A., & von Stockar, U. (2007). Packed-bed bioreactors for mammalian cell culture: Bioprocess and biomedical applications. *Biotechnology Advances*, *25*, 45–56. https://doi.org/10.1016/j.biotechadv.2006.08.004.

Michaels, J., Mallik, A. K., Nowak, J. E., Wasan, D. T., & Papoutsakis, E. T. (1994). Dynamic interfacial tension and rheological properties of cell culture medium with shear protectant additives. In R. E. Spier, J. B. Griffiths, & W. Berthold (Eds.), *Animal cell technology: Products of today, prospects for tomorrow* (pp. 389–391). Butterworth-Heinemann.

Mirro, R., & Voll, K. (2009). Which impeller is right for your cell line? A guide to impeller selection for stirred-tank bioreactors. *BioProcess International*, *7*, 52–57.

Monahan, P. B., & Holtzapple, M. T. (1993). Oxygen transfer in a pulse bioreactor. *Biotechnology and Bioengineering*, *42*, 724–728. https://doi.org/10.1002/bit.260420607.

Moreira, J. L., Cruz, P. E., Santana, P. C., & Feliciano, A. S. (1995). Influence of power input and aeration method on mass transfer in a laboratory animal cell culture vessel. *Journal of Chemical Technology & Biotechnology*, *62*, 118–131. https://doi.org/10.1002/jctb.280620203.

Nagel-Heyer, S., Leist, C., Lünse, S., Goepfert, C., & Pörtner, R. (2005). From biopsy to cartilage-carrier constructs by using microcarrier cultures as sub-process. *Proceedings of 19th ESACT meeting. European Society for Animal Cell Technology (ESACT), Harrogate, UK*.

Naghdi, M., Ghovvati, M., Rabiee, N., Ahmadi, S., Abbariki, N., Sojdeh, S., Ojaghi, A., Bagherzadeh, M., Akhavan, O., Sharifi, E., Rabiee, M., Saeb, M. R., Bolouri, K., Webster, T. J., Zare, E. N., & Zarrabi, A. (2022). Magnetic nanocomposites for biomedical applications. *Advances in Colloid and Interface Science*, *308*, 102771. https://doi.org/10.1016/j.cis.2022.102771.

Najafpour, G. D., Mehdi, T., Komeili, M., & Mohammadi, M. (2009). Kinetic model for an up-flow anaerobic packed bed bioreactor: Dairy wastewater treatment. *African Journal of Biotechnology*, *8*, 15. https://doi.org/10.4314/ajb.v8i15.61859.

Narciso-Ortiz, L., Vargas-García, K. A., Vázquez-Larios, A. L., Quiñones-Muñoz, T. A., HernándezMartínez, R., & Lizardi-Jiménez, M. A. (2020). Coral reefs and watersheds of the Gulf of Mexico in Veracruz: Hydrocarbon pollution data and bioremediation proposal. *Regional Studies in Marine Science*, *35*, 101155. https://doi.org/10.1016/j.rsma.2020.101155.

Nehring, D., Gonzales, R., Czermak, P., & Pörtner, R. (2004). Mathematical model of a membrane filtration process using ceramic membranes to increase retroviral pseudotype vector titer. *Journal of Membrane Science*, *237*, 25–38. https://doi.org/10.1016/j.memsci.2004.02.027.

Nelson, K. L. (1988a). Industrial scale mammalian cell culture part I: Bioreactor design considerations. *Biopharma Manufacturing*, *2*, 42–46.

Nelson, K. L. (1988b). Industrial scale mammalian cell culture part II: Design and scale up. *Biopharma Manufacturing*, *2*, 47–55.

Neuhaus, W., Lauer, R., Oelzant, S., Fringeli, U. P., Ecker, G. F., & Noe, C. R. (2006). A novel flow based hollow-fiber blood-brain barrier in vitro model with immortalised cell line PBMEC/C1-2. *Journal of Biotechnology*, *125*, 127–141. https://doi.org/10.1016/j.jbiotec.2006.02.019.

Ni, M., Teo, J. C. M., Bin Ibrahim, M. S., Zhang, K., Tasnim, F., Chow, P. Y., Zink, D., & Ying, J. Y. (2011). Characterization of membrane materials and membrane coatings for bioreactor units of bioartificial kidneys. *Biomaterials*, *32*, 1465–1476. https://doi.org/10.1016/j.biomaterials.2010.10.061.

Nikolova, M. P., & Chavali, M. S. (2019). Recent advances in biomaterials for 3D scaffolds: A review. *Bioactive Materials*, *4*, 271–292. https://doi.org/10.1016/j.bioactmat.2019.10.005.

Oo, Z. Y., Deng, R., Hu, M., Ni, M., Kandasamy, K., Bin Ibrahim, M. S., Ying, J. Y., & Zink, D. (2011). The performance of primary human renal cells in hollow fiber bioreactors for bioartificial kidneys. *Biomaterials*, *32*, 8806–8815. https://doi.org/10.1016/j.biomaterials.2011.08.030.

Palladino, F., Marcelino, P. R. F., Schlogl, A. E., José, Á. H. M., Rodrigues, R. D. C. L. B., Fabrino, D. L., Santos, I. J. B., & Rosa, C. A. (2024). Bioreactors: Applications and innovations for a sustainable and healthy future—A critical review. *Applied Sciences*, *14*, 9346. https://doi.org/10.3390/app14209346.

Pandey, A., Larroche, C., Soccol, C. R., & Dussap, C. G. (2008). *Advances in fermentation technology* (pp. 172–201). Asiatech Publishers.

Papoutsakis, E. T. (1991). Media additives for protecting freely suspended animal cells against agitation and aeration damage. *Trends in Biotechnology*, *9*, 316–397. https://doi.org/10.1016/0167-7799(91)90102-N.

Park, J. B. K., Craggs, R. J., & Shilton, A. N. (2011). Wastewater treatment high-rate algal ponds for biofuel production. *Bioresource Technology*, *102*, 35–42. https://doi.org/10.1016/j.biortech.2010.06.158.

Park, J.-K., & Lee, D.-H. (2005). Bioartificial liver systems: Current status and future perspective. *Journal of Bioscience and Bioengineering*, *99*, 311–319. https://doi.org/10.1263/jbb.99.311.

Pietig, G., Mehrens, T., Hirsch, J. R., Çetinkaya, I., Piechota, H., & Schlatter, E. (2001). Properties and regulation of organic cation transport in freshly isolated human proximal tubules. *Journal of Biological Chemistry*, *276*, 33741–33746. https://doi.org/10.1074/jbc.M104617200.

Pino, M. S., Rodríguez-Jasso, R. M., Michelin, M., Flores-Gallegos, A. C., Morales-Rodriguez, R., Teixeira, J. A., & Ruiz, H. A. (2018). Bioreactor design for enzymatic hydrolysis of biomass under the biorefinery concept. *Chemical Engineering Journal*, *341*, 119–136. https://doi.org/10.1016/j.cej.2018.04.057.

Prado, R. C., & Borges, E. R. (2018). Microbioreactors as engineering tools for bioprocess development. *Brazilian Journal of Chemical Engineering*, *35*, 1163–1182.

Putt, R., Manjinder, S., Chinnasamy, S., & Das, K. C. (2011). An efficient system for carbonation of high rate algae pond water to enhance CO_2 mass transfer. *Bioresource Technology*, *102*, 3240–3245. https://doi.org/10.1016/j.biortech.2010.11.029.

Racusen, L. C., Monteil, C., Sgrignoli, A., Lucskay, M., Marouillat, S., Rhim, J. G. S., & Morin, J.-P. (1997). Cell lines with extended in vitro growth potential from human renal proximal tubule: Characterization, response to inducers, and comparison with established cell lines. *The Journal of Laboratory and Clinical Medicine*, *129*(3), 318–329. https://doi.org/10.1016/S0022-2143(97)90180-3.

Radisic, M., Marsano, A., Maidhof, R., Wang, Y., & Vunjak-Novakovic, G. (2008). Cardiac tissue engineering using perfusion bioreactor systems. *Nature protocols*, *3*, 719–738. https://doi.org/10.1038/nprot.2008.40.

Rafa, N., Ahmed, S. F., Badruddin, I. A., Mofijur, M., & Kamangar, S. (2021). Strategies to produce cost-effective third-generation biofuel from microalgae. *Frontiers in Energy Research*, *946*, 174230. https://doi.org/10.3389/fenrg.2021.749968.

Rahman, U. T., Roy, H., Islam, M. R., Tahmid, M., Fariha, A., Mazumder, A., Tasnim, N., Pervez, M. N., Cai, Y., Naddeo, V., & Islam, M. S. (2023). The advancement in membrane bioreactor (MBR) technology toward sustainable industrial wastewater management. *Membranes*, *13*(2), 181. https://doi.org/10.3390/membranes13020181.

Rajarathinam, R., Rajarathinam, R., Kanagaraj, L. P., Ranganathan, R. V., Dhanasekaran, K., & Manickam, N. K. (2019). Evaluation and characterization of novel sources of sustainable lignocellulosic residues for bioethanol production using ultrasound-assisted alkaline pre-treatment. *Waste Management*, *87*, 368–374. https://doi.org/10.1016/j.wasman.2019.02.015.

Ranjbar, R., Inoue, R., Katsuda, T., Yamaji, H., & Katoh, S. (2008). High efficiency production of astaxanthin in an airlift photobioreactor. *Journal of Bioscience and Bioengineering*, *106*(2), 204–207. https://doi.org/10.1263/jbb.106.204.

Rasoul-Amini, S., Montazeri-Najafabady, N., Mobasher, M. A., Hoseini-Alhashemi, S., & Ghasemi, Y. (2011). *Chlorella* sp.: A new strain with highly saturated fatty acids for biodiesel production in bubble-column photobioreactor. *Applied Energy*, *88*(10), 3354–3356. https://doi.org/10.1016/j.apenergy.2010.12.040.

Ravi, R. K., Singh, M., & Fulekar, M. H. (2021). Bioconversion dynamics for bioremediation of chlorpyrifos pesticide using bacterial consortia in two immiscible liquid phase partitioning bioreactor. *Biomass Conversion and Biorefinery*, *12*(5), 1755–1769. https://doi.org/10.1007/s13399-021-01695-4.

Reisman, H. B. (1993). Problems in scale up of biotechnology production processes. *Critical Reviews in Biotechnology*, *13*(3), 195–253. https://doi.org/10.3109/07388559309041319.

Reuss, M. (1993). Oxygen transfer and mixing: Scale up implications. In G. Stephanopoulos (Ed.), *Biotechnology V 3: Bioprocessing* (pp. 186–217). VCH.

Reuveny, S., & Lazar, A. (1989). Equipment and procedures for production of monoclonal antibodies in culture. In A. R. Liss (Ed.), *Monoclonal antibodies and applications* (Vol. *11*, pp. 45–80).

Reyhanoglu, Y., & Gokturk, E. (2019). Synthesis of polyglycolic acid copolymers from cationic copolymerization of C1 feedstocks and long chain epoxides. *Journal of Saudi Chemical Society*, *23*(7), 879–886. https://doi.org/10.1016/j.jscs.2019.01.008.

Richmond, A., & Cheng-Wu, Z. (2001). Optimization of a flat plate glass reactor for mass production of *Nannochloropsis* sp. outdoors. *Journal of Biotechnology*, *85*(3), 259–269. https://doi.org/10.1016/S0168-1656(00)00353-9.

Rodríguez-Calvo, A., Silva-Castro, G. A., Olicón-Hernández, D. R., González-López, J., & Calvo, C. (2020). Biodegradation and absorption technology for hydrocarbon-polluted water treatment. *Applied Sciences, 10*(3), 841. https://doi.org/10.3390/app10030841.

Rosser, J., & Thomas, D. J. (2018). Bioreactor processes for maturation of 3D bioprinted tissue. In D. J. Thomas, Z. M. Jessop, & I. S. Whitaker (Eds.), *3D bioprinting for reconstructive surgery techniques and applications* (pp. 191–215). Elsevier. https://doi.org/10.1016/B978-0-08-101103-4.00010-7.

Ruiz-Molina, N., Parsons, J., Schroeder, S., Posten, C., Reski, R., & Decker, E. L. (2022). Process engineering of biopharmaceutical production in moss bioreactors via model-based description and evaluation of phytohormone impact. *Frontiers in Bioengineering and Biotechnology, 10*, 837965. https://doi.org/10.3389/fbioe.2022.837965.

Saito, A., Aung, T., Sekiguchi, K., & Sato, Y. (2006). Present status and perspective of the development of a bioartificial kidney for chronic renal failure patients. *Therapeutic Apheresis and Dialysis, 10*, 342–347. https://doi.org/10.1111/j.1744-9987.2006.00387.x.

Sánchez-Mirón, A., García, M. C., Camacho, F. G., Grima, E. M., & Chisti. (2002). Growth and biochemical characterization of microalgal biomass produced in bubble column and airlift photobioreactors: Studies in fed-batch culture. *Enzyme and Microbial Technology, 31*(7), 1015–1023. https://doi.org/10.1016/S0141-0229(02)00229-6.

Santos, A., Ma, W., & Judd, S. J. (2011). Membrane bioreactors: Two decades of research and implementation. *Desalination, 273*, 148–154. https://doi.org/10.1016/j.desal.2010.07.063.

Sauer, I. M., Neuhaus, P., & Gerlach, J. C. (2002). Concept for modular extracorporeal liver support for the treatment of acute hepatic failure. *Metabolic Brain Disease, 17*, 477–484. https://doi.org/10.1046/j.1525-1594.2002.06931_1.x.

Sharma, K. R. (2012). Design and operational key factors of bioreactor. *Biotech Articles, 4*(1), 55–60.

Singh, J., Kaushik, N., & Biswas, S. (2014). Bioreactors—Technology & design analysis. *The Scitech Journal, 1*(6), 28–36.

Smith, J. J., Lilly, M. D., & Fox, R. I. (1990). The effect of agitation on the morphology and penicillin production of *Penicillium chrysogenum*. *Biotechnology and Bioengineering, 35*, 1011. https://doi.org/10.1002/bit.260351009.

Snouber, L. C., Bunescu, A., Naudot, M., Legallais, C., Brochot, C., Dumas, M. E., Elena-Herrmann, B., & Leclerc, E. (2013). Metabolomics-on-a-chip of hepatotoxicity induced by anticancer drug flutamide and its active metabolite hydroxyflutamide using HepG2/C3a microfluidic biochips. *Toxicological Sciences, 132*, 8–20. https://doi.org/10.1093/toxsci/kfs230.

Sola, C., & Godia, F. (1995). Scale up. In J. A. Asenjo & J. C. Merchuk (Eds.), *Bioreactor system design* (pp. 511–552). Marcel Dekker.

Sonwani, R. K., Jaiswal, R. P., Rai, B. N., & Singh, R. S. (2022). Moving bed biofilm reactor-(MBBR-) based advanced wastewater treatment technology for the removal of emerging contaminants. In M. Shah, S. Rodriguez-Couto, & J. Biswas (Eds.), *Development in wastewater treatment research and processes* (pp. 349–370). Elsevier. https://doi.org/10.1016/B978-0-323-85583-9.00020-X.

Spier, M. R., Vandenberghe, L. P. D. S., Medeiros, A. B. P., & Soccol, C. R. (2011). Application of different types of bioreactors in bioprocesses. In P. G. Antolli & Z. Liu (Eds.), *Bioreactors: design, properties and applications* (pp. 55–90). Nova Science Publishers.

Srivastava, A. K., Singh, R. K., & Singh, D. (2021). Microbe-based bioreactor system for bioremediation of organic contaminants: Present and future perspective. In A. Kumar, V. K. Singh, & V. K. Mishra (Eds.), *Microbe mediated remediation of environmental contaminants* (pp. 241–253). Elsevier. https://doi.org/10.1016/B978-0-12-821199-1.00020-1.

Stanness, K. A., Westrum, L. E., Fornaciari, E., Mascagni, P., Nelson, J. A., Stenglein, S. G., Myers, T., & Janigro, D. (1997). Morphological and functional characterization of an in vitro blood–brain barrier model. *Brain Research*, *771*, 329–342. https://doi.org/10.1016/S0006-8993(97)00829-9.

Staub, J. M., Garcia, B., Graves, J., Hajdukiewicz, P. T. J., Hunter, P., Nehra, N., Paradakar, V., Schlittler, M., Carroll, J. A., Spatola, L., Ward, D., Ye, G., & Russell, D. A. (2000). High-yield production of a human therapeutic protein in tobacco chloroplasts. *Nature Biotechnology*, *18*, 333–338. https://doi.org/10.1038/73796.

Sun, M., Qian, K., Su, N., Chang, H., Liu, J., & Chen, G. (2003). Foot-and mouth disease virus VP1 protein fused with cholera toxin B subunit expressed in *Chlamydomonas reinhardtii* chloroplast. *Biotechnology Letters*, *25*, 1087–1092. https://doi.org/10.1023/A:1024140114505.

Taghizadeh-Alisaraei, A., Motevali, A., & Ghobadian, B. (2019). Ethanol production from date wastes: Adapted technologies, challenges, and global potential. *Renewable Energy*, *143*, 1094–1110. https://doi.org/10.1016/j.renene.2019.05.048.

Takagi, M., Ohara, K.-I., & Yoshida, T. (1995). Effect of hydrostatic pressure on hybridoma cell metabolism. *Journal of Fermentation and Bioengineering*, *80*(6), 619–621. https://doi.org/10.1016/0922-338X(96)87744-0.

Tamburic, B., Zemichael, F. W., Crudge, P., Maitland, G. C., & Hellgardt, K. (2011). Design of a novel flatplate photobioreactor system for green algal hydrogen production. *International Journal of Hydrogen Energy*, *36*, 6578–6591. https://doi.org/10.1016/j.ijhydene.2011.02.091.

Tec-Caamal, E. N., Rodríguez-Vázquez, R., Weijma, J., & Aguilar-López, R. (2021). Simulation platform for in-situ Fe (II) oxidation and bioscorodite crystallization in a one-step process for as (V) immobilization from acid wastewater. *Minerals Engineering*, *172*, 107170. https://doi.org/10.1016/j.mineng.2021.107170.

Tiwari, H., Sonwani, R. K., & Singh, R. S. (2023). Bioremediation of dyes: A brief review of bioreactor performance. *Environmental Technology Reviews*, *12*, 83–128. https://doi.org/10.1080/21622515.2023.2184276.

Townsend, T. G. (2018). *Landfill bioreactor design and operation* (1st ed.). Routledge. https://doi.org/10.1201/9780203749555.

Tramper, J. (1995). Oxygen gradients in animal cell bioreactors. In E. C. Beuvery, J. B. Griffiths, & W. P. Zeijlemaker (Eds.), *Animal cell technology: Developments towards the 21st century* (pp. 883–891). Kluwer Academic Publishers.

Tramper, J., de Gooijer, K. D., & Vlak, J. M. (1993). Scale up considerations and bioreactor development for animal cell cultivation. In J. Tramper, K. D. de Gooijer, & J. M. Vlak (Eds.), *Insect cell culture engineering: Bioprocess technology* (Vol. *17*, pp. 139–177). Marcel Dekker.

Tramper, J., Smit, D., Straatman, J., & Vlak, J. M. (1987). Bubble column design for growth of fragile insect cells. *Bioprocess Engineering*, *2*, 37–41. https://doi.org/10.1007/BF00372858.

Tredici, M. R. (2004). Mass production of microalgae: Photobioreactors. In A. Richmond (Ed.), *Handbook of microalgal culture: Biotechnology and applied phycology* (pp. 178–214). Blackwell Publishing.

Tregoning, J. S., Nixon, P., Kuroda, H., Svab, Z., Clare, S., Bowe, F., Fairweather, N., Ytterberg, J., van Wijk, K. J., Dougan, G., & Maliga, P. (2003). Expression of tetanus toxin Fragment C in tobacco chloroplasts. *Nucleic Acids Research*, *31*, 1174–1179. https://doi.org/10.1093/nar/gkg221.

Ugwu, C. U., Aoyagi, H., & Uchiyama, H. (2008). Photobioreactors for mass cultivation of algae. *Bioresource Technology*, *99*, 4021–4028. https://doi.org/10.1016/j.biortech.2007.01.046.

Ulvestad, M., Björquist, P., Molden, E., Åsberg, A., & Andersson, T. B. (2011). OATP1B1/1B3 activity in plated primary human hepatocytes over time in culture. *Biochemical Pharmacology*, *82*, 1219–1226. https://doi.org/10.1016/j.bcp.2011.07.076.

Van't Reit, K., & Tramper, J. (1991). *Basic bioreactor design*. Marcel Dekker, Inc. USA.

Vendruscolo, F., da Rocha, F. G. L., & Antoniosi, F. N. R. (2017). Biosorption of hexavalent chromium by microorganisms. *International Biodeterioration & Biodegradation*, *119*, 87–95. https://doi.org/10.1016/j.ibiod.2016.10.008.

Verdú-Navarro, F., Moreno-Cid, J. A., Weiss, J., & Egea-Cortines, M. (2023). The advent of plant cells in bioreactors. *Frontiers in Plant Science*, *14*, 1310405. https://doi.org/10.3389/fpls.2023.1310405.

Verhulst, A., Sayer, R., De Broe, M. E., D'Haese, P. C., & Brown, C. D. A. (2008). Human proximal tubular epithelium actively secretes but does not retain rosuvastatin. *Molecular Pharmacology*, *74*(4), 1084–1091. https://doi.org/10.1124/mol.108.047647.

Vinken, M., Vanhaecke, T., & Rogiers, V. (2012). Primary hepatocyte cultures as in vitro tools for toxicity testing: Quo vadis? *Toxicology in vitro*, *26*, 541–544. https://doi.org/10.1016/j.tiv.2012.01.002.

Votruba, J., & Sobotka, M. (1992). Physiological similarity and bioreactor scale up. *Folia Microbiologica*, *37*(5), 331–345. https://doi.org/10.1007/BF02815659.

Wang, E. Q., Li, S. Z., Tao, L., Geng, X., & Li, T. C. (2010). Modeling of rotating drum bioreactor for anaerobic solid-state fermentation. *Applied Energy*, *87*, 2839–2845. https://doi.org/10.1016/j.apenergy.2009.05.032.

Wang, G., Zhang, W., Freedman, D., Eppstein, L., & Kadouri, A. (1992). Animal cell technology: Developments: Process and Products. In R. E. Spier, G. Griffiths, & D. MacDonald (Eds.), *Continuous production of monoclonal antibodies in CelliGen packed bed reactor using Fibracel carrier* (pp. 460–464). Butterworth-Heinemann.

Wang, G., Zhang, W., Jacklin, C., Freedman, D., Eppstein, L., & Kadouri, A. (1992). Modified CelliGen-packed bed bioreactors for hybridoma cell cultures. *Cytotechnology*, *9*, 41–49. https://doi.org/10.1007/BF02521730.

Wayte, J., Boraston, R., Bland, H., Varley, J., & Brown, M. (1997). pH: Effects on growth and productivity of cell lines producing monoclonal antibodies: Control in large scale fermenters. *The Genetic Engineer and Biotechnologist*, *17*(2–3), 125–132.

Wei, M., Gao, Y., Li, X., & Serpe, M. J. (2017). Stimuli-responsive polymers and their applications. *Polymer Chemistry*, *8*(1), 127–143. https://doi.org/10.1039/C6PY01585A.

Wendt, D., Riboldi, S. A., Cioffi, M., & Martin, I. (2012). Bioreactors in tissue engineering: Scientific challenges and clinical perspectives. *Bioreactor Systems for Tissue Engineering*, *112*, 1–27. https://doi.org/10.1007/978-3-540-69357-4.

Werner, R. G. (1994). Potential and efficiency in the biotechnical process. *Pharmaceutical Technology Europe*, May, 20–28.

Wilkening, S., Stahl, F., & Bader, A. (2003). Comparison of primary human hepatocytes and hepatoma cell line Hepg2 with regard to their biotransformation properties. *Drug Metabolism and Disposition*, *31*, 1035–1042. https://doi.org/10.1124/dmd.31.8.1035.

Williams, J. A. (2002). *Keys to Bioreactor Selections*. Environmental & Production Solutions, LLC, cep magazine.

Xiao, K., Liang, S., Wang, X., Chen, C., & Huang, X. (2019). Current state and challenges of full-scale membrane bioreactor applications: A critical review. *Bioresource Technology*, *271*, 473–481. https://doi.org/10.1016/j.biortech.2018.09.061.

Xu, J., Ge, X., & Dolan, M. C. (2011). Towards high-yield production of pharmaceutical proteins with plant cell suspension cultures. *Biotechnology Advances*, *29*, 278–299. https://doi.org/10.1016/j.biotechadv.2011.01.002.

Yoshida, S., Hiraga, K., Takehana, T., Taniguchi, I., Yamaji, H., Maeda, Y., Toyohara, K., Miyamoto, K., Kimura, Y., & Oda, K. (2016). A bacterium that degrades and assimilates poly (ethylene terephthalate). *Science*, *351*(6278), 1196–1199. https://doi.org/10.1126/science.aad635.

Zaslavskaia, L. A., Lippmeier, J. C., Shih, C., Ehrhardt, D., Grossman, A. R., & Apt, K. E. (2001). Trophic conversion of an obligate photoautotrophic organism through metabolic engineering. *Science*, *292*, 2073–2075. https://doi.org/10.1126/science.160.

Zeilinger, K., Schreiter, T., Darnell, M., Söderdahl, T., Lübberstedt, M., Dillner, B., Knobeloch, D., Nüssler, A. K., Gerlach, J. C., & Andersson, T. B. (2011). Scaling down of a clinical three-dimensional perfusion multicompartment hollow fiber liver bioreactor developed for extracorporeal liver support to an analytical scale device useful for hepatic pharmacological in vitro studies. *Tissue Engineering*, *17*, 549–556. https://doi.org/10.1089/ten.tec.2010.0580.

Zhang, B., Matchinski, E. J., Chen, B., Ye, X., Jing, L., & Lee, K. (2019). Marine oil spills—Oil pollution, sources and effects. In C. Sheppard (Ed.), *World seas: An environmental evaluation* (pp. 391–406). Elsevier. https://doi.org/10.1016/B978-0-12-805052-1.00024-3.

Zhang, H., Li, Y., Liu, Y., Zhu, Q., Qi, X., Zhu, G., Wang, J., Jin, P., & Zeng, X. (2021). The effect of basal dislocation on {1121} twin boundary evolution in a Mg-Gd-Y-Zr alloy. *Journal of Materials Science and Technology*, *81*, 212–218. https://doi.org/10.1016/j.jmst.2020.11.062.

Zhang, H., Tasnim, F., Ying, J. Y., & Zink, D. (2009). The impact of extracellular matrix coatings on the performance of human renal cells applied in bioartificial kidneys. *Biomaterials*, *30*, 2899–2911. https://doi.org/10.1016/j.biomaterials.2009.01.046.

Zhang, H. Y., & Jackson, T. A. (2008). Autochthonous bacterial flora indicated by PCR-DGGE of 16S rRNA gene fragments from the alimentary tract of *Costelytra zealandica* (Coleoptera: Scarabaeidae). *Journal of Applied Microbiology*, *105*, 1277–1285. https://doi.org/10.1111/j.1365-2672.2008.03867.x.

Zhang, S., Handa-Corrigan, A., & Spier, R. E. (1992). Oxygen transfer properties of bubbles in animal cell culture media. *Biotechnology and Bioengineering*, *40*, 252–259. https://doi.org/10.1002/bit.260400209.

Zhang, Z. B., & Thomas, C. R. (1993). Micromanipulation—A new approach to studying animal cell damage in bioreactors. *The Genetic Engineer and Biotechnologist*, *13*(1), 19–29.

Zhong, J.-J. (2011). Bioreactor engineering. In M. Moo-Young (Ed.), *Comprehensive biotechnology* (pp. 257–269). Elsevier.

Zhu, X. G., Long, S. P., & Ort, D. R. (2008). What is the maximum efficiency with which photosynthesis can convert solar energy into biomass? *Current Opinion in Biotechnology*, *19*(2), 153–159. https://doi.org/10.1016/j.copbio.2008.02.004.

Zverlov, V. V., Oll, W. H., & Schwarz, W. H. (2003). Enzymes for digestion of cellulose and other polysaccharides in the gut of longhorn beetle larvae, *Rhagium inquisitor* L. (Col., Cerambycidae). *International Biodeterioration & Biodegradation*, *51*, 175–179. https://doi.org/10.1016/S0964-8305(02)00139-7.

10

Economic Analysis of Consolidated Bioprocess Technology in Green Fuel Production

Susheel Kumar Singh, Pradeep Kumar Mishra, and Ravi Prakash Jaiswal

Department of Chemical Engineering & Technology, IIT (BHU), Varanasi, Uttar Pradesh, India

10.1 Introduction

The world's population is increasing rapidly, and to sustain this large population, we need sustainable energy solutions. The global energy demand is projected to rise by almost 28% by 2040 compared to current levels (Osman et al., 2021). Limited fossil fuel reserves and the increasing necessity to combat climate change have encouraged efforts to develop cleaner, alternative renewable energy sources (Wang & Azam, 2024).

Biofuels have become a crucial component in shifting toward a low-carbon energy future. Second-generation, or cellulosic, biofuels, sourced from lignocellulosic biomass, including crop residues, forestry by-products, and dedicated energy crops, show tremendous potential. According to recent statistics, the annual lignocellulose biomass production is approximately 181.5 billion tons, of which only 8.2 billion tons of biomass are utilized in various sectors (Singh, Singhania, et al., 2022). These nonedible and readily available feedstocks leave a smaller footprint than first-generation biofuels, which are derived only from edible food crops. The utilization of lignocellulosic biomass also promotes a circular economy. However, only 2–5% of the total biomass is utilized in the biofuel production industry (Mukherjee et al., 2023; Shen & Sun, 2021).

Conversion of lignocellulosic biomass into biofuels presents an exciting opportunity, yet it is hindered by significant economic challenges due to its complex structure, which typically requires multistep processing. The cell wall of a plant mainly consists of lignin, cellulose, and hemicellulose, where lignin plays a vital role as a binder that unites cellulose and hemicellulose, offering necessary mechanical strength.

One of the primary cost drivers in conventional cellulosic biofuel production typically unfolds in multiple stages: biomass pretreatment, enzymatic hydrolysis, and microbial fermentation. Each step in this three-stage biofuel production needs a different reactor and a considerable amount of energy. Collectively, these factors increase the capital and operating expenditure, which hinders its economic availability in the market on a commercial scale (Li et al., 2024).

Consolidated bioprocessing (CBP) has emerged as a promising innovation in response to these challenges (Guo et al., 2022). CBP aims to simplify and streamline by integrating all three stages of conventional biofuel production, namely cellulase enzyme production, cellulose hydrolysis, and fermentation, into a single biological and operational step. This may be achieved using a single genetically engineered microorganism or a synergistic microbial consortium that can perform all necessary functions within one reactor system (Hossain et al., 2024). CBP offers significantly lower production costs by limiting the pretreatment processes and the addition of hydrolyzing enzymes externally. It also helps in improving energy efficiency (Tsai et al., 2022).

This chapter will assess the economic feasibility of CBP, both in terms of current market feasibility and sustainable potential to produce it on a large scale. The structure and fundamental principles of CBP technology are explored initially, followed by an investigation of the primary expenditure components. Additionally, we will also compare CBP with traditional biomass conversion methods through techno-economic assessments, outline key economic challenges, and evaluate CBP's long-term prospects within the evolving global energy context. Our aim is to understand and evaluate the role of CBP in biofuel production.

Considering the potential advantages, several challenges must be overcome before CBP can be applied on a large scale. These include the challenge of process scale-up, the requirement for stable and economical feedstock supply chains, and the development of strong microbial strains with both high cellulolytic activity and effective product manufacturing capabilities (Singhania et al., 2022). Apart from that, policy frameworks, technological maturity, and market conditions also affect CBP's economic performance.

10.2 Overview of CBP

CBP represents a next-generation approach in the biochemical conversion of lignocellulosic biomass into biofuels. Unlike conventional multistep processes that separate enzyme production, hydrolysis, and fermentation into distinct operations, CBP integrates all these functions into a single bioreactor (Hasunuma & Kondo, 2012). The goal of CBP is to streamline the production of biofuels, particularly ethanol and butanol, by employing a single microorganism or a consortium of microbes capable of producing cellulolytic enzymes, hydrolyzing biomass, and fermenting the resulting sugars into biofuels (Lynd et al., 2005).

Integrating several processes into one has substantially reduced production and increased energy efficiency. This approach is needed to achieve economic viability for second-generation biofuel production. It will also reduce the complexity of conventional biofuel production methods (Figure 10.1).

10.2.1 Conventional Bioconversion vs. CBP

Bioethanol production from lignocellulosic biomass typically involves multiple steps (Conde-Mejia et al., 2012). First, the biomass undergoes pretreatment to break down the rigid structure formed by cellulose, hemicellulose, and lignin. This step is followed by enzymatic hydrolysis, wherein cellulases and hemicellulases convert the polysaccharides into fermentable sugars such as glucose and xylose. The last step is fermentation, where microorganisms convert these sugars into biofuels, such as ethanol or butanol,

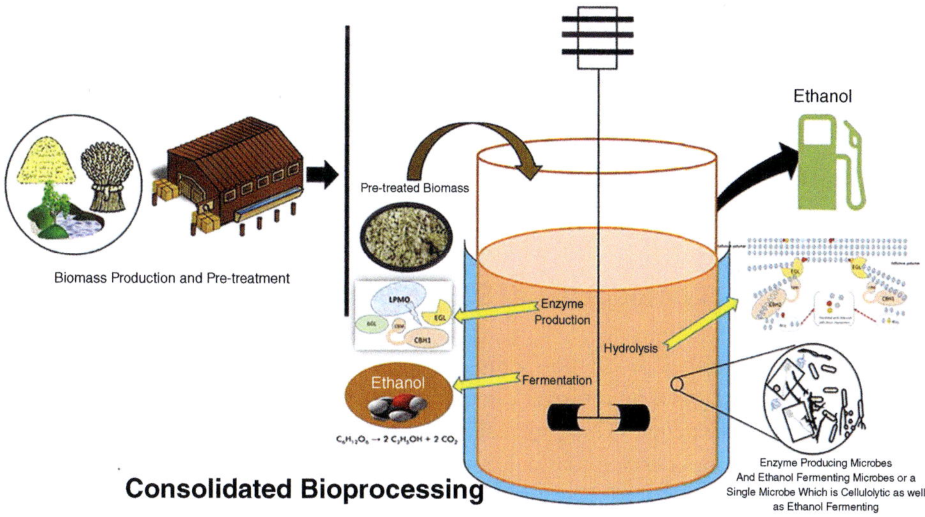

Figure 10.1 Consolidated bioprocessing for bioethanol production employing consortium (*Source:* Singhania et al. (2022) / with permission of Elsevier).

hydrogen, etc. These processes can be carried out using Separate Hydrolysis and Fermentation (SHF) or Simultaneous Saccharification and Fermentation (SSF). In SHF, enzymatic hydrolysis and fermentation occur in separate reactors, which provides optimal conditions for each stage. However, this approach requires time and increases equipment and enzyme costs (Phwan et al., 2018). SSF combines hydrolysis and fermentation in a single vessel, reducing process time and contamination risks (Phwan et al., 2018). However, it still depends on externally produced enzymes and challenges maintaining favorable enzymatic activity and microbial growth conditions (Ask et al., 2012).

CBP bypasses the need for exogenous enzyme addition by utilizing microorganisms that produce the necessary hydrolytic enzymes and perform fermentation. This single-step process eliminates one of the costliest and most technically demanding aspects of cellulosic biofuel production, which is external enzyme production, and thus has the potential to significantly lower overall production costs (Lynd et al., 2008).

10.2.2 Engineering Strategies and Synthetic Biology in CBP

Genetic biology and metabolic engineering advances are central to developing viable CBP systems (Ghim et al., 2010). These fields enable researchers to design and construct microbial strains with tailored functionalities to produce desired products very efficiently. Biotechnical engineering efforts focus on heterologous enzyme expression that can introduce genes encoded for cellulases (e.g., endoglucanases, cellobiohydrolases, and β-glucosidases) into fermentative microorganisms, rewiring metabolic pathways to maximize flux toward the desired product (e.g., ethanol or butanol) while minimizing by-product formation and stress tolerance engineering for enhancing tolerance to biofuel toxicity, pH variations, and inhibitors such as furfural and acetic acid, which are generated during biomass pretreatment.

For instance, *Saccharomyces cerevisiae* has been engineered to express a suite of cellulase genes from *Trichoderma reesei* and *Aspergillus niger*, enabling limited CBP activity (Yamada et al., 2013). Similarly, consortia-based approaches involve co-culturing multiple engineered strains, each performing a specialized function. One strain may hydrolyze cellulose in this setup, while another ferments sugars to bioethanol. Although such promising systems face challenges in maintaining population dynamics, along with preventing competition among the strains (Zuroff & Curtis, 2012).

10.3 Economic Components of Biofuel Production

The economics of biofuel production from lignocellulosic biomass depend upon raw material availability, capital investment, operational cost, process efficiency, market conditions, and regulatory policies (Kargbo et al., 2021). Each component must be optimized to ensure competitiveness with fossil fuels and first-generation biofuels. When analyzing the economic framework of CBP, these components provide the basis for comparison with conventional multi-step processes.

10.3.1 Feedstock Costs

Feedstock expenditures are widely recognized as the most significant contributor to the overall cost of biofuel production. It can be as high as 40–60% of the total production cost (Balan, 2014; Mujtaba et al., 2023). If we look among the various options for biofuel production, lignocellulose biomass, such as wheat straw, rice husk, and dedicated energy crops, like switchgrass, are widely available and low-cost options (Kargbo et al., 2021).

The economic advantage given by lignocellulosic biomass is somewhat reduced by its recalcitrant properties and complexities in its structure. These complexities lead to enzymatic resistance and difficulties in its degradation. To overcome this hindrance, pretreatment is needed, which again adds up in capital investment and increases the operational complexity (Figure 10.2).

Apart from the above challenges, such as seasonal availability and inconsistency in the supply chain of feedstock, difficulties associated with biomass harvesting, transportation, drying, and storage further exacerbate the economic burden.

CBP offers a promising alternative by potentially circumventing or minimizing many of these challenges (Singhania et al., 2021). The use of engineered or naturally capable microbial strains that can hydrolyze cellulose and hemicellulose and ferment the resulting sugars in a single step significantly reduces the need for extensive pretreatment. Mild or no pretreatment not only lowers chemical and thermal input requirements but also reduces the generation of inhibitory by-products, thereby improving overall process efficiency. Additionally, CBP systems can streamline feedstock handling by eliminating multiple intermediate steps, such as size reduction, drying, and enzyme addition, which are typically necessary in traditional multi-stage processing methods (John et al., 2025).

By simplifying the upstream and midstream processing chain, CBP not only reduces capital expenditure (CAPEX) and operating expenditure (OPEX) but also improves the economic viability of utilizing lower-cost, abundant lignocellulosic materials. This positions CBP as a cost-effective and scalable approach for next-generation biofuel production under resource-constrained and sustainability-driven scenarios (Wu et al., 2023).

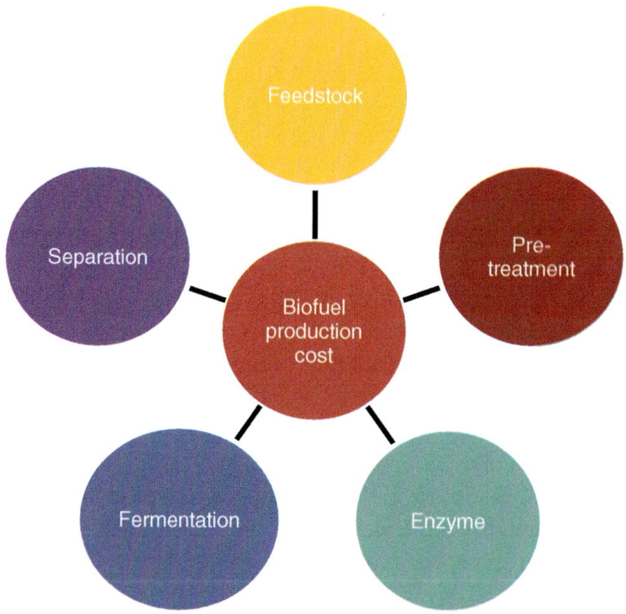

Figure 10.2 Components of biofuel production cost.

10.3.2 Capital Expenditures (CAPEX)

Capital expenditure in biofuel production includes the initial costs required for equipment, facility construction, and infrastructure. Traditional methods such as SHF and SSF involve multiple unit operations—pretreatment reactors, enzyme production units, hydrolysis tanks, and fermentation vessels. Each of these components contributes significantly to the total CAPEX, especially in large-scale facilities where scale-up must ensure adequate control and process separation.

In contrast, CBP reduces the number of required process units by integrating cellulase production, saccharification, and fermentation into a single reactor (Dempfle et al., 2021). This process integration results in a simplified design and lower installation costs. Fewer reactors mean reduced expenditure on piping, instrumentation, auxiliary systems, and construction labor. For instance, a techno-economic analysis estimated that CBP-based ethanol plants could reduce overall CAPEX by 15–25% compared to SHF, depending on the scale and feedstock used (Singhania et al., 2022).

Moreover, CBP is well-suited for modular and decentralized deployment, which avoids the high fixed capital costs often associated with large biorefineries. This flexibility allows small and medium-sized enterprises in rural areas to adopt the technology at a feasible investment scale, creating more distributed bioeconomy hubs (Sánchez-García et al., 2024).

Despite the potential savings, some CAPEX challenges persist for CBP, particularly the need for specially engineered bioreactors that can maintain optimal conditions for multifunctional microbial consortia or thermophilic organisms (Li et al., 2024). These specialized reactors may partially offset initial cost savings. However, the net capital advantage of CBP remains favorable in most economic models, especially as system integration technologies and reactor designs continue to evolve (Jaiswal et al., 2022).

10.3.3 Operating Expenditures (OPEX)

Operating expenditure is one of the main components in biofuel production, as it includes the cost of raw materials, labor cost, energy consumption, and maintenance of the production unit (Mohammadshirazi et al., 2014). In traditional biofuel production, a substantial portion of OPEX arises from the purchase and application of cellulolytic enzymes often accounting for 20–30% of total operating costs (Kwiatkowski et al., 2006). This is especially burdensome at industrial scales, where enzyme demand is high and must be continuously met.

CBP eliminates the need for exogenous enzyme addition by utilizing microorganisms that both produce cellulolytic enzymes and ferment sugars into ethanol or other target biofuels. The *in situ* production of enzymes not only reduces direct enzyme costs but also cuts down energy consumption associated with enzyme production, purification, storage, and dosing (Singh, Langyan, et al., 2022).

Moreover, CBP streamlines the logistics of production by minimizing material handling, decreasing the number of process steps, and improving the overall carbon and energy efficiency of the system. Labor and maintenance requirements are also reduced, as a single bioreactor system demands less oversight and upkeep compared to traditional multi-reactor setups.

However, CBP systems do have unique operational challenges. These include maintaining microbial viability and productivity over extended runs, controlling contamination risks in non-sterile lignocellulosic feedstocks, and managing co-product inhibition (Milessi et al., 2025). Despite these, advances in strain robustness, process monitoring, and continuous fermentation technologies are helping overcome many of these limitations.

Some studies suggest that CBP can reduce the operating cost by 20–40% in comparison to SHF and SSF (Lynd et al., 2017), especially when using engineered strains that have high conversion efficiency. This type of economic advancement is very significant during the scaling of production. This can also lead investors to apply and adopt CBP technology.

10.3.4 Product Yields and Conversion Efficiency

The functioning of a microorganism or an enzyme used during the fermentation is a critical factor in determining the viability of a process. In conventional systems, the use of external enzymes is the major cost component due to their large requirement for hydrolysis. Commercial enzymes are typically produced by filamentous fungi like *Trichoderma reesei*, and their cost can contribute up to $0.50 per gallon of ethanol produced (Rosas-Vega et al., 2025).

In CBP, high conversion ratios reduce the feedstock requirement and improve the cost effectiveness. CBP integrates both hydrolysis and fermentation, which improves the yield and minimizes sugar loss by optimizing the reaction conditions in a single bioreactor (Afedzi & Parakulsuksatid, 2023).

Efficiency of a system is primarily evaluated in terms of sugar conversion into ethanol or fuel production yield and overall production in comparison to feedstock. Conventional systems have less efficiency in comparison to CBP, due to the efficient coordination between enzymatic hydrolysis and fermentation (Ghinea et al., 2025). For example, enzymatic breakdown can be inhibited by the accumulation of carbohydrate or fermentation products, which can lead to a reduction in biomass breakdown and overall biofuel yield.

CBP addresses these limitations through process synergy. In CBP, hydrolysis and fermentation occur simultaneously within the same microbial system, allowing immediate consumption of released sugars. This continuous uptake helps prevent product inhibition and maintains favorable conditions for both saccharification and fermentation (Althuri et al., 2018). The thermodynamic and kinetic alignment of these reactions improves substrate utilization and reduces fermentation times.

Furthermore, CBP supports higher process intensification, where higher solids loading can be employed without drastically increasing viscosity or reducing mass transfer, as typically seen in separate hydrolysis. This increases ethanol concentration and reduces downstream separation costs.

That said, achieving high yields in CBP requires microbes that can degrade diverse plant polymers (cellulose, hemicellulose, and lignin) and convert resulting sugars into biofuels with minimal by-products (Le et al., 2024). Native cellulolytic microbes often lack the ethanol tolerance or productivity of industrial strains like *S. cerevisiae* (Tsegaye et al., 2024). To overcome this, synthetic biology approaches have been used to engineer microbes that combine both robust enzyme systems and high fermentation capabilities.

Reported CBP efficiencies have reached 70–85% of theoretical ethanol yields under optimized conditions, which is competitive with or better than SHF/SSF when considering the total input/output energy ratio (Ghazali & Mustafa, 2025). The integration thus not only enhances conversion efficiency but also improves energy and carbon balances, which are key metrics for sustainable biofuel systems.

However, as of now, many CBP strains under development do not yet match the ethanol productivity of conventional processes, particularly concerning pentose sugar utilization and tolerance to inhibitors (Lynd et al., 2002). The economic benefit of CBP thus depends on achieving commercial-level yields that can match or exceed the benchmarks set by existing technologies.

10.4 Integration with Circular Bioeconomy Models

The circular bioeconomy emphasizes resource efficiency, waste valorization, and closed-loop production systems. CBP can provide a streamlined approach by using a low-waste economy for biofuel and biochemical production (Ambaye et al., 2021). CBP aligns with these principles by offering a lower-waste, integrated pathway to biofuel and biochemical production. By minimizing chemical inputs, process steps, and energy use, CBP reduces the environmental and economic burdens of waste treatment and water use (Ambaye et al., 2021).

Moreover, CBP strains can be engineered not only for ethanol production but also to synthesize high-value co-products such as organic acids, bioplastic precursors (e.g., lactic acid and succinic acid), and enzymes (Patel & Shah, 2021). These co-product streams enhance the revenue potential of CBP-based biorefineries and increase their economic resilience in volatile fuel markets.

CBP and biorefineries, when combined with each other, can produce fuel and chemicals at the same time, which can lead the path toward a sustainable revenue system.

10.5 Scalability and Deployment Potential

CBP's economic viability is closely related to developments in synthetic biology and microbial engineering. A number of technical obstacles need to be removed before CBP can reach its full economic potential and compete with conventional biofuel production routes. These include improving resistance to inhibitory compounds produced during pretreatment, facilitating the efficient co-fermentation of both hexose and pentose sugars derived from lignocellulosic biomass, raising ethanol titers and tolerance in microbial hosts, and boosting the effectiveness of downstream product recovery and separation (Olguin-Maciel et al., 2020).

Innovations in genetic and metabolic engineering are very much beneficial, as they are accelerating the development of microbial strains that are more suitable for CBP, like CRISPER-Cas9-based genome and adaptive laboratory evolution, which have provided precise modification in the genome of microbes. This technique enables the optimization of metabolic pathways for improved yield of enzyme production along with sugar utilization for biofuel production (Jakočiūnas et al., 2016). Thanks to advanced engineering strategies, scientists are now able to design microbes that perform at an industrial level, delivering high product yields, fermenting quickly, and standing up to tough processing conditions.

In addition to strain development, parallel improvements in process engineering are contributing to enhanced economic viability. Innovations in bioreactor design, such as high-solids fermenters and continuous stirred tank reactors (CSTRs) (Deb et al., 2024), as well as consolidated high-gravity fermentation systems, allow for better mass transfer, temperature control, and scalability. Furthermore, the integration of real-time process monitoring technologies and automated control systems reduces labor costs and minimizes operational variability. Advanced analytics and artificial intelligence-driven process optimization enable dynamic adjustments to fermentation conditions, improving overall process consistency and yield.

Together, these developments in microbial strain engineering and process technology are gradually reducing both capital and operational expenditures associated with CBP. As these technologies mature and become widely adopted, CBP is expected to surpass traditional methods in economic performance, making it a more commercially viable and sustainable option for large-scale biofuel production (Figure 10.3).

10.6 Factors Influencing CBP in Scaling Up

10.6.1 Market Viability in the Low-Carbon Economy

As the world is adopting the strategies to decarbonize the energy sector, renewable technologies like CBP are most likely well positioned to benefit significantly from supportive policy of government policies and market-based environment incentives. CBP is significantly more advantageous due to its potential to produce biofuels in a single bioreactor and with lower to zero greenhouse gas (GHG) emissions in comparison to traditional fossil fuels and even other methods of biofuel production. This is largely due to CBP's integrated process design, which minimizes energy consumption, reduces the reliance on

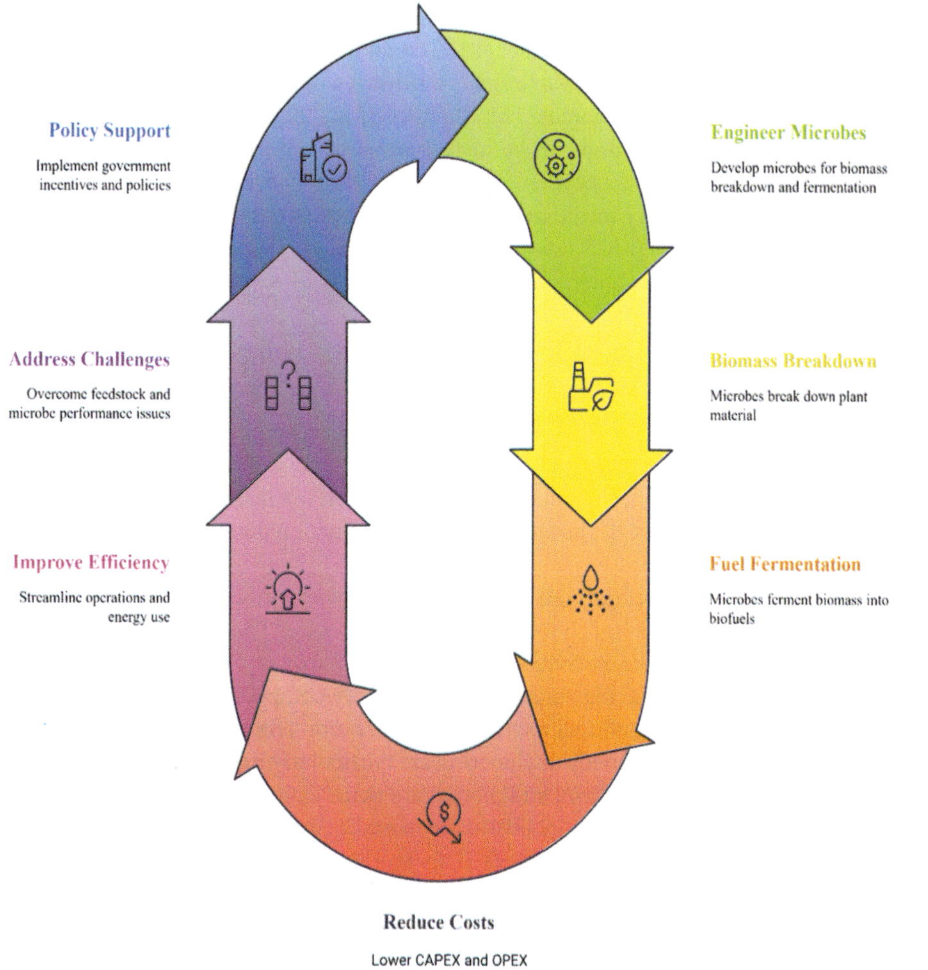

Figure 10.3 The CBP cycle for biofuel production.

externally produced enzymes and fossil-derived chemicals, and enhances overall carbon and energy efficiency (Olguin-Maciel et al., 2020).

These characteristics of CBP-derived biofuel qualify it for preferential treatment under various national and international regulatory mechanisms aimed at promoting low-carbon alternatives. In jurisdictions such as California, the Low Carbon Fuel Standard (LCFS) assigns carbon intensity (CI) scores to fuels, offering higher credit values for those that demonstrate significant GHG reductions (Axsen & Wolinetz, 2023). Fuels produced via CBP, with their inherently lower CI, can thus command higher market value through tradable LCFS credits.

Similarly, under the Renewable Fuel Standard (RFS2) in the United States, biofuels are categorized based on their lifecycle GHG emissions relative to a petroleum baseline (Stock, 2015). CBP-based ethanol, particularly from lignocellulosic feedstocks, can qualify as "cellulosic biofuel," which meets the highest reduction thresholds and receives the most favorable Renewable Identification Number (RIN) credit multipliers (McPhail et al., 2011).

In the European Union, the Renewable Energy Directive (RED II) prioritizes advanced biofuels derived from non-food biomass (Mai-Moulin et al., 2021). CBP aligns well with RED II's sustainability criteria and can be instrumental in helping member states meet renewable energy targets for transport.

Importantly, these regulatory incentives offer financial resilience against fluctuating oil prices. Even when fossil fuel prices are low, carbon credits and compliance subsidies can generate substantial supplementary revenues, enhancing CBP's economic viability and attractiveness to investors and fuel distributors alike.

10.6.2 Innovation and Strain Development Trajectory

The long-term economic sustainability of CBP is closely related to perpetual improvement in microbial engineering and synthetic biology. It combines the various platforms that are essential to enzyme production, hydrolysis, and fermentation within a single biological system. To perform these tasks, CBP heavily relies on the functional ability of its microbial agents.

To fulfill its economic potential and match or exceed the performance of traditional methods, several critical biological challenges must be addressed. These include improving ethanol tolerance, enabling the co-fermentation of both hexose and pentose sugars from lignocellulosic biomass, enhancing resistance to inhibitory compounds generated during pretreatment, and optimizing product recovery and separation efficiency.

Advances in metabolic and genetic engineering are particularly helpful because they are speeding up the creation of microbial strains that are better suited for CBP. For example, adaptive laboratory evolution and CRISPER-Cas9-based genomes have made it possible to precisely modify the genome of microbes. This method makes it possible to optimize metabolic pathways for increased enzyme production yield and sugar utilization for the production of biofuel. As these engineered microorganisms become more resilient and efficient, they are anticipated to outperform conventional fermentative organisms in terms of yield, process stability, and productivity, thereby enhancing the economic viability of CBP-based fuel production.

Along with the microbial modification, a significant modification in bioprocessing engineering is also needed. The development of a next-gen bioreactor system that can have high feedstock handling capabilities and improve the mass transfer, along with real-time monitoring tools and an automated control system, reduces the labor and energy costs. Technologies such as soft sensors, machine learning-based optimization, and digital twins can further streamline operations and minimize downtime.

Collectively, the synergistic progress in microbial strain engineering and process automation is driving CBP toward becoming a commercially viable, cost-effective, and sustainable solution for large-scale biofuel production.

10.7 Future Outlook and Strategic Recommendations

If we look forward, CBP's future success and economic sustainability will depend upon scientific development, consistent support from the government and society, and market integration. Scientists have reported significant development in microbial engineering and design, but further advancement is still necessary to fulfill the goals of commercializing CBP as a core technology in biofuel production. The path ahead requires coordinated action across multiple domains.

Strain development is a critical factor in CBP. It can be performed through the open-source platform and the involvement of researchers from around the world who can share the genetic tools, metabolic models, and performance data. Collaborative efforts can help overcome the limitations of proprietary development and enable broader, faster innovation.

The establishment of a transparent, techno-economic, and efficient model is essential to gain the belief of policymakers and private sector investment. These models must incorporate various factors that influence capital investment, like feedstock variability, efficiency of the process, and real-world variability.

The integration of CBP into existing biorefineries that produce biofuel and value-added chemicals can enhance the revenue generation and improve the economic resilience. This multifunctional approach aligns with the principles of circular bioeconomy and optimizes biomass utilization.

Finally, more investment in skill enhancement programs to operate the CBP models and in the education sector to ensure smooth operation. Training programs in bioprocess engineering, synthetic biology, and industrial automation will form the foundation of a competitive, future-ready bioeconomy.

Given the intensifying global focus on decarbonization, energy transition, and sustainable development, CBP is uniquely positioned to become a cornerstone technology. Its continued evolution will be instrumental in shaping a low-carbon, circular future for bio-based industries.

10.8 Conclusion

CBP emerges as an innovative approach in the field of sustainable biofuel production. Traditional methods require multiple stages, each with separate reactors and the use of costly enzymes that are produced externally. However, CBP simplifies the process by integrating all these steps and uses engineered microbes that can break down lignocellulosic material and ferment it into fuels like ethanol or hydrogen. This integrated design significantly reduces capital (CAPEX) and operating costs (OPEX). One of the biggest expenses in conventional systems is the use of externally produced enzymes to break down biomass. CBP avoids this by enabling microorganisms to produce these enzymes themselves, cutting down on both cost and complexity. It also streamlines operations by reducing the number of reactors and energy-intensive steps, making the process more efficient and reliable.

Still, there are hurdles to overcome. Many naturally occurring microbes don't produce fuel as efficiently as conventional fermenting microbes; combining them with other microbes to enhance their performance and biofuel production is a key focus of ongoing research. Another challenge is ensuring a stable, cost-effective supply of lignocellulosic feedstocks like agricultural waste or energy crops, along with efficient logistics.

Despite these challenges, CBP fits well with global trends for low-carbon technologies and circular bio-economies. It is also a way forward for producing biofuels along with other value-added items like bioplastics or organic acids. Decentralized CBP biorefineries could bring new economic opportunities to rural communities.

Government policies and incentives, such as carbon credits or renewable fuel mandates, will play a crucial role in bringing CBP to market. With continued innovation in synthetic biology and strong policy support, CBP could become a cornerstone of sustainable energy systems, helping the world transition to a cleaner, more circular bio-based economy.

References

Afedzi, A. E. K., & Parakulsuksatid, P. (2023). Recent advances in process modifications of simultaneous saccharification and fermentation (SSF) of lignocellulosic biomass for bioethanol production. *Biocatalysis and Agricultural Biotechnology, 54*, 102961. https://doi.org/10.1016/j.bcab.2023.102961.

Althuri, A., Chintagunta, A. D., Sherpa, K. C., & Banerjee, R. (2018). Simultaneous saccharification and fermentation of lignocellulosic biomass. In S. Kumar & R. K. Sani (Eds.), *Biorefining of biomass to biofuels: Opportunities* (pp. 265–285). Cham: Springer.

Ambaye, T. G., Vaccari, M., Bonilla-Petriciolet, A., Prasad, S., van Hullebusch, E. D., & Rtimi, S. (2021). Emerging technologies for biofuel production: A critical review on recent progress, challenges and perspectives. *Journal of Environmental Management, 290*, 112627. https://doi.org/10.1016/j.jenvman.2021.112627.

Ask, M., Olofsson, K., Di Felice, T., Ruohonen, L., Penttilä, M., Lidén, G., & Olsson, L. (2012). Challenges in enzymatic hydrolysis and fermentation of pretreated Arundo donax revealed by a comparison between SHF and SSF. *Process Biochemistry, 47*(10), 1452–1459. https://doi.org/10.1016/j.procbio.2012.05.016.

Axsen, J., & Wolinetz, M. (2023). What does a low-carbon fuel standard contribute to a policy mix? An interdisciplinary review of evidence and research gaps. *Transport Policy, 133*, 54–63. https://doi.org/10.1016/j.tranpol.2023.01.008.

Balan, V. (2014). Current challenges in commercially producing biofuels from lignocellulosic biomass. *International Scholarly Research Notices, 2014*(1), 463074. https://doi.org/10.1155/2014/463074.

Conde-Mejia, C., Jimenez-Gutierrez, A., & El-Halwagi, M. (2012). A comparison of pretreatment methods for bioethanol production from lignocellulosic materials. *Process Safety and Environment Protection, 90*(3), 189–202. https://doi.org/10.1016/j.psep.2011.08.004.

Deb, N., Rahman, T., Alam, M. Z., Miah, M. S., & Kamal, R. (2024). A single-reactor system for simultaneous pretreatment and fermentation of POME for bioethanol production. *International Journal of Polymeric Science, 2024*(1), 5264918. https://doi.org/10.1155/2024/5264918.

Dempfle, D., Kröcher, O., & Studer, M. H.-P. (2021). Techno-economic assessment of bioethanol production from lignocellulose by consortium-based consolidated bioprocessing at industrial scale. *New Biotechnology, 65*, 53–60. https://doi.org/10.1016/j.nbt.2021.07.005.

Ghazali, M. F. S. M., & Mustafa, M. (2025). Bioethanol as an alternative fuels: A review on production strategies and technique for analysis. *Energy Conversion and Management: X*, 100933. https://doi.org/10.1016/j.ecmx.2025.100933.

Ghim, C.-M., Kim, T., Mitchell, R. J., & Lee, S. K. (2010). Synthetic biology for biofuels: Building designer microbes from the scratch. *Biotechnology and Bioprocess Engineering, 15*, 11–21. https://doi.org/10.1007/s12257-009-3065-5.

Ghinea, C., Ungureanu-Comăniţă, E.-D., Ţâbuleac, R. M., Oprea, P. S., Coşbuc, E. D., & Gavrilescu, M. (2025). Cost-benefit analysis of enzymatic hydrolysis alternatives for food waste management. *Foods, 14*(3), 488. https://doi.org/10.3390/foods14030488.

Guo, Z.-W., Ni, Z.-F., Zong, M.-H., & Lou, W.-Y. (2022). Modular metabolic engineering of *Bacillus licheniformis* for efficient 2, 3-butanediol production by consolidated bioprocessing of Jerusalem Artichoke tubers. *ACS Sustainable Chemistry & Engineering, 10*(29), 9624–9634. https://doi.org/10.1021/acssuschemeng.2c02994.

Hasunuma, T., & Kondo, A. (2012). Consolidated bioprocessing and simultaneous saccharification and fermentation of lignocellulose to ethanol with thermotolerant yeast strains. *Process Biochemistry, 47*(9), 1287–1294. https://doi.org/10.1016/j.procbio.2012.05.004.

Hossain, G. S., Liang, Y., Foo, J. L., & Chang, M. W. J. B. N. (2024). Engineered microbial consortia for next-generation feedstocks. *Biotechnology Notes, 5*, 23–26. https://doi.org/10.1016/j.biotno.2024.01.002.

Jaiswal, K. K., Chowdhury, C. R., Yadav, D., Verma, R., Dutta, S., Jaiswal, K. S., & Karuppasamy, K. S. K. (2022). Renewable and sustainable clean energy development and impact on social, economic, and environmental health. *Energy Nexus, 7*, 100118. https://doi.org/10.1016/j.nexus.2022.100118.

Jakočiūnas, T., Jensen, M. K., & Keasling, J. D. (2016). CRISPR/Cas9 advances engineering of microbial cell factories. *Metabolic Engineering, 34*, 44–59. https://doi.org/10.1016/j.ymben.2015.12.003.

John, S. U., Onu, C. E., Ezechukwu, C. M.-J., Nwokedi, I. C., & Onyenanu, C. N. (2025). Multi-product biorefineries for biofuels and value-added products: Advances and future perspectives. *Academia Green Energy, 2*(1). https://doi.org/10.20935/AcadEnergy7605.

Kargbo, H., Harris, J. S., & Phan, A. N. (2021). "Drop-in" fuel production from biomass: Critical review on techno-economic feasibility and sustainability. *Renewable and Sustainable Energy Reviews, 135*, 110168. https://doi.org/10.1016/j.rser.2020.110168.

Kwiatkowski, J. R., McAloon, A. J., Taylor, F., & Johnston, D. B. (2006). Modeling the process and costs of fuel ethanol production by the corn dry-grind process. *Industrial Crops and Products, 23*(3), 288–296. https://doi.org/10.1016/j.indcrop.2005.08.004.

Le, Y., Zhang, M., Wu, P., Wang, H., & Ni, J. (2024). Biofuel production from lignocellulose via thermophile-based consolidated bioprocessing. *Engineering Microbiology, 4*, 100174. https://doi.org/10.1016/j.engmic.2024.100174.

Li, Z., Waghmare, P. R., Dijkhuizen, L., Meng, X., & Liu, W. (2024). Research advances on the consolidated bioprocessing of lignocellulosic biomass. *Engineering Microbiology, 4*(2), 100139. https://doi.org/10.1016/j.engmic.2024.100139.

Lynd, L. R., Laser, M. S., Bransby, D., Dale, B. E., Davison, B., Hamilton, R., Sheehan, J. (2008). How biotech can transform biofuels. *Nature Biotechnology, 26*(2), 169–172. https://doi.org/10.1038/nbt0208-169.

Lynd, L. R., Liang, X., Biddy, M. J., Allee, A., Cai, H., Foust, T., Michael E. Himmel, M.S. Laser, M. Wang Wyman, C. E. (2017). Cellulosic ethanol: Status and innovation. *Current Opinion in Biotechnology, 45*, 202–211. https://doi.org/10.1016/j.copbio.2017.03.008.

Lynd, L. R., Van Zyl, W. H., McBride, J. E., & Laser, M. (2005). Consolidated bioprocessing of cellulosic biomass: An update. *Current Opinion in Biotechnology, 16*(5), 577–583. https://doi.org/10.1016/j.copbio.2005.08.009.

Lynd, L. R., Weimer, P. J., Van Zyl, W. H., & Pretorius, I. S. (2002). Microbial cellulose utilization: Fundamentals and biotechnology. *Microbiology and Molecular Biology Reviews, 66*(3), 506–577. https://doi.org/10.1128/mmbr.66.3.506-577.2002.

Mai-Moulin, T., Hoefnagels, R., Grundmann, P., & Junginger, M. (2021). Effective sustainability criteria for bioenergy: Towards the implementation of the European Renewable Directive II. *Renewable and Sustainable Energy Reviews, 138*, 110645. https://doi.org/10.1016/j.rser.2020.110645.

McPhail, L. L., Westcott, P., & Lutman, H. (2011). *The renewable identification number system and US biofuel mandates*: SSRN. Economic Research Service, U.S. Department of Agriculture.

Milessi, T. S., Reis, C. E., Ramos, M. D., Carvalho, A. K. F., Aminabhavi, T., & Chandel, A. K. (2025). Native and tailor-made microbial cell factories for economic production of bio-renewables: A paradigm shift in industrial biotechnology. *Biomass Bioenergy, 201*, 108097. https://doi.org/10.1016/j.biombioe.2025.108097.

Mohammadshirazi, A., Akram, A., Rafiee, S., & Kalhor, E. B. (2014). Energy and cost analyses of biodiesel production from waste cooking oil. *Renewable and Sustainable Energy Reviews, 33*, 44–49. https://doi.org/10.1016/j.rser.2014.01.067.

Mujtaba, M., Fraceto, L. F., Fazeli, M., Mukherjee, S., Savassa, S. M., de Medeiros, G. A., Anderson do Espírito Santo Pereira, Sandro Donnini Mancini, Juha Lipponen Vilaplana, F. (2023). Lignocellulosic biomass from agricultural waste to the circular economy: A review with focus on biofuels, biocomposites and bioplastics. *Journal of Cleaner Production, 402*, 136815. https://doi.org/10.1016/j.jclepro.2023.136815.

Mukherjee, P. K., Das, B., Bhardwaj, P. K., Tampha, S., Singh, H. K., Chanu, L. D., Nanaocha Sharma, Devi, S. I. (2023). Socio-economic sustainability with circular economy—An alternative approach. *Science of the Total Environment, 904*, 166630. https://doi.org/10.1016/j.scitotenv.2023.166630.

Olguin-Maciel, E., Singh, A., Chable-Villacis, R., Tapia-Tussell, R., & Ruiz, H. A. (2020). Consolidated bioprocessing, an innovative strategy towards sustainability for biofuels production from crop residues: An overview. *Agronomy, 10*(11), 1834. https://doi.org/10.3390/agronomy10111834.

Osman, A. I., Mehta, N., Elgarahy, A. M., Al-Hinai, A., A. H. Al-Muhtaseb, Rooney, D. W. (2021). Conversion of biomass to biofuels and life cycle assessment: A review. *Environmental Chemistry, 19*, 4075–4118. https://doi.org/10.1007/s10311-021-01273-0.

Patel, A., & Shah, A. R. (2021). Integrated lignocellulosic biorefinery: Gateway for production of second generation ethanol and value added products. *Journal of Bioresources and Bioproducts, 6*(2), 108–128. https://doi.org/10.1016/j.jobab.2021.02.001.

Phwan, C. K., Ong, H. C., Chen, W.-H., Ling, T. C., Ng, E. P., & Show, P. L. (2018). Overview: Comparison of pretreatment technologies and fermentation processes of bioethanol from microalgae. *Energy Conversion and Management, 173*, 81–94. https://doi.org/10.1016/j.enconman.2018.07.054.

Rosas-Vega, F. E., Pozzan, R., Martínez-Burgos, W. J., Letti, L. A. J., de Mattos, P. B. G., Ramos-Neyra, L. C., Gabriel Spinillo Dudeque, Gustavo Amaro Bittencourt, Gabriela dos S. Costa, Porto de Souza Vandenberghe, L. (2025). Enzymes produced by the genus *Aspergillus* integrated into the biofuels industry using sustainable raw materials. *Fermentation, 11*(2), 62. https://doi.org/10.3390/fermentation11020062.

Sánchez-García, E., Martínez-Falcó, J., Marco-Lajara, B., & Manresa-Marhuenda, E. (2024). Revolutionizing the circular economy through new technologies: A new era of sustainable progress. *Environmental Technology and Innovation, 33*, 103509. https://doi.org/10.1016/j.eti.2023.103509.

Shen, X., & Sun, R. (2021). Recent advances in lignocellulose prior-fractionation for biomaterials, biochemicals, and bioenergy. *Carbohydrate Polymers, 261*, 117884. https://doi.org/10.1016/j.carbpol.2021.117884.

Singh, N., Singhania, R. R., Nigam, P. S., Dong, C.-D., Patel, A. K., & Puri, M., (2022). Global status of lignocellulosic biorefinery: Challenges and perspectives. *Bioresource Technology, 344*, 126415. https://doi.org/10.1016/j.biortech.2021.126415.

Singh, R., Langyan, S., Rohtagi, B., Darjee, S., Khandelwal, A., Shrivastava, M., Kothari, R., Mohan, H., Raina, S., Kaur, J., & Singh, A., (2022). Production of biofuels options by contribution of effective and suitable enzymes: Technological developments and challenges. *Materials Science for Energy Technologies, 5*, 294–310. https://doi.org/10.1016/j.mset.2022.05.001.

Singhania, R. R., Patel, A. K., Singh, A., Haldar, D., Soam, S., Chen, C.-W., Dong, C.-D. (2022). Consolidated bioprocessing of lignocellulosic biomass: Technological advances and challenges. *Bioresource Technology, 354*, 127153. https://doi.org/10.1016/j.biortech.2022.127153.

Singhania, R. R., Ruiz, H. A., Awasthi, M. K., Dong, C.-D., Chen, C.-W., & Patel, A. K. (2021). Challenges in cellulase bioprocess for biofuel applications. *Renewable and Sustainable Energy Reviews, 151*, 111622. https://doi.org/10.1016/j.rser.2021.111622.

Stock, J. H. (2015). *The renewable fuel standard: A path forward.* Center on Global Energy Policy, Columbia University.

Tsai, S.-L., Sun, Q., & Chen, W., (2022). Advances in consolidated bioprocessing using synthetic cellulosomes. *Current Opinion in Biotechnology, 78*, 102840. https://doi.org/10.1016/j.copbio.2022.102840.

Tsegaye, K. N., Alemnew, M., & Berhane, N. (2024). Saccharomyces cerevisiae for lignocellulosic ethanol production: A look at key attributes and genome shuffling. *Frontiers in Bioengineering and Biotechnology, 12*, 1466644. https://doi.org/10.3389/fbioe.2024.1466644.

Wang, J., & Azam, W. (2024). Natural resource scarcity, fossil fuel energy consumption, and total greenhouse gas emissions in top emitting countries. *Geoscience Frontiers, 15*(2), 101757. https://doi.org/10.1016/j.gsf.2023.101757.

Wu, B., Lin, R., Bose, A., Huerta, J. D., Kang, X., Deng, C., & Murphy, J. D. (2023). Economic and environmental viability of biofuel production from organic wastes: A pathway towards competitive carbon neutrality. *Energy, 285*, 129322. https://doi.org/10.1016/j.energy.2023.129322.

Yamada, R., Hasunuma, T., & Kondo, A. (2013). Endowing non-cellulolytic microorganisms with cellulolytic activity aiming for consolidated bioprocessing. *Biotechnology Advances, 31*(6), 754–763. https://doi.org/10.1016/j.biotechadv.2013.02.007.

Zuroff, T. R., & Curtis, W. R. (2012). Developing symbiotic consortia for lignocellulosic biofuel production. *Applied Microbiology and Biotechnology, 93*, 1423–1435. https://doi.org/10.1007/s00253-011-3762-9.

11

Frequency Adaptation of Consolidated Bioprocess Technology in Green Fuel Production

Susheel Kumar Singh, Pradeep Kumar Mishra, and Ravi Prakash Jaiswal

Department of Chemical Engineering & Technology, IIT (BHU), Varanasi, Uttar Pradesh, India

11.1 Introduction

The global transition from conventional to sustainable green fuel production systems has intensified in recent years. In particular, second-generation biofuel is derived from lignocellulosic biomass or dedicated fuel crops (Singhania et al., 2022). However, traditional bioprocessing approaches, such as simultaneous saccharification and fermentation (SSF) or separate hydrolysis and fermentation (SHF), necessitate the addition of external enzymes and complex multi-stage processing, which substantially raises costs and complicates the process (Liu et al., 2021).

Consolidated bioprocessing (CBP) is an advanced biotechnological approach in biofuel production that combines all the stages, like enzyme synthesis, lignocellulosic hydrolysis, and fermentation, of conventional biofuel production into a single microbial reactor process and reduces the overall production cost (Davison et al., 2020). Current research suggests the CBP's increasing relevance in the increasing global energy demands (Wall et al., 2023). Many scientists conceptualize CBP within broader biomass conversion strategies, stating its role in second-generation biorefineries that valorize agricultural residues while mitigating greenhouse gas emissions and enhancing rural economies.

While CBP remains at an emergent stage, substantial progress has been made toward its adoption, particularly through research into cellulolytic and fermentative microbial platforms, including both naturally capable organisms like *Clostridium thermocellum* and engineered hosts such as *Saccharomyces cerevisiae* endowed with heterologous cellulase systems (Mbaneme-Smith & Chinn, 2015). In a different study, scientists developed a more efficient strain of *Neurospora crassa* that has elevated cellulase activity, enabling the direct conversion of lignocellulosic biomass into itaconic acid without the need for an external enzyme, which demonstrates the CBP's potential beyond biofuels. The integration significantly improves itaconic acid yield and simplifies the production process (Zhao et al., 2024). CBP can also offer a cost-effective and sustainable approach toward producing a valuable chemical. Similarly, recombinant *S. cerevisiae* strains capable of secreting multiple lignocellulolytic enzymes can be used in fermenting the glucose and xylose. They produce up to 4.46 g.L/h in synthetic media and up to 1.86 g/L/h using industrial sugarcane bagasse feedstock (Perez et al., 2023).

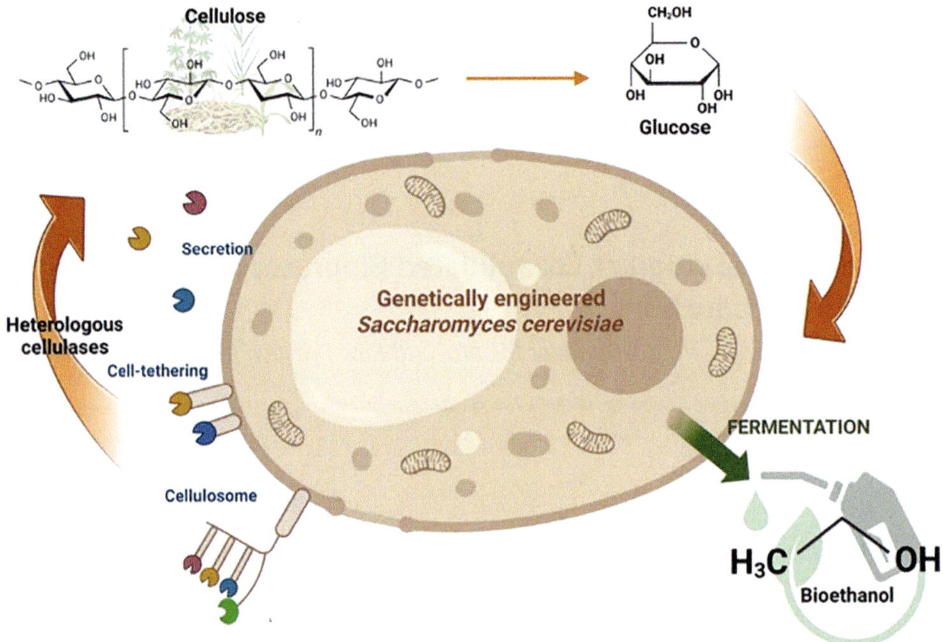

Figure 11.1 Genetically engineered *S. cerevisiae* for consolidated bioprocessing (*Source:* Adapted from Fortuin et al. (2025)).

Despite the encouraging benefits, including process intensification, reduced enzyme cost, and improved yield, CBP remains a pilot- or academic-scale project. In this chapter, we will explore the adaptation of CBP and engineering strategies and how to make CBP more viable in market as a commercial project (Figure 11.1).

11.2 Technological Basis of Consolidated Bioprocessing (CBP)

11.2.1 Core Components of CBP

CBP integrates three sequential biochemical operations enzymes, synthesis, saccharification, and fermentation, within a single bioreactor. This configuration reduces capital and operational expenditures inherent to SHF or SSF processes (Mbaneme-Smith & Chinn, 2015). By reducing the dependency on external addition of cellulase enzymes, CBP enables continuous consumption of liberated sugars, thereby mitigating inhibitory effects associated with sugar accumulation (Cunha et al., 2020). Early techno-economic models, such as those by the National Renewable Energy Laboratory, indicate that CBP has the greatest potential for cost reduction among all biomass conversion improvements (Larnaudie et al., 2019).

11.2.2 Microorganisms in CBP: Native to Engineered

Development of viable CBP strains remains the central challenge. Two principal engineering strategies have been explored:

(i) Transforming native cellulolytic organisms (e.g., *C. thermocellum*) into ethanol-producing hosts.
(ii) Modification of ethanologenic and hydrogen-producing microbes (e.g., *S. cerevisiae*, *Zymomonas mobilis*, and *Escherichia coli*) with cellulolytic capabilities (den Haan et al., 2021; Vasan et al., 2011).

While significant progress has been made, heterologous expression of multiple cellulases often impairs host viability or enzyme folding and secretion efficiency (Muzaffar et al., 2025).

Notably, research on engineered industrial *S. cerevisiae* strains has demonstrated direct fermentation of corn cob-derived hemicellulose into ethanol. These strains display hemicellulolytic enzymes on the cell surface and assimilate xylose, yielding up to 11.1 g/L ethanol without external enzymes, surpassing SSF with commercial hydrolases in efficiency (Cunha et al., 2020; Moysés et al., 2016).

11.2.3 Feedstock Compatibility and Pretreatment Considerations

CBP broadly targets lignocellulosic biomass, including agricultural residues such as wheat straw, rice straw, corn stover, and sugarcane bagasse, which globally accumulate in the hundreds of millions of tons annually (Srivastava et al., 2023). Combining both the processes, enzyme production and hydrolysis, reduces or even eliminates the need for energy-intensive pretreatment, although some optimized biomass processing is still required to enhance digestibility and lignin accessibility (Devi et al., 2022).

11.2.4 Advanced CBP Strategies: Thermophiles, Consortia, and Hybrid Processes

Emerging approaches leverage thermophilic organisms such as *C. thermocellum* and *Myceliophthora thermophila*, which possess innate cellulolytic activity and tolerance to high-temperature operation, reducing the risk of contamination and improving reaction kinetics (da Rosa-Garzon et al., 2022). For instance, *M. thermophila* was engineered to incorporate CO_2 fixation pathways (Calvin–Benson–Bassham and PYC modules), enabling direct conversion of plant biomass and captured CO_2 into malic acid at yields reaching ~0.53 g/g and titers above 70 g/L (Li et al., 2021).

Synthetic microbial consortia also present promising hybrid strategies. Division of metabolic labor enables cellulolysis and product fermentation in separate strains, facilitating optimized performance and mitigating metabolic burden on a single host. Though outside the scope of native-strain systems, such consortia are instrumental where a fully integrated strain is not yet achievable (Jiang et al., 2020).

11.3 Drivers of CBP Adoption

11.3.1 Economic Drivers

- Use of external cellulase enzymes forms a substantial portion of production costs in traditional bioprocessing. CBP eliminates the need for exogenous enzymes by integrating enzyme synthesis into the fermentation stage, thereby substantially reducing operational costs. One techno-economic study reports that CBP could lower ethanol production costs from approximately \$0.19 to \$0.04 per gallon compared to conventional saccharification and fermentation routes (Afedzi et al., 2025; Olguin-Maciel et al., 2020).

- CBP integrates different stages of the conventional process into a single reactor, decreasing infrastructure, utility demand, operational complexity, and staffing needs (Patel et al., 2025). Such simplifications translate into reduced capital expenditure, overall operational cost, and streamlined facilities designed for continuous operation or modular deployment.
- CBP enables simultaneous enzymatic hydrolysis and fermentation, which means that the produced sugars are instantly consumed by microorganisms. This dynamic counters end-product inhibition, maximizing sugar utilization and improving conversion efficiencies (Paulova et al., 2015).
- CBP enables the use of agricultural residues and non-food lignocellulosic feedstocks (e.g., wheat straw and rice straw) that mostly remain underutilized. This valorization aligns with circular bioeconomy goals and minimizes competition with food systems.

11.3.2 Technological and Innovation Drivers

- Modern world bioengineering tools, including CRISPR/Cas genome editing, are facilitating the construction of CBP-capable microbial hosts. Particularly, *S. cerevisiae* and other ethanologens are being engineered to express cellulases and hemicellulases while retaining high ethanol productivity (Garg et al., 2023). Despite persisting challenges like low titer or metabolic burden, recent progress is promising in developing yeast-based CBP platforms (Sharma et al., 2022).
- Multi-species systems enable partitioning of biochemical tasks such as cellulolysis, sugar consumption, and fermentation across specialized strains. This reduces metabolic trade-offs and enhances functional performance compared to single-organism systems (Li et al., 2024).

11.3.3 Policy and Regulatory Drivers

- Biofuel production policy, renewable fuel standards, and direct subsidies from the government have incentivized the development and piloting of advanced bioprocesses, including CBP, globally. For example, the US Renewable Fuel Standard and tax credits like 45 V under the Inflation Reduction Act directly support technologies that lower carbon intensity, including lignocellulosic biofuels.
- Many nations, including India, are investing in second-generation (2G) biofuel capacity, with major oil companies planning multiple 2G ethanol plants. Such infrastructure projects elevate CBP's potential as a cost-effective conversion technology for lignocellulosic feedstocks like sugarcane bagasse and rice residues (Shahid et al., 2021).

11.4 Technological Enhancements and Digestibility Improvements

Technological improvement is needed to improve the overall adaptability of CBP. Here we try to analyze recent innovations in genetic engineering, pretreatment strategies, thermophilic hosts, and microbial consortia that have driven significant gains in the digestibility and performance of CBP systems for green fuel production.

11.4.1 Engineered *Clostridium thermocellum* for Hemicellulose Co-Utilization

A groundbreaking study genetically evolved *C. thermocellum* to assimilate hemicellulose, overcoming a long-standing limitation in thermophilic CBP systems (An et al., 2020). The evolved strain KJC19-9 showed a ~ threefold increase in specific growth rate on xylose, and through recombinant expression of β-xylosidase genes, the resulting KJCBXint strain co-utilized cellulose and xylan simultaneously during fermentation of pretreated corn stover, markedly improving xylan deconstruction and sugar utilization compared to earlier strains (Chou et al., 2024). This co-utilization capability directly enhances digestibility and substrate conversion efficiency within a single CBP organism.

11.4.2 Discovery of Novel Thermophiles for Hydrogen Production

In a different domain of green fuels, *Thermoanaerobacterium thermosaccharolyticum* M18, a newly isolated moderately thermophilic bacterium, demonstrated high cellulolytic activity and hydrogen production directly from raw biomass (Cao et al., 2014). It achieved degradation rates of 56–63% for untreated corn cob, stalk, and wheat straw, yielding ~3.2–3.5 mmol H_2/g substrate (≈10.9 mmol H_2/g Avicel) and rates up to 2.05 mmol/L/h. Such single-step conversion capabilities make it a compelling candidate for hydrogen-focused CBP systems with robust digestibility (Cao et al., 2014).

11.4.3 Thermophile-Based CBP Platforms and Synthetic Biology

A recent review highlights the growing potential of thermophile-based CBP hosts. Scientists emphasize CRISPR/Cas genome editing and Design–Build–Test–Learn (DBTL) synthetic biology pipelines to engineer microbes like *C. thermocellum* and *M. thermophila* for consolidated enzyme production, biomass hydrolysis, and fermentation (Liao et al., 2022). These platforms are being optimized for thermostable enzyme expression, contamination resistance, and process scalability—all contributing to improved digestibility and product yield in lignocellulosic biorefineries (Kothari et al., 2018).

11.4.4 Co-Culture and Semi-Fed-Batch Strategies for Enhanced Sugar Release

An innovative semi-fed-batch saccharification approach combining *C. thermocellum* and *Thermobrachium celere* A9 offer improved biomass conversion efficiency, not strictly CBP, but conceptually related (Nhim et al., 2022). By adding feedstock in stages, enzyme inhibition is reduced, and higher substrate loadings are enabled. The resulting cellulose conversion yields are increased, although saccharification and fermentation remain temporally separated. This hybrid strategy highlights that digestibility challenges in CBP, particularly solid loading and inhibition, can be addressed with process ingenuity (Nhim et al., 2025).

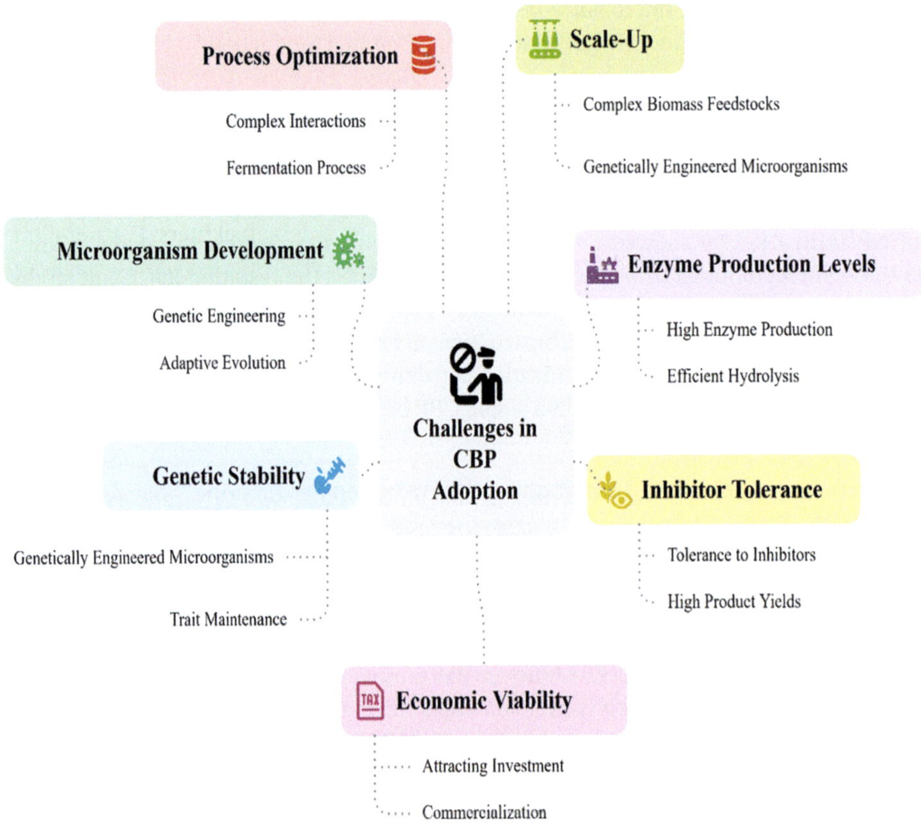

Figure 11.2 Depicting the challenges in CBP adoption.

11.5 Challenges in the Adoption of CBP Technology

Despite its numerous advantages, CBP technology faces a range of scientific, technical, and commercial challenges that hinder its widespread adoption in green fuel industries and other industries that can valorize the waste to produce some useful chemicals. To know the limitations of CBP, we have to critically examine these limitations, supported by recent research, and highlight areas where innovation is needed to enable large-scale implementation (Figure 11.2).

11.5.1 Genetic Instability and Limited Host Diversity

One of the primary drawbacks of CBP lies in the scarcity of microbial strains that possess all three essential capabilities: effective biomass degradation, robust fermentation, and tolerance to process inhibitors (Olguin-Maciel et al., 2020). Most naturally cellulolytic organisms, such as *C. thermocellum* and *Caldicellulosiruptor* species, lack the genetic tools and pathway efficiencies required for high-yield product formation (Cha et al., 2024; Lynd et al., 2017). On the contrary, traditional fermentative microbes like *S. cerevisiae* lack cellulolytic capabilities, requiring either complex genetic engineering or the creation of co-culture systems (Jeon et al., 2009).

11.5.2 Enzyme Activity and Product Yield

In CBP systems, in situ enzyme production can result in trade-offs between enzyme expression and fermentation performance (Han & Zhang, 2020). Overexpression of cellulases or hemicellulases can burden host metabolism, reducing biomass and biofuel titers. Furthermore, cellulosic hydrolysis under industrial solid loadings (>15% w/v) remains suboptimal, as high substrate concentrations increase viscosity, mass transfer limitations, and product inhibition (Datta, 2016).

11.5.3 Feedstock Variability and Pretreatment Dependence

CBP is often promoted as a low-pretreatment alternative; however, practical implementations still depend on some form of pretreatment to improve biomass accessibility (Agbor et al., 2014). The inconsistency of lignocellulosic feedstock composition due to plant type, harvest conditions, and storage also complicates process standardization.

11.5.4 Lack of Scalable Bioreactor Designs

CBP requires unique reactor designs capable of maintaining anaerobic or thermophilic conditions, handling high solid content, and enabling effective mixing despite non-Newtonian rheology. Few commercial bioreactors are optimized for CBP processes, and scale-up from lab to pilot-scale introduces new challenges in heat transfer, foam control, and pH stability (Akinosho et al., 2014).

In the US DOE Bioenergy Technologies Office, scale-up studies for CBP-based ethanol production remain in the early stages, with high capital costs and risk factors delaying industrial adoption (Dempfle et al., 2021).

11.5.5 Regulatory and Market Uncertainty

The absence of mature regulatory frameworks, feed-in tariffs, and market incentives for advanced biofuels, especially in developing countries, slows investment in CBP R&D and infrastructure. Furthermore, the lack of standard metrics to evaluate CBP system performance across different feedstocks and configurations creates uncertainty in commercial decision-making.

11.6 Case Studies and Industrial Applications of CBP

11.6.1 Industrial Actors Leading CBP Implementation

Firms like Mascoma Corporation and Qteros in the United States have spearheaded early entrepreneurship and innovation within CBP. Mascoma successfully engineered *C. thermocellum* to degrade cellulose and (Bomgardner, 2013) produce ethanol with high efficiency and ethanol tolerance, earning an $80 million Department of Energy award in 2011 to build a planned commercial-scale CBP facility in Kinross, Michigan (Mbaneme-Smith & Chinn, 2015; Mossberg et al., 2021). Similarly, Qteros developed the Q Microbe®, a strain of *Clostridium phytofermentans* capable of saccharifying and fermenting a variety of feedstocks (e.g., wheat straw, sugarcane bagasse, corn stover) in a single microbial platform with consistent saccharification performance (Bomgardner, 2013).

Another example is the SBIR-supported C-CBP (co-treatment-CBP) projects undertaken by Terragia Biofuel and Enchi Corporation, targeting cash-positive conversion of corn stover co-located with existing ethanol mills. Phase I trials reported nearly doubled ethanol titers with simultaneous C5 and C6 sugar conversion, and techno-economic modeling suggested an eightfold shorter payback period for plants using C-CBP relative to conventional pretreatment-based designs.

11.6.2 Case Studies and Pilot Facilities

While many academic studies advance understanding of CBP, industrial-scale deployment remains in pilot and demonstration phases. For instance, Global Bioenergies partnered with Fraunhofer CBP to develop a pilot plant for isobutene production via CBP, integrating fermenters with purification to replicate commercial pipelines, albeit not focused solely on fuels but also chemical precursors to drop-in fuel components such as isooctane (Smit et al., 2022).

Similarly, Fraunhofer CBP has facilitated the scale-up of various biochemical processes, such as the extraction of xylan and lignin from wood at pilot scales using their 500 L facilities, demonstrating their ability to host CBP processes toward commercial viability (Najjoum et al., 2025).

11.6.2.1 Case Studies on the Mascoma Corporation Pilot Plant and Industrial Applications of CBP

To better understand the practical impact and translational potential of CBP technology, it is essential to examine how it has been applied in industrial and pilot-scale settings. One of the earliest industrial-scale attempts to apply CBP for bioethanol production came from the Mascoma Corporation, which constructed a pilot facility in New York, United States. The company engineered *C. thermocellum* strains capable of expressing high levels of cellulolytic enzymes and optimized fermentation pathways for ethanol production. Mascoma's CBP platform aimed to reduce operational costs by eliminating externally added enzymes and simplifying process steps.

While the technology showed promise at the demonstration scale, the process faced scalability issues such as low ethanol titers (~25–30 g/L) and sensitivity to substrate variation. In the year 2014, the project merged with another industry, Lallemand Biofuels & Distilled Spirits, indicating a shift toward hybrid enzyme-assisted systems. Nonetheless, this case marked a significant milestone in adapting CBP from lab to pilot and industrial scale.

11.6.3 Geographic and Sectoral Trends

Efforts in CBP adoption are largely concentrated in North America and Europe, with the US Department of Energy playing a catalytic role through grants to firms like Mascoma and Qteros. European pilot plants, though older, include initiatives like those by Global Bioenergies at Fraunhofer CBP in Germany (Mbaneme-Smith & Chinn, 2015). The Swedish biorefinery landscape highlights many pilot/demonstration projects using biochemical and gasification routes, though CBP-specific efforts appear less mature or still emerging in the current planning waves (Mossberg et al., 2021).

In emerging economies like India, while CBP-specific implementations remain nascent, rapid expansion in ethanol and biomass gasification indicates a growing platform in which CBP may eventually be incorporated in biorefineries exploring circular valorization of crop residues such as rice straw and sugarcane bagasse (Madhuvanthi et al., 2022).

11.7 Emerging Trends in CBP Technology

Recent advances in omics technologies, genome-scale modeling, and metabolic pathway engineering are expected to enhance the performance and stability of CBP microorganisms. For instance, CRISPR/Cas9-mediated gene editing now enables more precise metabolic rewiring in *C. thermocellum* and *Thermoanaerobacterium saccharolyticum*, improving ethanol tolerance and reducing by-product formation.

Machine learning-based predictive modeling is also being applied to optimize enzyme expression profiles and cofactor balances in real-time during fermentation. This approach was illustrated in 2024 research by Banerjee et al., where AI-driven adaptive control improved ethanol yields by 15% compared to static systems in a CBP platform.

Another emerging trend is the design of synthetic microbial consortia that distribute the CBP tasks across multiple engineered strains, allowing for greater process flexibility and resilience. This modular strategy has shown potential for multi-product generation, including ethanol, hydrogen, and organic acids, from the same biomass input.

11.8 Integration with Circular Bioeconomy

CBP technology fulfills the principles of circular bioeconomy by converting low-value agricultural and lignocellulosic biomass into high-value fuels and other chemical products with minimal or no waste produced during the process. Integration of CBP with waste valorization systems, such as biochar production, biogas, and chemical recovery from fermentation residues, could enhance energy return on investment (EROI) and reduce the lifecycle of greenhouse gas emissions.

In 2016, the International Renewable Energy Agency (IRENA) conducted a techno-economic analysis of CBP and concluded that integrated CBP systems could reduce the minimum ethanol selling price to below \$0.60/L by 2045 with appropriate scale and policy support, making it competitive with fossil fuels in many regions (IRENA, 2016).

11.9 Future Prospects and Policy Recommendations in Sustainable Biofuel Production

As the global climate is changing and nations seek urgent alternatives to fossil-based fuels and conventional energy systems, CBP is emerging as a highly promising, next-generation bioconversion technology. By integrating enzyme production, substrate hydrolysis, and fermentation into a single reactor, CBP offers high efficiency in converting lignocellulosic and other agricultural biomass into valuable biofuels like butanol, ethanol, biogas, and hydrogen. While the technology remains in its developmental and pilot phases, advancements in synthetic biology, microbial engineering, and systems-level process design are rapidly filling the gap between research innovation and industrial adoption with its benefits.

11.9.1 Vision for the Future

If supported by aligned policy and investment strategies, CBP could become the cornerstone of advanced biorefineries by 2040. These facilities would not only produce clean fuels but also contribute to decentralized, carbon-neutral economic models, especially in regions rich in agricultural waste, such as India, Brazil, and Southeast Asia. Importantly, the future of CBP will extend beyond biofuel production. In a fully developed biorefinery ecosystem, CBP can integrate the co-production of chemicals, biomaterials, and other bio-based products, combining the principles of a circular, zero-waste bioeconomy.

11.9.2 Strategic Policy Recommendations

Moving CBP from the lab to large-scale/industrial deployment will require a coordinated, multi-stakeholder approach involving government, industry, and academia. The following policy actions are proposed to catalyze this transition:

1) **Investment in Pilot and Demonstration Facilities**
   Public and government funding and infrastructure support from the established institute for pilot-scale CBP plants are essential to validate process performance at scale. These facilities will serve as innovation hubs for optimizing feedstock utilization, microbial co-cultures, and bioreactor design. Establishing such testbeds will also reduce perceived financial risk for private investors evaluating the viability of CBP technologies.
2) **Incentivizing Advanced Biofuel Integration through Blending Mandates**
   Implementing or expanding biofuel blending targets, such as transitioning from E10 to E15 or E20 ethanol-gasoline mixtures, can create guaranteed market pull for cellulosic biofuels. Government mandates should prioritize advanced biofuels sourced from non-food biomass, thus reinforcing long-term sustainability objectives while stimulating demand for CBP-derived products.
3) **Establishing Certification and Quality Standards for CBP-Based Fuels**
   Developing internationally recognized certification frameworks will be vital to ensure the environmental and performance credentials of CBP fuels. Clear standards can enhance market trust, enable cross-border trade, and ensure compliance with global sustainability benchmarks such as the EU RED II regulations or carbon credit initiatives.
4) **Workforce Development and Skill Building**
   Improving CBP technologies will require a new generation of skilled professionals and researchers trained in areas such as fermentation science, bioprocess engineering, synthetic biology, and industrial microbiology. Public and private initiatives should support vocational programs, university curricula, and hands-on training experiences to build this future-ready workforce, particularly in rural and agri-industrial regions.

5) **Fostering Collaborative Innovation through Public–Private Partnerships (PPPs)**
   Encouraging collaborative platforms where universities, startups, government labs, and industrial partners can co-develop and pilot CBP solutions will accelerate innovation. These partnerships can serve as accelerators for technology transfer, commercialization support, and open data sharing on feedstock viability, microbial performance, and life-cycle assessments.

11.9.3 A Path Toward a Carbon-Neutral Future

The role of CBP in building resilient and decentralized bio-economies cannot be overstated. By valorizing lignocellulosic agricultural waste, a resource often underutilized or discarded, CBP has the potential to reduce greenhouse gas emissions, enhance energy security, and promote rural income generation through biomass-based industries.

Looking ahead, the goal is not merely to scale a single technology, but to meaningfully embed CBP into a broader green energy matrix. In such a system, diverse products from fuels and fertilizers to bioplastics and platform chemicals can be manufactured through carbon-neutral, zero-waste processes, advancing both climate and development agendas.

11.10 Conclusion

The transition from conventional to sustainable and low-carbon energy sources is no longer a future ambition, but an urgent global requirement. Among various emerging fields of biotechnologies, CBP offers a promising path toward cost-effective and scalable production of green fuels from lignocellulosic and agricultural biomass. By integrating enzymatic hydrolysis, fermentation, and microbial enzyme production into a single reactor, CBP significantly reduces process complexity and production expenditure when compared to traditional methods.

In this chapter, we explored the frequency and nature of CBP adaptation across research and industrial areas. From the early pilot-scale trials by Mascoma Corporation to recent AI-integrated fermentation systems and synthetic consortia, the steady rise in technological maturity and commercial interest highlights growing confidence in the CBP approach. Importantly, CBP has evolved from being an experimental curiosity to a near-commercially viable solution, especially with support from synthetic biology, process intensification, and thermophilic microbial engineering.

Case studies from the United States underline the spread and versatility of CBP in producing diverse biofuels such as ethanol, butanol, and hydrogen. These real-world applications demonstrate not only technical feasibility but also economic and environmental benefits when CBP is integrated with the circular bioeconomy framework.

However, some challenges remain; issues such as low product titers, organism robustness under industrial conditions, and biomass variability need to be addressed through continued interdisciplinary collaboration and sustained policy support. If governments and biomass-based industrial sectors act side by side by investing in pilot-scale facilities, setting strong biofuel mandates, and establishing clear regulatory frameworks, CBP will become a central pillar in achieving net-zero emissions targets.

In conclusion, the future of CBP is intrinsically tied to the global bioenergy transition. With strategic investment, innovation, and policy alignment, CBP has the potential to transform agricultural waste into a cornerstone resource for the sustainable fuels of tomorrow. Thereby empowering both energy security and environmental stewardship in a world rapidly moving away from fossil dependence.

References

Afedzi, A. E. K., Afrakomah, G. S., Gyan, K., Khan, J., Seidu, R., Baidoo, T., Afrasiab, I., Tareen, A. K., & Parakulsuksatid, P. (2025). Enhancing economic and environmental sustainability in lignocellulosic bioethanol production: Key factors, innovative technologies, policy frameworks, and social considerations. *Sustainability*, *17*(2), 499. https://doi.org/10.3390/su17020499.

Agbor, V., Carere, C., Cicek, N., Sparling, R., & Levin, D. (2014). Biomass pretreatment for consolidated bioprocessing (CBP). *Advances in Biorefineries*, 234–258. https://doi.org/10.1533/9780857097385.1.234.

Akinosho, H., Yee, K., Close, D., & Ragauskas, A. (2014). The emergence of *Clostridium thermocellum* as a high utility candidate for consolidated bioprocessing applications. *Frontiers in Chemistry*, *2*, 66. https://doi.org/10.3389/fchem.2014.00066.

An, Q., Bu, J., Cheng, J.-R., Hu, B.-B., Wang, Y.-T., & Zhu, M.-J. (2020). Biological saccharification by *Clostridium thermocellum* and two-stage hydrogen and methane production from hydrogen peroxide-acetic acid pretreated sugarcane bagasse. *International Journal of Hydrogen Energy*, *45*(55), 30211–30221. https://doi.org/10.1016/j.ijhydene.2020.08.069.

Bomgardner, M. (2013). *REBORN Qteros, developer of one-step biofuels microbe, is revived by founders*. American Chemical Society. 1155 16TH ST, NW, WASHINGTON, DC 20036 USA. DOI: 10.1021/cen-09101-notw5.

Cao, G.-L., Zhao, L., Wang, A.-J., Wang, Z.-Y., & Ren, N.-Q. (2014). Single-step bioconversion of lignocellulose to hydrogen using novel moderately thermophilic bacteria. *Biotechnology for Biofuels*, *7*(1), 82. https://doi.org/10.1186/1754-6834-7-82.

Cha, M., Kim, J. K., Lee, W.-H., Song, H., Lee, T.-G., Kim, S.-K., & Kim, S.-J. (2024). Metabolic engineering of *Caldicellulosiruptor bescii* for hydrogen production. *Applied Microbiology and Biotechnology*, *108*(1), 65. https://doi.org/10.1007/s00253-023-12974-7.

Chou, K. J., Croft, T., Hebdon, S. D., Magnusson, L. R., Xiong, W., Reyes, L. H., Chen, X., Miller, E. J., Riley, D. M., Dupuis, S., Laramore, K. A., Keller, L. M., Winkelman, D., & Maness, P.-C. (2024). Engineering the cellulolytic bacterium, *Clostridium thermocellum*, to co-utilize hemicellulose. *Metabolic Engineering*, *83*, 193–205. https://doi.org/10.1016/j.ymben.2024.03.008.

Cunha, J. T., Romaní, A., Inokuma, K., Johansson, B., Hasunuma, T., Kondo, A., & Domingues, L. (2020). Consolidated bioprocessing of corn cob-derived hemicellulose: Engineered industrial *Saccharomyces cerevisiae* as efficient whole cell biocatalysts. *Biotechnology for Biofuels*, *13*(1), 138. https://doi.org/10.1186/s13068-020-01780-2.

da Rosa-Garzon, N. G., Laure, H. J., Rosa, J. C., & Cabral, H. (2022). Valorization of agricultural residues using *Myceliophthora thermophila* as a platform for production of lignocellulolytic enzymes for cellulose saccharification. *Biomass and Bioenergy*, *161*, 106452. https://doi.org/10.1016/j.biombioe.2022.106452.

Datta, S. (2016). Recent strategies to overexpress and engineer cellulases for biomass degradation. *Current Metabolomics*, *4*(1), 14–22. https://doi.org/10.2174/2213235X03666150702155845.

Davison, S., Den Haan, R., & Van Zyl, W. (2020). Exploiting strain diversity and rational engineering strategies to enhance recombinant cellulase secretion by *Saccharomyces cerevisiae*. *Applied Microbiology and Biotechnology*, *104*(12), 5163–5184. https://doi.org/10.1007/s00253-020-10602.

Dempfle, D., Kröcher, O., & Studer, M. H.-P. (2021). Techno-economic assessment of bioethanol production from lignocellulose by consortium-based consolidated bioprocessing at industrial scale. *New Biotechnology*, *65*, 53–60. https://doi.org/10.1016/j.nbt.2021.07.005.

den Haan, R., Rose, S. H., Cripwell, R. A., Trollope, K. M., Myburgh, M. W., Viljoen-Bloom, M., & van Zyl, W. H. (2021). Heterologous production of cellulose-and starch-degrading hydrolases to expand *Saccharomyces cerevisiae* substrate utilization: Lessons learnt. *Biotechnology Advances*, *53*, 107859. https://doi.org/10.1016/j.biotechadv.2021.107859.

Devi, A., Bajar, S., Kour, H., Kothari, R., Pant, D., & Singh, A. (2022). Lignocellulosic biomass valorization for bioethanol production: A circular bioeconomy approach. *Bioenergy Research*, *15*(4), 1820–1841. https://doi.org/10.1007/s12155-022-10401-9.

Fortuin, J., Hoffmeester, L. J., Minnaar, L. S., & den Haan, R. (2025). Advancing cellulose utilization and engineering consolidated bioprocessing yeasts: Current state and perspectives. *Applied Microbiology and Biotechnology*, *109*(1), 43.

Garg, D., Samota, M. K., Kontis, N., Patel, N., Bala, S., & Rosado, A. S. (2023). Revolutionizing biofuel generation: Unleashing the power of CRISPR-Cas mediated gene editing of extremophiles. *Microbiological Research*, *274*, 127443. https://doi.org/10.1016/j.micres.2023.127443.

Han, Y., & Zhang, F. (2020). Control strategies to manage trade-offs during microbial production. *Current Opinion in Biotechnology*, *66*, 158–164. https://doi.org/10.1016/j.copbio.2020.07.004.

IRENA. (2016). *Innovation outlook advanced liquid biofuels*. Retrieved from USA.

Jeon, E., Hyeon, J.-E., Suh, D. J., Suh, Y.-W., Kim, S. W., Song, K. H., & Han, S. O. (2009). Production of cellulosic ethanol in *Saccharomyces cerevisiae* heterologous expressing *Clostridium thermocellum* endoglucanase and *Saccharomycopsis fibuligera* β-glucosidase genes. *Molecules and Cells*, *28*(4), 369–374. https://doi.org/10.1007/s10059-009-0131-y.

Jiang, Y., Dong, W., Xin, F., & Jiang, M. (2020). Designing synthetic microbial consortia for biofuel production. *Trends in Biotechnology*, *38*(8), 828–831. https://doi.org/10.1016/j.tibtech.2020.02.002.

Kothari, N., Holwerda, E. K., Cai, C. M., Kumar, R., & Wyman, C. E. (2018). Biomass augmentation through thermochemical pretreatments greatly enhances digestion of switchgrass by *Clostridium thermocellum*. *Biotechnology for Biofuels*, *11*(1), 219. https://doi.org/10.1186/s13068-018-1216-7.

Larnaudie, V., Ferrari, M. D., & Lareo, C. (2019). Techno-economic analysis of a liquid hot water pretreated switchgrass biorefinery: Effect of solids loading and enzyme dosage on enzymatic hydrolysis. *Biomass and Bioenergy*, *130*, 105394. https://doi.org/10.1016/j.biombioe.2019.105394.

Li, J., Chen, B., Gu, S., Zhao, Z., Liu, Q., Sun, T., Zhang, Y., Wu, T., Liu, D., Sun, W., & Tian, C. (2021). Coordination of consolidated bioprocessing technology and carbon dioxide fixation to produce malic acid directly from plant biomass in *Myceliophthora thermophila*. *Biotechnology for Biofuels*, *14*(1), 186. https://doi.org/10.1186/s13068-021-02042-5.

Li, Z., Waghmare, P. R., Dijkhuizen, L., Meng, X., & Liu, W. (2024). Research advances on the consolidated bioprocessing of lignocellulosic biomass. *Engineering Microbiology*, *4*(2), 100139. https://doi.org/10.1016/j.engmic.2024.100139.

Liao, X., Ma, H., & Tang, Y. J. (2022). Artificial intelligence: A solution to involution of design–build–test–learn cycle. *Current Opinion in Biotechnology*, *75*, 102712. https://doi.org/10.1016/j.copbio.2022.102712.

Liu, Y., Tang, Y., Gao, H., Zhang, W., Jiang, Y., Xin, F., & Jiang, M. (2021). Challenges and future perspectives of promising biotechnologies for lignocellulosic biorefinery. *Molecules*, *26*(17), 5411. https://doi.org/10.3390/molecules26175411.

Lynd, L. R., Guss, A. M., Himmel, M. E., Beri, D., Herring, C., Holwerda, E. K., Murphy, S. J., Olson, D. G., Paye, J., Rydzak, T., Shao, X., Tian, L., & Worthen, R. (2017). Advances in consolidated bioprocessing using *Clostridium thermocellum* and *Thermoanaerobacter saccharolyticum*. *Industrial Biotechnology: Microorganisms*, *1*, 365–394. https://doi.org/10.1002/9783527807796.ch10.

Madhuvanthi, S., Jayanthi, S., Suresh, S., & Pugazhendhi, A. (2022). Optimization of consolidated bioprocessing by response surface methodology in the conversion of corn stover to bioethanol by thermophilic *Geobacillus thermoglucosidasius*. *Chemosphere*, *304*, 135242. https://doi.org/10.1016/j.chemosphere.2022.135242.

Mbaneme-Smith, V., & Chinn, M. S. (2015). Consolidated bioprocessing for biofuel production: Recent advances. *Energy Emission Control Technologies*, *3*, 23–44. https://doi.org/10.2147/EECT.S63000.

Mossberg, J., Söderholm, P., & Frishammar, J. (2021). Challenges of sustainable industrial transformation: Swedish biorefinery development and incumbents in the emerging biofuels industry. *Biofuels, Bioproducts and Biorefining*, *15*(5), 1264–1280. https://doi.org/10.1002/bbb.2249.

Moysés, D. N., Reis, V. C. B., de Almeida, J. R., de Moraes, L. M., & Torres, F. A., (2016). Xylose fermentation by *Saccharomyces cerevisiae*: Challenges and prospects. *International Journal of Molecular Sciences, 17*(3), 207. https://doi.org/10.3390/ijms17030207.

Muzaffar, N., Raziq, A., Khan, M. W., Khan, N. M., Shahid, B., Gul, A., & Ullah, H. (2025). Recent developments in heterologous expression of cellulases using the Pichia pastoris expression system: A comprehensive literature review. *Applied Microbiology*, *5*(1), 22. https://doi.org/10.3390/applmicrobiol5010022.

Najjoum, N., Grimi, N., Benali, M., Chadni, M., & Castignolles, P. (2025). Extraction and chemical features of wood hemicelluloses: A review. *International Journal of Biological Macromolecules*, *311*, 143681. https://doi.org/10.1016/j.ijbiomac.2025.143681.

Nhim, S., Baramee, S., Tachaapaikoon, C., Pason, P., Ratanakhanokchai, K., Uke, A., Ceballos, R. M., Kosugi, A., & Waeonukul, R. (2025). Effective semi-fed-batch saccharification with high lignocellulose loading using co-culture of *Clostridium thermocellum* and *Thermobrachium celere* strain A9. *Frontiers in Microbiology*, *15*, 1519060. https://doi.org/10.3389/fmicb.2024.1519060.

Nhim, S., Waeonukul, R., Uke, A., Baramee, S., Ratanakhanokchai, K., Tachaapaikoon, C., Pason, P., Liu, Y. J., & Kosugi, A. (2022). Biological cellulose saccharification using a coculture of *Clostridium thermocellum* and *Thermobrachium celere* strain A9. *Applied Microbiology and Biotechnology*, *106*(5), 2133–2145. https://doi.org/10.1007/s00253-022-11818-0.

Olguin-Maciel, E., Singh, A., Chable-Villacis, R., Tapia-Tussell, R., & Ruiz, H. A. (2020). Consolidated bioprocessing, an innovative strategy towards sustainability for biofuels production from crop residues: An overview. *Agronomy, 10*(11), 1834. https://doi.org/10.3390/agronomy10111834.

Patel, R., Rajaraman, T., Rana, P., Ambegaonkar, N., & Patel, S. (2025). A review on techno-economic analysis of lignocellulosic biorefinery producing biofuels and high-value products. *Results in Chemistry, 13*, 102052. https://doi.org/10.1016/j.rechem.2025.102052.

Paulova, L., Patakova, P., Branska, B., Rychtera, M., & Melzoch, K. (2015). Lignocellulosic ethanol: Technology design and its impact on process efficiency. *Biotechnology Advances, 33*(6), 1091–1107. https://doi.org/10.1016/j.biotechadv.2014.12.002.

Perez, C. L., Milessi, T. S., Sandri, J. P., Ramos, M. D., Carvalho, B. T., Claes, A., Demeke, M., Thevelein, J., & Zangirolami, T. C. (2023). Evaluation of consolidated bioprocessing of sugarcane biomass by a multiple hydrolytic enzyme producer Saccharomyces yeast. *Bioenergy Research, 16*(4), 1973–1989. https://doi.org/10.1007/s12155-023-10607-5.

Shahid, M. K., Batool, A., Kashif, A., Nawaz, M. H., Aslam, M., Iqbal, N., & Choi, Y. (2021). Biofuels and biorefineries: Development, application and future perspectives emphasizing the environmental and economic aspects. *Journal of Environmental Management, 297*, 113268. https://doi.org/10.1016/j.jenvman.2021.113268.

Sharma, J., Kumar, V., Prasad, R., & Gaur, N. A. (2022). Engineering *of Saccharomyces cerevisiae* as a consolidated bioprocessing host to produce cellulosic ethanol: Recent advancements and current challenges. *Biotechnology Advances, 56*, 107925. https://doi.org/10.1016/j.biotechadv.2022.107925.

Singhania, R. R., Patel, A. K., Singh, A., Haldar, D., Soam, S., Chen, C.-W., Tsai, M. L., & Dong, C.-D. (2022). Consolidated bioprocessing of lignocellulosic biomass: Technological advances and challenges. *Bioresource Technology, 354*, 127153. https://doi.org/10.1016/j.biortech.2022.127153.

Smit, A. T., Verges, M., Schulze, P., van Zomeren, A., & Lorenz, H. (2022). Laboratory-to pilot-scale fractionation of lignocellulosic biomass using an acetone organosolv process. *ACS Sustainable Chemistry & Engineering, 10*(32), 10503–10513. https://doi.org/10.1021/acssuschemeng.2c01425.

Srivastava, N., Singh, R., Singh, P., Ahmad, I., Singh, R. P., Rai, A. K., Asiri, M., & Gupta, V. K. (2023). Recent advances on lignocellulosic bioresources and their valorization in biofuels production: Challenges and viability assessment. *Environmental Technology and Innovation, 29*, 103037. https://doi.org/10.1016/j.eti.2023.103037.

Vasan, P. T., Piriya, P. S., Prabhu, D. I. G., & Vennison, S. J. (2011). Cellulosic ethanol production by *Zymomonas mobilis* harboring an endoglucanase gene from *Enterobacter cloacae. Bioresource Technology, 102*(3), 2585–2589. https://doi.org/10.1016/j.biortech.2010.09.110.

Wall, A. A., Rabemanolontsoa, H., & Venus, J. J. A. S. (2023). Bioprocessing and fermentation technology for biomass conversion. *Applied Sciences, 14*, 5. In (Vol. 14, pp. 5): Applied Sciences. https://doi.org/10.3390/app14010005.

Zhao, C., Zhao, J., Han, J., Mei, Y., & Fang, H. (2024). Improved consolidated bioprocessing for itaconic acid production by simultaneous optimization of cellulase and metabolic pathway of *Neurospora crassa. Biotechnology for Biofuels and Bioproducts, 17*(1), 57. https://doi.org/10.1186/s13068-024-02505-5.

12

Cost Analysis of Consolidated Bioprocess Technology for Green Energy Production and Environmental Sustainability Impact

Vikram Singh[1], Neeraj Gupta[2], Deepti Singh[2], Gyanendra Tripathi[1], Ashish[1], and Fredrick Kayusi[3,4]

[1] *Department of Bioengineering, Integral University, Lucknow, Uttar Pradesh, India*
[2] *Institute of Biosciences and Technology, Shri Ramswaroop Memorial University, Lucknow, Uttar Pradesh, India*
[3] *Department of Environmental Studies, Geography and Planning, Maasai Mara University, Narok, Kenya*
[4] *Department of Environmental and Earth Sciences, Pwani University, Kilifi, Kenya*

12.1 Introduction

The increasing global energy demand, coupled with mounting environmental concerns such as climate change, fossil fuel depletion, and greenhouse gas emissions, has necessitated a transition toward cleaner, renewable energy systems (Palwan et al., 2025) pollution. The current global energy matrix is heavily dependent on fossil fuels, which account for over 80% of primary energy consumption, contributing substantially to CO_2 emissions and environmental degradation (Kumar et al., 2025; IEA, 2021). In response, renewable energy technologies—particularly solar, wind, hydropower, and bioenergy—are being promoted as viable alternatives. Among these, bioenergy holds a unique and strategic position due to its dual role in waste valorization and sustainable energy generation (IRENA, 2020).

The bioeconomy, defined as the use of renewable biological resources to produce food, materials, and energy, has emerged as a central strategy for climate-resilient development. Bioenergy plays a key role within this framework by enabling the conversion of lignocellulosic biomass (LCB), agricultural residues, and organic wastes into value-added products like biofuels, biochemicals, and biomaterials, thereby reducing environmental pressure while promoting rural economic development (European Commission, 2018).

Biofuels such as bioethanol, biodiesel, and biogas have become important alternatives to fossil fuels, particularly in the transportation sector. However, the economic viability and environmental sustainability of biofuels remain key challenges (Alvira et al., 2010). First-generation biofuels, derived from edible crops like corn and sugarcane, have led to land-use concerns and food-vs.-fuel debates. Consequently, there has been a shift toward second-generation (2G) biofuels produced from non-food LCB and third-generation (3G) biofuels from algae and microbial sources. Despite their promise, these advanced biofuels face high production costs, complex processing steps, and low conversion efficiencies, primarily due to the need for expensive enzyme cocktails and multi-stage processing (Humbird et al., 2011; Klein-Marcuschamer et al., 2012).

To overcome these limitations, consolidated bioprocessing (CBP) has been proposed as a transformative approach in the field of industrial biotechnology. CBP is an integrated, one-pot bioprocessing strategy that combines cellulase enzyme production, biomass hydrolysis, and microbial fermentation into a single step within one bioreactor. This eliminates the need for external enzyme addition, significantly reducing both capital (CAPEX) and operational expenditures (OPEX) while simplifying process logistics (Lynd et al., 2005; Olson et al., 2012).

CBP relies on either naturally occurring microorganisms with native cellulolytic and fermentative abilities (e.g., *Clostridium thermocellum* and *Thermoanaerobacterium saccharolyticum*) or genetically engineered strains (e.g., *Saccharomyces cerevisiae* and *Zymomonas mobilis*) tailored to perform all necessary bioconversion tasks. These microorganisms are capable of *in situ* enzyme production, followed by the breakdown of complex polysaccharides (cellulose and hemicellulose) into fermentable sugars, and their subsequent conversion into bioethanol, butanol, or other advanced biofuels (Hasunuma & Kondo, 2012; Stephanopoulos, 2007).

Furthermore, CBP aligns well with the principles of green chemistry and sustainable process engineering, as it minimizes waste, water usage, and energy input. It supports circular bioeconomy models by enabling the valorization of agro-industrial waste streams, forest residues, and energy crops into high-value products (Rashed et al., 2025; Zhou et al., 2016). With recent advancements in synthetic biology, metabolic pathway engineering, and AI-based process optimization, CBP is moving closer to commercial-scale deployment (Iruj et al., 2025; Lee et al., 2020).

This chapter aims to critically evaluate the economic feasibility, environmental sustainability, and technical viability of CBP in the context of green energy production. It covers the fundamental principles, key process steps, cost components, and comparative benefits of CBP over conventional multistep bioprocessing. The goal is to provide an in-depth understanding of how CBP can be leveraged as a cost-effective, simplified, and scalable solution for sustainable biofuel production in both developed and developing countries.

12.1.1 Scope and Objectives of the Chapter

This chapter explores the economic and environmental potential of CBP for green energy production. The scope encompasses both the technological fundamentals and the economic implications of deploying CBP at a commercial scale. Specific objectives include:

- To define the principles and advantages of CBP over conventional bioprocessing approaches.
- To analyze the cost components associated with CBP, including feedstock, microbial strain development, equipment, and operational inputs.
- To provide a detailed techno-economic assessment (TEA), including key financial indicators such as CAPEX, OPEX, return on investment (ROI), net present value (NPV), and internal rate of return (IRR).
- To evaluate the environmental sustainability benefits of CBP through reduced greenhouse gas emissions and waste generation.
- To identify technical and economic challenges and to propose future directions involving synthetic biology, AI integration, and circular bioeconomy models.

Through this comprehensive review, the chapter aims to present CBP as a viable and forward-looking solution for sustainable biofuel production and the energy transition.

12.2 Fundamentals of Consolidated Bioprocessing (CBP)

12.2.1 Definition and Evolution of CBP

CBP is a next-generation bioconversion strategy that integrates enzyme production, biomass hydrolysis, and microbial fermentation into a single, streamlined step within one bioreactor. Unlike traditional multi-stage processes that require separate vessels and the external addition of expensive enzyme cocktails, CBP utilizes robust microbial strains capable of carrying out all key functions in one system. This "one-pot" biorefinery model offers improved process efficiency, simplified operations, and substantial cost savings (Lynd et al., 2005; Olson et al., 2012).

The conceptual framework for CBP was initially proposed to address the major cost bottlenecks associated with cellulosic ethanol production—particularly the high cost of cellulases and the complexity of biomass-to-biofuel conversion. With advances in synthetic biology, metabolic engineering, and systems biology, researchers have since developed various microbial strains (native and engineered) that can both degrade complex polysaccharides and convert monomeric sugars into ethanol, butanol, and other biofuels (Hasunuma & Kondo, 2012; Wyman et al., 2005).

CBP has evolved from simple co-culture systems to sophisticated genetically engineered organisms capable of secreting high-yield enzyme systems, tolerating inhibitors, and utilizing pentose and hexose sugars simultaneously. Its evolution reflects a significant milestone in the shift toward low-input, high-output biorefineries within a circular and sustainable bioeconomy model.

CBP reduces technical complexity and improves the feasibility of scaling up advanced biofuel production in both industrial and decentralized contexts. Figure 12.1 shows a comparison of traditional bioprocessing vs. CBP. The technology also aligns with the UN Sustainable Development Goals (SDGs), promoting clean energy, industrial innovation, and sustainable resource use (UN, 2022), Table 12.1 shows the key features of CBP vs. traditional multistep bioprocessing.

12.2.2 Microorganisms Used in CBP (Native and Engineered Strains)

The success of CBP depends largely on the performance of microbial strains capable of simultaneous saccharification and fermentation (SSF). These strains are classified into native and engineered categories:

12.2.2.1 Native CBP Microorganisms

These organisms naturally exhibit both cellulolytic activity and fermentation capability. *C. thermocellum* is a thermophilic, anaerobic bacterium known for its efficient cellulose degradation via cellulosome complexes. It produces ethanol, acetate, and other volatile products (Lynd et al., 2002). *T. saccharolyticum* works synergistically with *C. thermocellum* to ferment both hexose and pentose sugars. *Trichoderma reesei* is a filamentous fungus recognized for its robust cellulase secretion system.

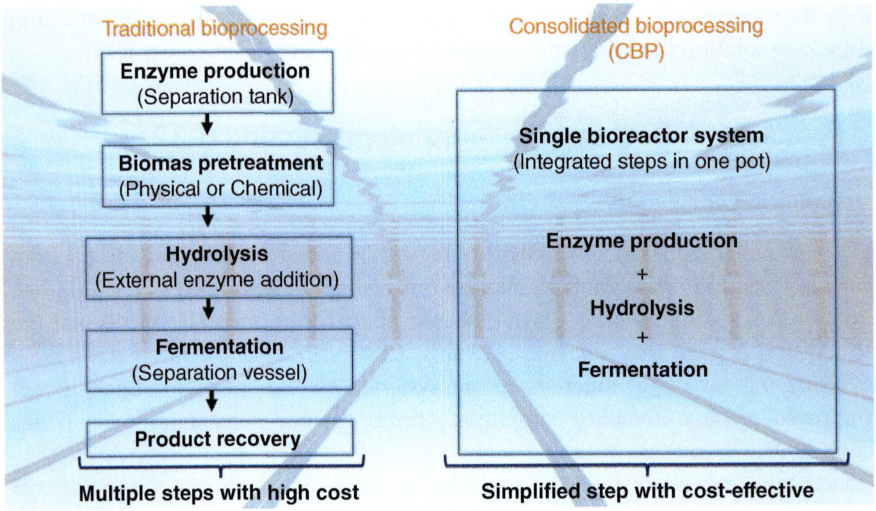

Figure 12.1 Comparison of traditional bioprocessing vs. consolidated bioprocessing (CBP).

Table 12.1 Key Features of CBP vs. Traditional Multistep Bioprocessing.

Feature	Traditional Bioprocessing	Consolidated Bioprocessing (CBP)
Process stages	Separate: enzyme production, hydrolysis, and fermentation	All steps integrated into one system
Enzyme requirement	Requires external enzyme production and addition	Enzymes produced *in situ* by microbes
Microbial systems	Different microbes or stages are used sequentially	Single multifunctional strain or consortium
Operational complexity	High, requires multiple units and sterile transfers	Low, single vessel reduces risk and handling
Capital and operating costs	High due to multiple process units and control systems	Lower due to integration and simplification
Energy and water use	Intensive	More efficient; minimal reuse requirements
Environmental impact	Moderate; generates more waste and emissions	Lower; supports green chemistry and zero-waste approaches

12.2.2.2 Engineered CBP Microorganisms

Engineered strains are developed via gene editing and synthetic biology to perform complete biomass-to-fuel conversion. *S. cerevisiae*, the most studied yeast in biotechnology, has been engineered to express cellulases (endoglucanase, β-glucosidase, and cellobiohydrolase) on the cell surface or via secretion (Hasunuma & Kondo, 2012). *Escherichia coli* is modified to produce butanol, ethanol, or lactate by co-expressing cellulases and fermentation pathways. *Z. mobilis* is enhanced to metabolize pentoses like xylose and arabinose, making it suitable for lignocellulosic hydrolysates (Zhang et al., 2007).

These microbial systems are increasingly optimized through metabolic flux analysis, adaptive evolution, and Clustered Regularly Interspaced Short Palindromic Repeats (CRISPR)-based editing, enabling greater resilience to inhibitors, higher substrate loading, and improved product titers.

12.2.3 Substrates Used: Lignocellulosic Biomass and Agricultural Waste

CBP systems are especially designed to process renewable, non-food biomass sources such as:

Agricultural residues: Rice husk, wheat straw, sugarcane bagasse, and corn stover.
Forest biomass: Sawdust, wood chips, and bark.
Energy crops: Switchgrass, miscanthus, and sorghum.
Agro-industrial waste: Fruit peels, vegetable waste, spent grains, and coffee pulp.

These materials are abundant, cost-effective, and rich in polysaccharides. LCB is primarily composed of cellulose (30–50%)—a linear glucose polymer, hemicellulose (20–35%)—a heteropolymer of pentoses/hexoses, and lignin (10–25%)—an aromatic, recalcitrant polymer providing structural rigidity.

Pretreatment (physical, chemical, or biological) is often used to increase enzyme accessibility by disrupting lignin and exposing cellulose fibers. However, in CBP systems, milder pretreatments are sufficient since cellulolytic microbes or enzymes are already present, reducing the need for harsh chemicals (Zhou et al., 2016). Valorizing such biomass through CBP addresses waste management, reduces landfill usage, and contributes to carbon neutrality, particularly when integrated with carbon capture or energy recovery systems (Rodrigues et al., 2025).

12.3 Process Flow and System Design of Consolidated Bioprocessing (CBP)

12.3.1 One-Pot Conversion: Hydrolysis and Fermentation

A core innovation of CBP is the one-pot bioconversion system, which integrates enzymatic hydrolysis and fermentation within a single bioreactor. In traditional lignocellulosic biofuel production, the process is segmented into multiple stages: (i) pretreatment, (ii) enzymatic saccharification, and (iii) fermentation, each requiring different process conditions, microbial strains, and often separate equipment. This segmentation not only increases capital and operating costs but also introduces a greater risk of contamination and longer processing times (Lynd et al., 2002).

In contrast, CBP enables simultaneous hydrolysis and fermentation (SHF) using a single organism or microbial consortium capable of secreting hydrolytic enzymes and metabolizing the released sugars. This consolidation of steps eliminates the need for externally produced cellulases and facilitates continuous sugar-to-product conversion, thereby avoiding end-product inhibition, a common bottleneck in separate hydrolysis and fermentation systems (Zhang et al., 2006).

Additionally, CBP systems can be engineered to operate under thermophilic conditions (~50–60 °C), which not only enhances enzyme activity but also reduces the risk of contamination by mesophilic organisms. The one-pot design also allows for process intensification, wherein higher substrate loadings can be tolerated, improving productivity per unit volume (Olson et al., 2012). This makes CBP particularly advantageous for rural, small-scale, or modular biorefineries, where cost, footprint, and robustness are key performance indicators.

12.3.2 Role of Enzyme Production *In Situ*

One of the transformative aspects of CBP is *in situ* enzyme production—the endogenous synthesis of hydrolytic enzymes by the same microorganism that ferments the sugars into biofuels or chemicals. This biological self-sufficiency removes the need for purchasing or producing enzymes externally, which traditionally accounts for 30–50% of the bioethanol production cost (Klein-Marcuschamer et al., 2012).

The enzymes involved include cellulases (endoglucanases, cellobiohydrolases, and β-glucosidases), which target crystalline cellulose; hemicellulases (xylanases and mannanases), which degrade hemicellulose into pentoses; and accessory enzymes (laccases and esterases), which aid lignin modification and sugar release.

Microorganisms like *C. thermocellum* naturally form cellulosomes, large multi-enzyme complexes anchored to their cell surface, which synergistically degrade cellulose in close proximity to the microbial uptake systems (Bayer et al., 2004). This proximity effect improves hydrolysis efficiency and ensures that sugars are rapidly consumed, minimizing feedback inhibition. In engineered systems, *S. cerevisiae* and *Z. mobilis* have been modified to express cellulases using surface display systems, secretion via signal peptides, and cell fusion constructs for enzyme clustering (Hasunuma & Kondo, 2012). The goal is to optimize the enzyme-to-substrate ratio, synchronize sugar release with fermentation kinetics, and maintain high product yields without intermediate processing steps. This self-contained system improves the economic viability and scalability of CBP platforms.

12.3.3 Integrated Bioreactor Design

The bioreactor is central to CBP implementation and must support both enzymatic activity and microbial fermentation. These processes often have different optimal conditions; for example, hydrolysis may favor acidic pH and higher temperatures, whereas fermentation tends to perform best at moderate pH and mesophilic/thermophilic conditions. Therefore, reactor design must ensure thermochemical compatibility between hydrolysis and fermentation, efficient heat and mass transfer, and sterility and containment, especially in anaerobic systems.

Stirred tank reactors (STRs) are commonly used due to their excellent mixing and control capabilities, while packed-bed, fluidized-bed, and air-lift bioreactors are also being explored for continuous or semi-continuous CBP operations (van Dijken & Pronk, 2005). Key features include automated feed systems to regulate substrate concentration and avoid sugar accumulation, agitation and aeration controls for aerobic or anaerobic conditions, and in-line probes for real-time monitoring (e.g., pH, dissolved oxygen, and redox potential).

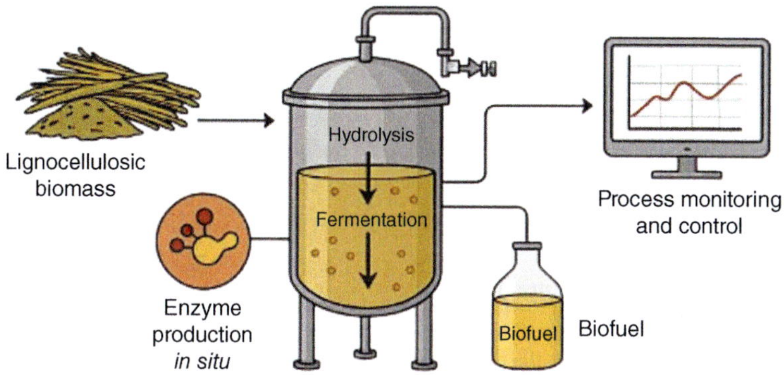

Figure 12.2 Schematic representation of a consolidated bioprocessing (CBP) system.

Newer systems incorporate membrane-integrated bioreactors that enable product separation during fermentation, thereby reducing inhibition and improving yields (Raghavendran et al., 2017). A modular CBP reactor design is being increasingly favored for decentralized, small-footprint biorefineries, particularly in developing countries or remote areas where infrastructure and supply chain access are limited.

12.3.4 Process Monitoring and Control

CBP systems must be precisely monitored to ensure consistent and efficient performance. Unlike conventional batch processing, CBP requires a holistic process control strategy that simultaneously tracks enzymatic conversion, microbial growth, sugar utilization, and product formation. Figure 12.2 shows a schematic representation of a CBP system. Key parameters to monitor include (a) substrate and sugar concentrations (e.g., cellulose, glucose, and xylose); (b) fermentation metabolites (ethanol, organic acids, and gases); (c) dissolved oxygen, redox potential, pH, and temperature; and (d) cell biomass concentration and viability.

Modern CBP facilities use real-time sensors and feedback control loops to regulate these variables, reducing downtime and improving product quality. Advanced monitoring systems may also include biosensors for detecting specific metabolic products, spectrophotometric or electrochemical probes for sugar and alcohol concentrations, and machine learning algorithms that predict process deviations based on historical data (Lee et al., 2020).

Integrating digital twins and AI-driven control platforms is becoming more feasible with the advent of Industry 4.0. These technologies offer predictive maintenance, intelligent optimization of process parameters, and real-time decision-making, which are essential for large-scale, economically competitive CBP deployment.

12.4 Cost Components of CBP

12.4.1 Feedstock Collection and Pretreatment

The selection of feedstock plays a crucial role in determining the overall cost-effectiveness of CBP. LCB, particularly high-density sources, offers economic advantages by reducing transportation costs, which are a significant factor in biorefinery feasibility. Mechanized pretreatment strategies, such as steam explosion and ammonia fiber expansion, enhance biomass accessibility while minimizing the formation of chemical inhibitors that could hinder enzymatic hydrolysis. These methods improve the efficiency of downstream processing, ensuring higher yields and reducing operational bottlenecks. Recent studies highlight integrated pretreatment approaches, such as hydrothermal processing combined with enzymatic hydrolysis, which have demonstrated improved sugar release efficiency and reduced energy consumption (Konda et al., 2014).

For effective enzymatic saccharification of lignocellulose, pretreatment of the feedstock is a necessary step. The primary goal of this pretreatment process is to separate the main components of LCB, with a particular focus on removing lignin from cellulose and hemicellulose. Additionally, it disrupts hydrogen bonds and van der Waals interactions in cellulose microfibrils, thereby loosening the rigid structure of lignocellulose. Various pretreatment methods—physical, chemical, and biological—have been developed to modify the structure of LCB. Physical techniques primarily aim to reduce particle size and increase the specific surface area through mechanical crushing and high-temperature hydrothermolysis. In recent years, irradiation-based approaches such as microwave and ultrasonic treatments have gained popularity for their ability to quickly release intracellular cellulose under high-energy radiation. Despite their effectiveness, these physical methods often require high energy input, posing challenges for large-scale application. Chemical pretreatment strategies utilize specific chemicals to selectively dissolve lignocellulose components, altering their structural composition. Approaches such as acid, alkali, oxidation, and organic solvent pretreatment significantly enhance biomass surface area and improve lignocellulose breakdown, thus facilitating enzymatic conversion. The optimization of pretreatment conditions remains a key research focus to enhance CBP viability. By refining lignocellulose pretreatment techniques, biomass accessibility is enhanced, ultimately leading to more efficient enzymatic saccharification and improved bioprocessing results (Li, et al., 2024).

12.4.2 Microbial Strain Development and Inoculum Preparation

Microorganisms from diverse environments inherently synthesize a wide range of bioactive compounds, which serve as essential sources for fuels, pharmaceuticals, and other valuable chemicals. The development of robust microbial strains capable of producing both hydrolytic enzymes and fermentative products is fundamental to CBP success. For example, genetically modified *S. cerevisiae* strains designed to produce cellulases have demonstrated significant potential in fermenting complex substrates like sweet potato residue (Li et al., 2024). Advances in synthetic biology and genetic engineering have enabled the creation of highly efficient microbial consortia that streamline the conversion of lignocellulosic feedstocks into biofuels. Investments in strain engineering and inoculum preparation are essential, but these costs are offset by the elimination of separate enzyme production and purification steps. The use of CRISPR-based genetic modifications has further improved microbial efficiency, allowing for enhanced substrate utilization and higher fermentation yields (Nagarajan et al., 2019).

12.4.3 Capital Costs: Equipment, Bioreactor, and Utilities

CBP systems often employ single-use bioreactors, which provide advantages over traditional stainless-steel systems, including reduced cleaning and sterilization demands. These systems can lead to cost savings of more than 60% compared to fixed-asset bioreactors. Additionally, the integration of CBP processes simplifies facility design, potentially lowering capital expenditures. The global fermenters and bioreactors market is projected to grow from $20.3 billion in 2024 to $55.45 billion by 2034, reflecting the increasing adoption of bioprocessing technologies (Uniyal & Bijalwan, 2024). Recent advancements in bioreactor design focus on continuous processing, which enhances productivity while reducing operational costs. The shift toward modular bioreactor systems has also improved scalability, making CBP more adaptable to industrial applications (Venkatraman et al., 2024).

12.4.4 Operating Costs: Enzymes, Media, Energy, and Labor

The majority of approaches for transforming LCB into biofuels depend on the enzymatic depolymerization of polysaccharides. However, the high cost of these enzymes remains a major challenge in establishing a financially viable lignocellulose-based biofuel industry. One of the most significant advantages of CBP is the reduction in enzyme costs. By utilizing microorganisms that produce their own hydrolytic enzymes, CBP eliminates the need for externally produced enzymes, leading to substantial operational cost savings. However, media costs remain a considerable expense. Strategies to reduce media consumption and optimize fermentation conditions are ongoing to further decrease operating costs (Chen et al., 2024). The global industrial enzyme market, valued at $7.5 billion in 2024, is expected to grow at a Compounded Annual Growth Rate (CAGR) of 5.23%, driven by increased demand for bio-based catalysts. Efforts to enhance enzyme efficiency through protein engineering and directed evolution have led to improved catalytic activity and stability under industrial conditions (Bangaru et al., 2022).

12.4.5 Downstream Processing and Product Recovery

CBP simplifies downstream processing by combining hydrolysis and fermentation into a single step, reducing the need for complex separation and purification processes. This integration not only lowers equipment and labor costs but also minimizes the risk of contamination between stages. Furthermore, the immediate consumption of sugars by microorganisms during fermentation prevents the accumulation of inhibitory by-products, enhancing overall process efficiency. Advances in membrane filtration, chromatography, and crystallization techniques have improved the recovery of biofuels and biochemicals from fermentation broths. Recent studies emphasize the importance of process optimization in downstream recovery, particularly in the purification of high-value bio-based chemicals. The development of automated separation technologies has further streamlined product recovery, reduced energy consumption, and improved yield (Jungbauer et al., 2024).

12.5 Economic Feasibility and Techno-Economic Assessment

CBP is gaining traction as a cost-effective and sustainable approach to biofuel production. Evaluating its economic feasibility requires analyzing key financial metrics, including CAPEX, OPEX, ROI, NPV, and IRR (Jalaei et al., 2025).

- CAPEX refers to the initial investment required for infrastructure, equipment, and facility setup. CBP facilities often require advanced bioreactors, microbial strain development, and integrated processing units, leading to high upfront costs. However, single-use bioreactors have been shown to reduce CAPEX by up to 60% compared to traditional stainless-steel systems (Bauer et al., 2020).
- OPEX includes ongoing operational costs, such as enzyme production, feedstock procurement, labor, and utilities. CBP significantly reduces enzyme costs by utilizing microorganisms that self-produce hydrolytic enzymes, eliminating the need for externally sourced enzymes (Berrada et al., 2025).
- ROI measures the profitability of CBP investments. Studies indicate that CBP-based ethanol production achieves a higher ROI compared to conventional biofuel processes due to lower enzyme costs and streamlined processing (Berrada et al., 2025).
- NPV assesses the long-term profitability of CBP facilities by discounting future cash flows. A study on CBP ethanol production found that NPV increases with scale, as larger facilities benefit from economies of scale and reduced feedstock transport costs (Hoey et al., 2025).
- IRR represents the rate of return on CBP investments. A techno-economic analysis of CBP-based lignocellulosic ethanol production reported an IRR of 16.8%, indicating strong investment viability (Shafinas Muhammad & Rosentrater, 2021).

12.5.1 Case Study: Cost–Benefit Analysis of a CBP Facility

CBP is an advanced biofuel production method that integrates enzyme production, hydrolysis, and fermentation into a single step, reducing costs and improving efficiency. A cost–benefit analysis of a CBP facility evaluates its economic feasibility by comparing initial investment costs with long-term sustainability benefits.

12.5.1.1 Cost Considerations

The primary costs associated with CBP facilities include infrastructure development, operational expenses, and microbial strain optimization. Recent studies highlight that capital investment in CBP facilities ranges between $50 and 100 million, depending on scale and technology integration. Additionally, feedstock costs contribute significantly to overall expenses, with fluctuations in biomass availability affecting production efficiency. CBP facilities require specialized microbial strains capable of efficiently converting LCB into ethanol. The development and optimization of these strains involve significant research and development costs. However, advancements in genetic engineering have led to the creation of robust microorganisms that enhance ethanol yield while minimizing processing steps.

CBP facilities contribute to environmental sustainability by utilizing agricultural residues and other LCB sources. This reduces reliance on fossil fuels and minimizes waste generation. The adoption of CBP technology aligns with global efforts to transition toward renewable energy sources and mitigate climate change effects. Economically, CBP facilities offer long-term cost savings by streamlining the biofuel production process. The reduction in enzyme costs and improved conversion efficiency translates to lower production expenses and higher profitability. Governments and private investors are increasingly supporting CBP initiatives due to their potential to enhance energy security and promote sustainable development (Olguin-Maciel et al., 2020).

12.5.1.2 Benefit Analysis

CBP facilities offer multiple advantages, including reduced enzyme costs, higher conversion efficiency, and lower energy consumption. The integration of genetically engineered microorganisms enhances ethanol yield while minimizing processing steps. Furthermore, CBP significantly reduces greenhouse gas emissions compared to traditional bioprocessing methods. Recent studies indicate that CBP can achieve ethanol yields of up to 90% of the theoretical maximum, making it a highly efficient process. Additionally, CBP eliminates the need for separate enzyme production, reducing overall operational costs and simplifying the production process (Olguin-Maciel et al., 2019).

12.5.2 Sensitivity Analysis: Effect of Feedstock Price and Yield

Sensitivity analysis is a crucial tool in evaluating the economic feasibility of biofuel production. The price of feedstock and its yield significantly impact the overall cost-effectiveness of bioprocessing. Variability in feedstock availability and quality can lead to fluctuations in sugar yields and production costs, affecting the sustainability of biofuel production. Studies indicate that single-pass harvested corn stover yields higher sugar content and lower production costs compared to multiphase harvested stover. Additionally, blending different feedstocks, such as switchgrass and sorghum, can enhance availability and sustainability while maintaining cost efficiency (Hemmelgarn et al., 2024).

The cost-based pricing of sustainable aviation fuel (SAF) is highly sensitive to feedstock prices, demonstrating a direct correlation between feedstock costs and SAF market dynamics. Over the past five years, feedstock price volatility has influenced SAF pricing, with reduced fluctuations observed after initial declines. This highlights the importance of robust price assessments in ensuring market stability and advancing green fuel technologies (Wang et al., 2024).

12.5.3 Comparison with Traditional Bioprocessing and Other Green Technologies

Traditional bioprocessing methods often rely on linear economic models, which contribute to environmental degradation through excessive waste generation. In contrast, green technologies emphasize circular bioeconomy principles, promoting waste valorization and sustainable resource utilization. The transition from conventional refineries to biorefineries has facilitated the production of bio-based products, including next-generation biofuels, enzymes, and vitamins. Emerging bioprocessing technologies, such as continuous biomanufacturing and digitalization, are revolutionizing the industry

(Ramchuran et al., 2023). Continuous biomanufacturing enhances productivity and reduces operational footprints, while artificial intelligence-driven automation optimizes process control. Sustainable bioprocessing practices, including single-use technologies and bio-renewable feedstocks, minimize environmental impact and improve efficiency. Life cycle assessments (LCAs) provide valuable insights into the sustainability of single-use technologies in biopharmaceutical manufacturing. By evaluating environmental stress factors, LCA helps industries identify areas for improvement and implement strategies for reducing carbon footprints. The adoption of circular economy principles in bioprocessing is essential for achieving long-term sustainability and mitigating climate change. (Sodhi et al., 2025).

12.6 Challenges and Opportunities

CBP provides a unique path for green energy production and sustainable development; however, it encounters difficulties related to genetically designing ideal microorganisms and boosting processes on an industrial scale.

12.6.1 Technical Barriers in Microbial Engineering and Scale-Up

It is a great challenge to design a single strain of microorganism that is capable of all types of CBP, including enzyme production, hydrolysis, and fermentation processes. Some microorganisms can perform one or two of these processes, but all three of these processes are required for green energy production and sustainable development. The second major challenge is to enhance the production from laboratory scale to pilot scale while maintaining product yield and quality (Rizvi & Ashish, 2024).

12.6.2 Economic Viability for Developing Countries

Many developing countries depend on their energy on non-renewable energy resources, which are limited and harmful to the environment. CBP is not only eco-friendly but also promotes economic development, reduces infrastructure costs, and minimizes waste.

12.6.3 Opportunities in Synthetic Biology and AI-Driven Process Control

CBP combines numerous steps into a single process. Nowadays, AI-based control processes provide significant opportunities in synthetic biology. Genetically engineered microbes are used in CBP to decrease costs and improve efficiency by removing steps like enzyme production and feedstock pre-processing and the fermentation process. AI is helpful to optimize these processes and enhance product yields through data analysis and machine learning algorithms.

12.6.4 Future Directions for Sustainable and Circular CBP Models

The future model of bioprocessing is mainly focused on green technology and sustainability by utilizing biological and agricultural waste. CBP can contribute to waste minimization, decrease environmental impact, and reduce carbon by converting CO_2 into important, valuable products like ethanol.

References

Alvira, P., Tomás-Pejó, E., Ballesteros, M., & Negro, M. J. (2010). Pretreatment technologies for an efficient bioethanol production process based on enzymatic hydrolysis: A review. *Bioresource Technology, 101*(13), 4851–4861. https://doi.org/10.1016/j.biortech.2009.11.093.

Bangaru, A., Sree, K. A., Kruthiventi, C., Banala, M., Shreya, V., Vineetha, Y., Shalini, A., Mishra, B., Yadavalli, R., Chandrasekhar, K., & Reddy, C. N. (2022). Role of enzymes in biofuel production: Recent developments and challenges. *Bio-Clean Energy Technologies, 1*, 81–112. https://doi.org/10.1007/978-981-19-0188-7_4.

Bauer, I., Dreher, T., Eibl, D., Glöckler, R., Husemann, U., John, G. T., Kaiser, S. C., Kampeis, P., Kauling, J., Kleebank, S., & Kraume, M. (2020). *Recommendations for process engineering characterisation of single-use bioreactors and mixing systems by using experimental methods.* DECHEMA Gesellschaft für Chemische Technik und Biotechnologie eV. https://doi.org/10.13140/RG.2.2.26390.06726.

Bayer, E. A., Belaich, J. P., Shoham, Y., & Lamed, R. (2004). The cellulosomes: Multienzyme machines for degradation of plant cell wall polysaccharides. *Annual Review of Microbiology, 58*, 521–554. https://doi.org/10.1146/annurev.micro.57.030502.091022.

Berrada, A., Sanjari, M. J., & El Mrabet, R. (2025). Techno-economic assessment of hydrogen production: Comparative analysis of electrolyser technologies in a hybrid PV/wind system. *International Journal of Hydrogen Energy, 141*, 193–211. https://doi.org/10.1016/j.ijhydene.2024.01.095.

Chen, Y. W., Feng, G., Hong, X., Wang, S., Dong, H., Yin, F., & Tang, Y.-Q. (2024). Effects of high solid content and straw proportion on volatile fatty acids production from straw, sludge and food wastes: performance and microbial community characteristics. *Appl Biol Chem., 67*, 83. https://doi.org/10.1186/s13765-024-00935-1.

European Commission. (2018). *A sustainable bioeconomy for Europe: Strengthening the connection between economy, society and the environment*. https://ec.europa.eu.

Hasunuma, T., & Kondo, A. (2012). Development of yeast cell factories for consolidated bioprocessing of lignocellulose to bioethanol through cell surface engineering. *Biotechnology Advances, 30*(6), 1207–1218. https://doi.org/10.1016/j.biotechadv.2011.10.011.

Hemmelgarn, A. B., Shinners, K. J., Timm, A. J., & Digman, M. F. (2024). Single-pass, single-stream harvest of corn grain and stover: A comparison of two harvest methods. *BioEnergy Research, 17*(4), 2158–2163. https://doi.org/10.1007/s12155-024-10638-5.

Hoey, M., Clark, R., Webb, E., & Lynd, L. (2025). Scale sensitivity of ethanol production via consolidated bioprocessing with consideration of feedstock cost. *Biofuels, Bioproducts and Biorefining, 19*(1), 9–17. https://doi.org/10.1002/bbb.2581.

Humbird, D., Davis, R., Tao, L., Kinchin, C., Hsu, D., Aden, A., & Dudgeon, D. (2011). *Process design and economics for biochemical conversion of lignocellulosic biomass to ethanol.* NREL Report. https://doi.org/10.2172/1013269.

International Energy Agency (IEA). (2021). *Renewables 2021: Analysis and forecast to 2026.* https://www.iea.org/reports/renewables-2021.

International Renewable Energy Agency (IRENA). (2020). *Bioenergy for the transition: Ensuring sustainability and overcoming barriers*. https://www.irena.org.

Iruj, M., Khan, D. M., Yaqoob, S., & Iqbal, Z. (2025). Sustainable Modeling of the Urban Air Quality in Abu Dhabi Using Machine Learning and Open-Source Satellite Data. *Sustainable Processes Connect, 1*(1), 1–10. https://doi.org/10.69709/SusProc.2025.188432.

Jalaei, F., Gourabpasi, A. H., & Mehran, D. (2025). Data delivery for standardizing sustainable whole-building lifecycle assessment using the proposed openBIM framework. *Sustainable Processes Connect*, *1*(1), 1–10. https://doi.org/10.69709/SusProc.2025.155435.

Jungbauer, A., Satzer, P., Duerauer, A., Azevedo, A., Aires-Barros, R., Nilsson, B., Farid, S., Goldrick, S., Ottens, M., Sponchioni, M., & Lahore, H. M. (2024). Continuous downstream processing. *Separation and Purification Technology*, *338*, 126439. https://doi.org/10.1016/j.seppur.2024.126439.

Klein-Marcuschamer, D., Oleskowicz-Popiel, P., Simmons, B. A., & Blanch, H. W. (2012). The challenge of enzyme cost in the production of lignocellulosic biofuels. *Biotechnology and Bioengineering*, *109*(4), 1083–1087. https://doi.org/10.1002/bit.24370.

Klein-Marcuschamer, D., Simmons, B. A., & Blanch, H. W. (2012). Techno-economic analysis of a lignocellulosic ethanol biorefinery with ionic liquid pretreatment. *Biofuels, Bioproducts and Biorefining*, *6*(3), 246–256. https://doi.org/10.1002/bbb.1318.

Konda, N. M., Shi, J., Singh, S., Blanch, H. W., Simmons, B. A., & Klein-Marcuschamer, D. (2014). Understanding cost drivers and economic potential of two variants of ionic liquid pretreatment for cellulosic biofuel production. *Biotechnology for Biofuels*, *7*, 86. https://doi.org/10.1186/1754-6834-7-86.

Kumar, A., Devnani, G. L., & Pal, D. B. (2025). Sustainable degradation of organic pollutants using various treatment processes: a review. *Sustainable Processes Connect*, *1*(1), 1–10. https://doi.org/10.69709/SusProc.2025.111130.

Lee, J. W., Wood, D. N., & Trinh, C. T. (2020). Towards sustainable and automated microbial biomanufacturing for biochemicals. *Current Opinion in Biotechnology*, *64*, 48–60. https://doi.org/10.1016/j.copbio.2019.08.003.

Li, J., Wang, X., Zhou, Y., & Chen, L. (2024). Engineering *Saccharomyces cerevisiae* for consolidated bioprocessing of sweet potato residue. *Journal of Industrial Microbiology & Biotechnology*, *51*(2), 123–132. https://doi.org/10.1093/jimb/kuad080.

Li, Z., Waghmare, P. R., Dijkhuizen, L., Meng, X., & Liu, W. (2024). Research advances on the consolidated bioprocessing of lignocellulosic biomass. *Engineering Microbiology*, *4*(2), 100139. https://doi.org/10.1016/j.engmic.2024.100139.

Lynd, L. R., van Zyl, W. H., McBride, J. E., & Laser, M. (2005). Consolidated bioprocessing of cellulosic biomass: An update. *Current Opinion in Biotechnology*, *16*(5), 577–583. https://doi.org/10.1016/j.copbio.2005.08.009.

Lynd, L. R., Weimer, P. J., van Zyl, W. H., & Pretorius, I. S. (2002). Microbial cellulose utilization: Fundamentals and biotechnology. *Microbiology and Molecular Biology Reviews*, *66*(3), 506–577. https://doi.org/10.1128/MMBR.66.3.506-577.2002.

Nagarajan, D., Lee, D. J., & Chang, J. S. (2019). Recent insights into consolidated bioprocessing for lignocellulosic biohydrogen production. *International Journal of Hydrogen Energy*, *44*(28), 14362–14379. https://doi.org/10.1016/j.ijhydene.2019.04.182.

Olguin-Maciel, E., Larqué-Saavedra, A., Lappe-Oliveras, P. E., Barahona-Pérez, L. F., Alzate-Gaviria, L., Chablé-Villacis, R., Domínguez-Maldonado, J., Pacheco-Catalán, D., Ruíz, H. A., & Tapia-Tussell, R. (2019). Consolidated bioprocess for bioethanol production from raw flour of *Brosimum alicastrum* seeds using the native strain of *Trametes hirsuta* bm-2. *Microorganisms*, *7*(11), 483. https://doi.org/10.3390/microorganisms7110483.

Olguin-Maciel, E., Singh, A., Chable-Villacis, R., Tapia-Tussell, R., & Ruiz, H. A. (2020). Consolidated bioprocessing, an innovative strategy towards sustainability for biofuels production from crop residues: An overview. *Agronomy, 10*(11), 1834. https://doi.org/10.3390/agronomy10111834.

Olson, D. G., McBride, J. E., Shaw, A. J., & Lynd, L. R. (2012). Recent progress in consolidated bioprocessing. *Current Opinion in Biotechnology, 23*(3), 396–405. https://doi.org/10.1016/j.copbio.2011.11.026.

Palwan, E., Debbarma, A., Debbarma, S., & Chaudhuri, S. R. (2025). Enhancing Fiber Yield and Environmental Sustainability through Integrated Fertilization Strategies and waste valorization of Ramie (Boehmeria nivea L). *Sustainable Processes Connect, 1*(1), 1–10. https://doi.org/10.69709/SusProc.2025.101023.

Raghavendran, V., Gummadi, S. N., & Sridhar, S. (2017). Membrane bioreactors for bioethanol production: Recent advances and challenges. *Biotechnology Journal, 12*(10), 1600452.

Ramchuran, S. O., O'Brien, F., Dube, N., & Ramdas, V. (2023). An overview of green processes and technologies, biobased chemicals and products for industrial applications. *Current Opinion in Green and Sustainable Chemistry, 41*, 100832. https://doi.org/10.1016/j.cogsc.2023.100832.

Rashed, M. S., Fakhry, S., Satour, R., Sobhy, M., & Haghi, A. K. (2025). Water Footprint and Water Resources Sustainability: A Review. *Sustainable Processes Connect, 1*(1), 1–10. https://doi.org/10.69709/SusProc.2025.188432.

Rodrigues, B. A., Machado, C. H. D. L., Mendonça, C. M., Prudente, E. S., Furtado, A. C. B., Fontanari, G. G., Komesu, A., & Martins, L. H. (2025). Drying Kinetics and Thermodynamic Properties of Cassava (Manihot esculenta Crantz) Peel as a Sustainable Agro-Industrial By-Product. *Sustainable Processes Connect, 1*(1), 1–10. https://doi.org/10.69709/SusProc.2025.1993765.

Rizvi, H., & Ashish. (2024 Jul 26). A critical review on scale-up strategies of biosurfactant production and its applications. *Tenside Surfactants Detergents., 61*(4), 297–309. https://doi.org/10.1515/tsd-2023-2567.

Shafinas Muhammad, N. I., & Rosentrater, K. A. (2021). Economic evaluation of combined heat and power integrated with food waste-based ethanol production. *Frontiers in Energy Research, 9*, 682609. https://doi.org/10.3389/fenrg.2021.6826.

Sodhi, A. S., Bhatia, S., & Batra, N. (2025). Advances in green technologies and global circular bioeconomy framework for a sustainable society. *Circular Economy and Sustainability, 5*, 2991–3016. https://doi.org/10.1007/s43615-025-00536-0.

Stephanopoulos, G. (2007). Challenges in engineering microbes for biofuels production. *Science, 315*(5813), 801–804. https://doi.org/10.1126/science.1137266.

UN. (2022). *United Nations sustainable development goals*. https://sdgs.un.org/goals.

Uniyal, S., & Bijalwan, P. K. (2024). Scaling up bioreactor systems for sustainable biorefinery: A crucial step in advancing the green economy. In K. Agrawal & P. Verma (Eds.), *Biotechnological Advances in Biorefinery* (pp. 231–248). Singapore: Springer. https://doi.org/10.1007/978-981-97-5544-8.

van Dijken, J. P., & Pronk, J. T. (2005). Sustainable microbial production of fuels, chemicals, and pharmaceuticals from biomass. *Current Opinion in Biotechnology, 16*(3), 254–259. https://doi.org/10.1016/j.copbio.2005.04.004.

Venkatraman, S., Rajendran, D. S., & Vaidyanathan, V. K. (2024). Development and scale-up of the bioreactor system in biorefinery: A significant step toward a green and bio-based economy. In S. Uniyal & P. K. Bijalwan (Eds.), *Biotechnological advances in biorefinery* (pp. 207–230). Springer Nature Singapore. https://doi.org/10.1007/978-981-97-5544-8_10.

Wang, B., Ting, Z. J., & Zhao, M. (2024). Sustainable aviation fuels: Key opportunities and challenges in lowering carbon emissions for aviation industry. *Carbon Capture Science & Technology*, *13*, 100263. https://doi.org/10.1016/j.ccst.2024.100263.

Wyman, C. E., Dale, B. E., & Elander, R. T. (2005). Coordinated development of leading biomass pretreatment technologies. *Bioresource Technology*, *96*(18), 1959–1966. https://doi.org/10.1016/j.biortech.2005.01.010.

Zhang, M., Eddy, C., Deanda, K., Finkelstein, M., & Picataggio, S. (2007). Metabolic engineering of a pentose metabolism pathway in ethanologenic *Zymomonas mobilis*. *Science*, *276*(5318), 1485–1487. https://doi.org/10.1126/science.267.5195.240.

Zhang, Y. H. P., Ding, S. Y., Mielenz, J. R., Cui, J. B., Elander, R. T., Laser, M., Himmel, M. E., McMillan, J. R., & Lynd, L. R. (2006). Fractionating recalcitrant lignocellulose at modest reaction conditions. *Biotechnology and Bioengineering*, *94*(5), 851–861. https://doi.org/10.1002/bit.20906.

Zhou, Y., Zhang, X., & Tan, T. (2016). Integration of alkaline hydrogen peroxide pretreatment with simultaneous saccharification and fermentation for ethanol production from corn stover. *Industrial Crops and Products*, *92*, 99–104. https://doi.org/10.1016/j.indcrop.2016.08.005.

13

Biochemical Enzymes to Influence the Bioprocess of Lignocellulosic Biomass for Bioethanol Production Process and Cost: An Analysis

Nihan Arabacı

Department of Biology, Faculty of Arts and Sciences, University of Cukurova, Adana, Türkiye

13.1 Introduction

Various environmental problems, such as excess waste, environmental pollution, global warming, climate change, depletion of fossil fuels, and the energy crisis that we face due to the increase in the world population, have been the driving forces for researchers to develop new methods. The energy crisis, one of the biggest problems, has led to a shift toward renewable and natural energy sources as an alternative to fossil fuels (Nargotra et al., 2022).

Fossil fuels such as oil, natural gas, and coal are primary energy sources, but their formation processes in nature are quite long. Moreover, CO_2 emissions from burning fossil fuels pose a danger in many situations, including global warming, environmental pollution, and a decrease in biodiversity (Periyasamy et al., 2023; Broda et al., 2022). The global energy demand, increasing continuously in recent years, indicates that fossil fuel reserves will be depleted, and the energy crisis may become even more inextricable (Broda et al., 2022). The price increase, parallel to the decrease in fossil fuels, has caused an oil crisis. For these reasons, alternative energy and fuel sources have been sought, and the production of various biofuels has gained priority globally. The production processes of fuels such as bioethanol, biobutanol, biogas, and biodiesel, included under the title of biofuels, have been developed and produced in a way that can solve the energy crisis. These fuels can be obtained from natural resources and waste, are renewable, nature-friendly, low-cost, and are quality energy sources that contribute to the country's economy (Periyasamy et al., 2023). Among these fuel types, bioethanol attracts attention especially because it can be used instead of gasoline.

In response to the depletion of energy resources in the world, interest in obtaining sustainable energy from renewable energy sources, especially from waste biomass, which is abundant on Earth, has increased (Lin & Tanaka, 2006). In recent years, the use of biomass formed by agricultural wastes as a raw material in biorefinery processes has come to the fore (Rajesh & Gummadi, 2023). In this context, bioethanol production has become very important worldwide. This bioenergy offers some significant advantages, such as evaluating waste biomass as alternative substrates in fuel production, solving the issue of waste disposal, using enzyme and fermentation technology, and minimizing greenhouse gas emissions by using fossil fuels (Lin & Tanaka, 2006).

Bioethanol, which can be used instead of petroleum-derived fossil fuels, can be produced from first-, second-, and third-generation raw materials (Sharma, Xu, et al., 2019). Lignocellulose is one of the most advantageous of the second-generation raw materials. It is more environmentally friendly, economical, and sustainable compared to fossil resources. When used as a raw material in bioethanol production, both the production cost of the fuel decreases and the waste in nature is recycled. The fact that it is not food and is abundant in nature makes it a preferred choice in bioethanol production. This is because when plant products that are also consumed as food are used in first-generation bioethanol production, problems occur in food supply. However, the production of second-generation bioethanol from lignocellulosic wastes is a solution to these problems and is commercially very valuable (Siqueira et al., 2020). Lignocellulose biomass is inexpensive, abundantly available, and can be supplied as an agricultural by-product or waste, as it is derived from plant raw material (Saini et al., 2022). It has a durable structure consisting of cellulose, hemicellulose, and lignin. Therefore, as a raw material, lignocellulose must first be hydrolyzed into the carbohydrates in its structure. For this purpose, physical, chemical, and biological processes are applied. Biological methods are environmentally friendly approaches, as they allow the use of cellulases, hemicellulases, and ligninolytic enzymes obtained from microorganisms for the hydrolysis of lignocellulose. They require less energy, are low-cost, and do not produce inhibitory by-products compared to chemical and physical methods (Nargotra et al., 2022; Prasad et al., 2019). In this context, microbial cellulases and lignocellulosic raw materials used in bioethanol production will be discussed in this section.

13.2 Bioethanol Production

One of the biofuels that can provide a solution to the environmental problems caused by petroleum-based fuels worldwide is bioethanol (Periyasamy et al., 2023). It is valuable in the biofuel sector both because it reduces greenhouse gas emissions and is an economical alternative to petroleum (Bala & Singh, 2019).

First-generation bioethanol is produced from starchy foods containing fermentable sugars such as sugar beet, corn, rice, and sugar cane. Biofuel production by direct use of these food products negatively affects the industry as it can endanger the supply of food and feed resources (Periyasamy et al., 2023). In order not to negatively affect the food sector, the use of abundant, economical, and renewable second-generation biomass lignocellulosic plant raw material sources has become important globally, and bioethanol production from this biomass provides an effective solution to food supply problems (Areeshi, 2022).

Second-generation biofuels, including cellulosic bioethanol, are environmentally friendly fuel types obtained from lignocellulosic biomass consisting of herbaceous and woody forest and agricultural wastes, which are abundant in nature, do not affect the food supply, and are known as cheap raw materials for fuel production (Prasad et al., 2019). Bioethanol production from renewable resources, especially lignocellulosic biomass, reveals that the industry has an eco-friendly approach. The conversion of this biomass into simple sugars using microbial enzymes (e.g., cellulases) and ethanol production by microbial fermentation is a highly value-added procedure for the industry (Bala & Singh, 2019). The durable cellulosic polysaccharide in the structure of this raw material can be hydrolyzed to fermentable sugars with various cellulosic enzymes. The efficient saccharification that occurs as a result of this hydrolysis with enzymes offers a more efficient bioconversion approach due to its environmental friendliness (Bhardwaj et al., 2021). In the production of second-generation bioethanol, two methods are preferred: simultaneous saccharification and fermentation (SSF) and separate hydrolysis and fermentation (SHF). In the SHF method, hydrolysis and fermentation are carried out separately at optimum pH and temperature. However, the sugar accumulated during hydrolysis can inhibit cellulase activity. Since better hydrolysis occurs in the SSF method, the ethanol yield is higher. However, since SSF does not occur under optimum conditions, the hydrolysis stage takes longer, and more enzymes may be needed. Thus, the production cost of bioethanol increases (Areeshi, 2022).

13.3 Biomass Feedstocks in the Manufacturing of Bioethanol

Bioethanol feedstocks can be broadly divided into three categories:

13.3.1 First-Generation Feedstocks

In first-generation bioethanol production, sugar plants such as sugar beet and sugar cane and plants with high starch content such as grain, corn, maize, cassava, potato, and wheat are used (Prasad et al., 2019; Soccol et al., 2019). In this production process, sugar conversion to bioethanol is very simple because only the mechanical process (grinding/crushing), like pretreatment, is required to obtain sugar from these plants. The starch of the starchy raw materials is gelatinized by cooking and then subjected to enzymatic hydrolysis, and the released glucose monomers are obtained and subjected to fermentation. The biggest drawback of first-generation bioethanol production is that the use of first-generation raw materials consumed as food in fuel production creates problems in terms of food demand and food safety in the food sector. When these raw materials, which are food products, are used in bioethanol production, problems arise in the food industry in terms of food supply. To eliminate these problems, very large agricultural lands are needed. Biodiversity will be negatively affected by meeting this need. For this reason, there has been a trend in second-generation biofuel production (Periyasamy et al., 2023; Prasad et al., 2019).

13.3.2 Second-Generation Feedstocks

Second-generation biofuels are obtained from renewable biomass that cannot be used as food. This biomass includes raw materials such as straw, rice husk, rice straw, corn cob, cornhusk, corn stover, and sugarcane bagasse. In the second-generation biofuel production process, the most abundant lignocellulosic biomass on Earth is evaluated as agricultural waste and used as raw material (like substrate) in biofuel production (Soccol et al., 2019). Bioethanol is produced by microbial fermentation of simple sugars obtained from this biomass with the help of enzymatic hydrolysis. Using lignocellulose is very valuable regarding food supply and security (Sharma, Xu, et al., 2019). Second-generation biofuels help to reduce carbon emissions, decrease fossil fuel use, and increase energy efficiency compared to the first-generation biofuels. It is predicted that approximately two-thirds of the world's energy needs will be provided by second-generation biomass by 2050. Enzymatic hydrolysis of cellulose and hemicellulose polysaccharides in lignocellulose composition is the most important stage in bioethanol production. However, the handicap in this production is that it requires a hard pretreatment process to separate lignin in the structure of lignocellulose from other components (Periyasamy et al., 2023). This situation is also overcome thanks to the activity of various ligninolytic enzymes.

13.3.3 Third-Generation Feedstocks

Algal biomass and seaweeds are the raw materials of third-generation biofuels. Carbohydrates, proteins, and lipids of microalgae have significant potential in the production of liquid biofuels such as bioethanol and biodiesel. Algae are highly efficient in biofuel production due to their ability to easily adapt to environmental conditions and to convert light into chemical energy thanks to their chlorophyll pigment. Third-generation raw materials have various advantages, such as being suitable for continuous production, being able to grow in water, not requiring land, having a high biofuel production capacity, being able to purify agricultural waste, producing by-products such as protein, and being used as feed and fertilizer (Periyasamy et al., 2023; Prasad et al., 2019). Considering these features, these raw materials have the potential to take place in the biorefinery industry as valuable biofuel biomass in the future due to their ability to produce sustainable green energy. They can be used as a huge bioethanol source that can meet the needs of the increasing population (Demirbas, 2010). However, in commercial bioethanol production, lignocellulosic biomass, an agricultural waste, is still preferred as a raw material compared to algae.

13.4 Lignocellulosic Biomass in the Bioethanol Process

In order to meet the energy demand that comes with population growth, sustainable resources that will be an alternative to limited fossil fuels have been sought. In this context, there has been a demand for biofuel production from lignocellulosic raw materials (Saini et al., 2022). This raw material is a carbon source that can be obtained from approximately more than 200 billion tons of plant waste (rice straw, corn straw, wheat straw, wood, agricultural wastes, grasses, trees, and forestry wastes) annually and contains abundant polysaccharides (Ilić et al., 2023; Broda et al., 2022; Prasad et al., 2019). It is an advantageous raw material in that it does not require extra agricultural land, is easy to obtain, and does not hinder the food and feed sector since it is not consumed as food. Its use in the bioethanol production process contributes positively to the environment, solves the problem of waste biomass, and provides economic benefits (Broda et al., 2022).

Lignocellulosic biomass is less harmful to the environment compared to coal and petroleum-derived resources. During the production processes of fossil fuels, by-products and hazardous gases that cause environmental pollution are released. However, biofuels, alternatives to fossil fuels, play a vital role in protecting the environment by reducing greenhouse gas emissions (Shahzadi et al., 2014). In addition, lignocellulose found in plants has a very durable structure. To break down the complex structure of lignocellulose, consisting of cellulose, hemicellulose, and lignin, the biomass must be subjected to various pretreatments. The pretreatment process should be environmentally friendly, easy to apply, have low inhibitory by-product production, be low-cost, not damage lignocellulosic raw materials, and increase bioethanol yield. For this purpose, physical, chemical, physicochemical, and biological (enzymatic) methods are applied (Broda et al., 2022). This biopolymer attracts great attention due to its ability to be used in bioethanol production by hydrolysis with microbial methods rather than chemical and physical pretreatment. Various hydrolytic enzymes are needed to hydrolyze this complex substrate. These enzymes are also obtained using microbial sources. Enzymatic hydrolysis is more eco-friendly and economical because it can be obtained by microbial means. The conversion of cellulose into simple sugars by enzymatic hydrolysis facilitates bioethanol fermentation and reduces the process time and cost (Periyasamy et al., 2023). Cellulose and hemicellulose, structural polymers of lignocellulose, are used in bioethanol production because they are easily hydrolyzable polysaccharides. However, since the other component of lignocellulose, lignin, has a more stubborn structure, lignocellulose must first be pretreated in the bioconversion process to ethanol. The production process is generally completed by fermenting the simple sugars released as a result of hydrolysis and obtaining ethanol (Broda et al., 2022).

In enzymatic processes, the cross-links between the carbohydrates in the structure and the lignin are broken, and the hydrogen bonds between the cellulose chains are broken, thus reducing cellulose's degree of polymerization and crystallinity. Therefore, the surface area of the lignocellulose increases, and the enzymes have easy access to the carbohydrates in the structure, facilitating enzymatic hydrolysis. As a result, access to the fermentable sugars required for the fermentation stage is facilitated (Broda et al., 2022). In the second stage, these simple sugars released due to hydrolysis are converted into bioethanol in the microbial fermentation stage. Yeast or bacterial strains are responsible for microbial fermentation (Periyasamy et al., 2023; Figure 13.1).

Lignocellulose is the main constituent of plant cell walls and composes the dry matter of all plants worldwide. It is structurally composed of 35–50% cellulose, 20–35% hemicellulose, and 15–30% lignin as its main components (Figure 13.2). In addition to these basic polymers, it contains some compounds such as lipids, proteins, pectins, phenols, vitamins and cyclic hydrocarbons, ash, wax, and resin (Guo et al., 2023; Ilić et al., 2023). Cellulose, which forms the backbone of the lignocellulose, is surrounded by hemicellulose and lignin. Lignin is a protective and durable structure that provides chemical and molecular resistance of biomass against microbial or enzymatic degradation of cellulose and hemicellulose. It is found in layers connected by hydrogen and anisole bonds on the cellulosic structure. This prevents hydrolytic enzymes from reaching cellulose and hemicellulose and the enzymatic saccharification of biomass (Guo et al., 2023; Zhu & Pan, 2022). Enzymatic hydrolysis of biomass is facilitated by pretreatments that remove lignin (e.g., ligninolytic enzymes). As cellulases and hemicellulases hydrolyze the structure, sugars that will be used in fermentation, the final stage of bioethanol production, are released (Nargotra et al., 2022).

Cellulose, a linear glucose polymer consisting of crystalline and amorphous regions linked by β-1,4 glycosidic bonds, constitutes 30–50% of lignocellulose and is hydrolyzed

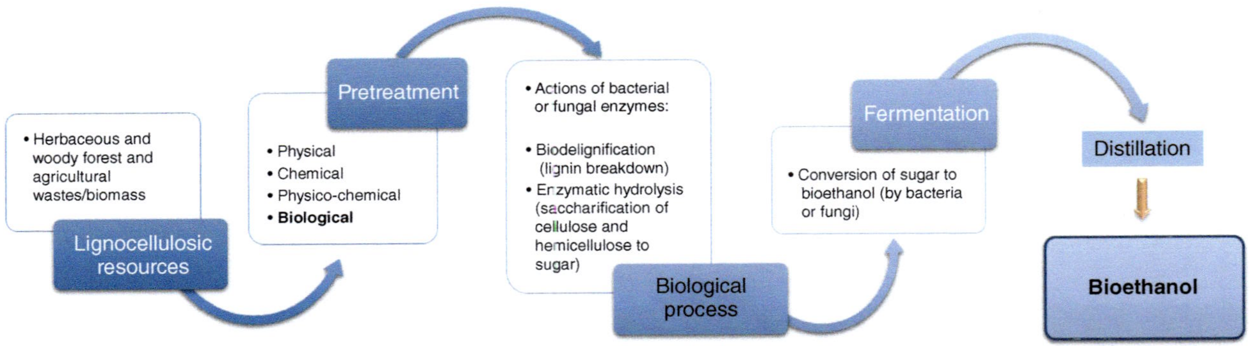

Figure 13.1 Biological conversion steps of lignocellulosic biomass to bioethanol (*Source:* Designed by the author).

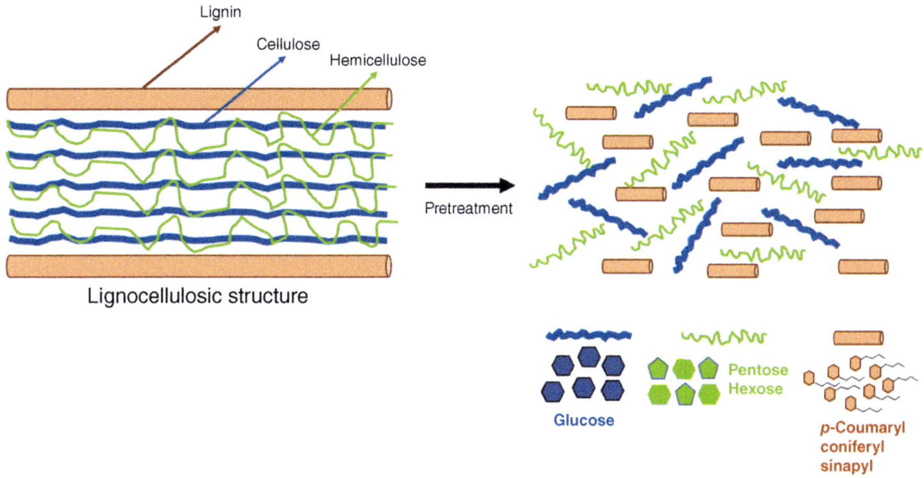

Figure 13.2 Lignocellulose structure after pretreatment (*Source:* Designed by the author).

by cellulases (endoglucosidases, exoglucosidases, and β-1,4-glucosidases) (Guo et al., 2023; Ilić et al., 2023; Nargotra et al., 2022). Cellulose fibrils connected by hydrogen bonds form the microfibrils of the plant cell wall (Ilić et al., 2023).

Hemicellulose is an amorphous polysaccharide formed by the establishment of β-1,3 and β-1,4 glycosidic bonds between various pentoses and hexoses. They constitute 20% of the lignocellulosic structure (Saini et al., 2022; Sharma, Xu, et al., 2019). Hemicellulose hydrolyzing enzymes are hemicellulases (Vasić et al., 2021).

Lignin, which consists of *p*-coumaryl, coniferyl, and sinapyl phenylpropanoids, is an aromatic heteropolymer and a very resistant structure that constitutes 15–40% of lignocellulose (Zoghlami & Paës, 2019). Thanks to the phenylpropane units it contains, it has gained a durable, permeable, and microorganism-resistant form. In this way, lignin degradation is quite difficult (Sharma, Xu, et al., 2019). This polymer is covalently bonded to hemicellulose and cellulose, providing rigidity and hydrophobicity to the plant cell wall, thus resisting oxidative stress and microbial attacks (Ilić et al., 2023). Lignin hydrolytic enzymes are lignin-degrading enzymes (peroxidase and phenol oxidase) and lignin-degrading coenzymes (phenyl coumarin and superoxide dismutase) (Guo et al., 2023).

There are some obstacles to the use of lignocellulosic biomass, such as the particle size of the biomass, the crystalline structure of the cellulose wrapped with hemicellulose, the degree of polymerization, and the recalcitrance of lignin, which reduces the enzymatic hydrolysis area of the lignocellulosic structure (Zoghlami & Paës, 2019). Therefore, to break down cellulose and hemicellulose, pretreatment methods must be applied to remove lignin from the biomass. For this purpose, physical pretreatments such as grinding, crushing, milling, ultrasonic, and chemical pretreatments with alkali, acids, or oxidative delignification are applied in the industry. However, disadvantages such as environmental damage and the high cost of chemical processes bring forward biological pretreatment methods that have less pollution, do not contain chemicals, require low energy, and have higher efficiency. In biological processes, enzymes mediate the conversion of raw materials into biofuels by hydrolyzing them into 5- and 6-carbon sugars. These enzymes are obtained from bacteria and fungi (Sharma, Xu, et al., 2019).

13.5 Microbial Cellulases in Hydrolyzing Lignocellulose: Biological Pretreatment

Hydrolysis of the complex polysaccharide structure of the inexhaustible waste lignocellulosic biomass with various commercial enzymes is a costly process. For this reason, converting this valuable waste raw material into bioethanol, a value-added product, using microbial enzymes provides a sustainable approach for the environment and national economies (Rajesh & Gummadi, 2023). Environmentally friendly cellulases, hemicellulases, and ligninolytic enzymes used in biological pretreatment with microbial enzymes play an active role in the conversion process of waste lignocellulosic biomass into bioethanol in the biorefinery industry as "green biocatalysts" (Nargotra et al., 2022; Figure 13.3).

A large amount of lignocellulosic biomass is released as waste worldwide and causes environmental pollution. Bioethanol production by microbial hydrolysis of this waste

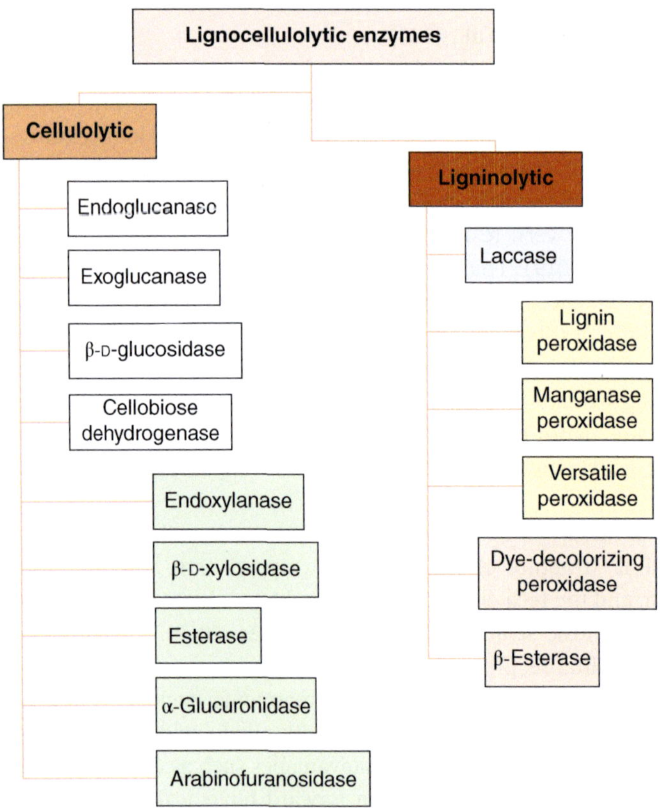

Figure 13.3 Various lignocellulolytic enzymes involved in the pretreatment of lignocellulose (*Source:* Designed by the author).

biomass and fermentation of the sugars obtained from hydrolysis is an important strategy for the fuel industry. Microorganisms and microbial enzymes contribute to the increase in the hydrolysis efficiency of lignocellulosic raw materials (Chukwuma et al., 2020). Thanks to the microbial enzymes, they indirectly contribute to the carbon cycle by hydrolyzing this biomass obtained from plants into sugars that help convert it into biofuel during fermentation (Samuel et al., 2022).

Biological pretreatments have significant advantages, such as being environmentally friendly, not containing chemicals, requiring less energy, and having low processing costs. They also have some disadvantages, such as continuous control of microbial growth and activity and slow progress of the process (Broda et al., 2022; Figure 13.4).

To facilitate the biological transformation of lignocellulosic raw materials, biomass is hydrolyzed using microorganisms (bacteria, fungi) and microbial enzymes, yielding compounds that can be utilized in ethanol production. Cellulolytic enzymes (cellulase, hemicellulase, and ligninase) produced by bacteria and fungi play an active role in the biological conversion of lignocellulosic biomass into biofuels when applied in appropriate combinations (Chukwuma et al., 2020). While lignin and hemicellulose in the composition of lignocellulose can be degraded during various pretreatments, cellulose is

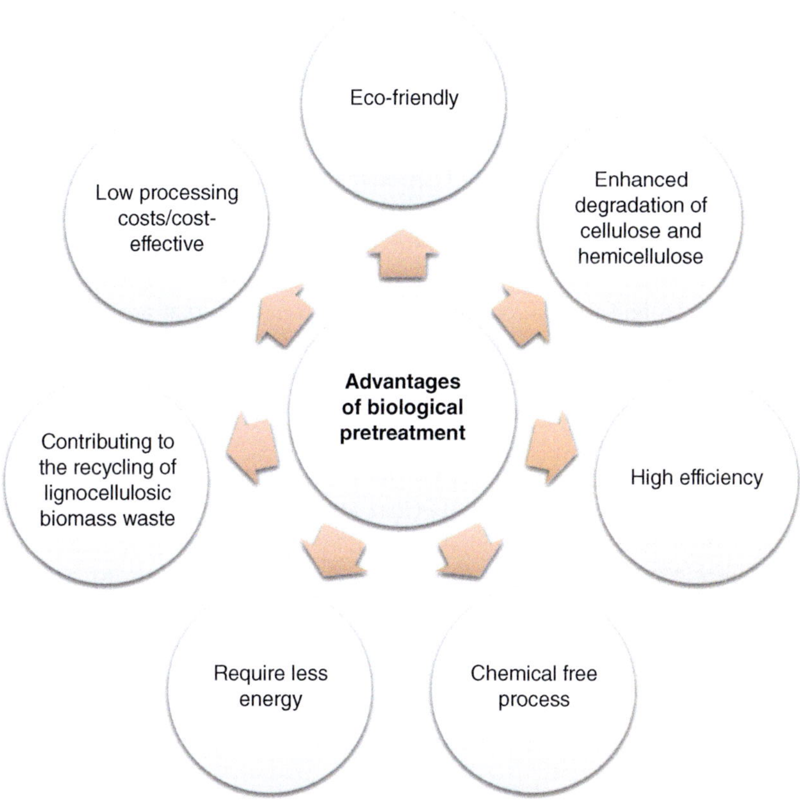

Figure 13.4 Advantages of pretreatment of lignocellulosic biomass with biological methods (*Source:* Designed by the author).

hydrolyzed by various microbial cellulases for use in energy conversion. Sugars released by hydrolysis are produced as bioethanol as a result of fermentation reactions in the fuel industry (Samuel et al., 2022).

Various white-rot fungi, such as *Trametes versicolor*, *Ceriporiopsis subvermispora*, *Phanerochaete chrysosporium*, and *Pleurotus ostreatus*, which can synthesize ligninolytic enzymes (lignin peroxidase, laccase, and manganese peroxidase), are quite effective in lignin degradation (Millati et al., 2011). Bacterial laccases, peroxidases, and β-etherases are also involved in delignification (De Gonzalo et al., 2016).

Cellulases, which release glucose by hydrolyzing the β-1,4 glycosidic bonds of cellulose, have a large share in the production process of high-value-added bioethanol thanks to their ability to convert waste lignocellulose into fermentable sugars (Rajesh & Gummadi, 2023; Chukwuma et al., 2020). They perform the saccharification of the cellulose component of lignocellulose and make it available to ethanologenic microorganisms responsible for bioethanol production (Bhardwaj et al., 2021; Kumar et al., 2018). Considering that microorganisms are potential producers of cellulases, there is a continuing need for the isolation and production of microorganisms. Therefore, producing microbial cellulases and their inclusion in bioethanol production processes is an important industrial step in biofuel production from both environmental and economic perspectives. Various bacterial strains such as *Bacillus* sp., *Paenibacillus* sp., *Aeromonas* sp., *Cellulomonas* sp., *Streptomyces* sp., *Clostridium* sp., *Escherichia* sp., and *Pseudomonas* sp. have the capacity to produce cellulase (Ilić et al., 2023; Sharma, Xu, et al., 2019).

As mentioned, cellulolytic enzymes are microbial catalyzers synthesized by various aerobic/anaerobic, thermophilic, mesophilic, and extremophilic microorganisms (fungi, bacteria, and actinomycetes) (Table 13.1). Among these groups, bacteria, more resistant to high temperatures and extreme pH conditions, produce enzymes that are more resistant to harsh industrial process conditions (Areeshi, 2022). The enzymatic hydrolysis conditions of cellulose and hemicellulose are environments affected by external factors such as temperature, pH, and incubation time. Therefore, enzymes that are resistant to these environments and can maintain their hydrolysis ability and ensure continuity of process efficiency are preferred in industrial processes (Prasad et al., 2019).

Cellulases are multi-enzyme complexes composed of three enzymes that can hydrolyze cellulose to glucose monomers. They are divided into three groups according to their catalytic action mechanisms (Figure 13.5): endoglucanases (EC 3.2.1.4), exoglucanases (EC 3.2.1.91), and β-glucosidases (EC 3.2.1.21) (Ejaz et al., 2021). They function synergistically in the hydrolysis of cellulose. First, endo-1,4-β-D-glucanases randomly break intermolecular glycosidic bonds in the amorphous regions of cellulose to produce oligomers. Second, exo-1,4-β-D-glucanase/exo-cellobiohydrolases remove dimers and monomers from the end of the glucose chain. Third, β-glucosidases hydrolyze glucose dimers, cellobiose, and other short cellulose oligomers to glucose monomers. While endo- and exo-glucanase reduce the degree of polymerization of cellulose during the liquefaction phase and release cellobiose, β-glucosidase is involved in the conversion of cellobiose into glucose (Siqueira et al., 2020).

Table 13.1 Some Microorganisms that Produce Lignocellulosic Enzymes.

	Microorganism	Lignocellulosic Enzyme	Lignocellulosic Feedstock	References
Bacteria	*Bacillus* sp. PM06	Cellulase	Wheat bran, sago waste, and rice bran	Rajesh and Gummadi (2023)
	Bacillus ligniniphilus L1	Laccase	Rice straw	Nazar et al. (2022)
	Bacillus subtilis CD001	Cellulase	Sugarcane bagasse and wheat straw	Malik and Javed (2021)
	Streptomyces fulvissimus CKS7	Endoglucanase and exoglucanase	Rye bran	Mihajlovski et al. (2021)
	Geobacillus sp.	Laccase	Corn stover and bagasse	Rai et al. (2019)
	Paenibacillus illinoisensis CX11	Cellulase and xylanase	Wheat straw	Ahmed et al. (2018)
	Streptomyces ipomoeae (SilA)	Laccase	Wheat straw	De La Torre et al. (2017)
	Streptomyces sp. MDS	Cellulase	Rice waste	Saratale et al. (2017)
	Bacillus sp. BS-5	Xylanase and endoglucanase	Corn cob	Xu et al. (2017)
	Bacillus licheniformis AMF-07	Cellulase	Rice bran and wheat bran	Azadian et al. (2016)
	Ruminiclostridium thermocellum M3	Endoglucanase, exoglucanase, β-glucosidase, and xylanase	Corn straw, corn cobs, rice straw, and poplar sawdust	Sheng et al. (2016)
Fungi	*Penicillium aurantiogriseum*	Endoglucanase, cellobiohydrolase, and β-glucosidase	Corn stover	Kovács et al. (2022)
	Aspergillus niger ITV02	Cellulase (β-glucosidase and endoglucanase)	Wheat straw	Infanzón-Rodríguez et al. (2022)
	Schizophyllum commune NAIMCC-F-03379	Cellulase	Wheat bran	Kumar et al. (2018)
	Trichoderma harzianum strain HZN10	Laccase	Sugarcane bagasse	Bagewadi et al. (2017)
	Trametes villosa	Laccase	Wheat straw	De La Torre et al. (2017)
	Pycnoporus sanguineus	Laccase	Sugarcane bagasse	Hernández et al. (2017)
	Aspergillus niger NRRL3	Endoglucanase, exoglucanase, and β-glucosidase	Soybean hulls	Julia et al. (2016)
	Aspergillus niger	Cellulase	Wheat bran	Dhillon et al. (2011)
	Aspergillus assiutensis VS34	Cellulase and xylanase	Sugarcane bagasse	Sharma, Nargotra, et al. (2019)
	Trichoderma citrinovirdae	Cellulase	Sugarcane straw	Guerra et al. (2006)

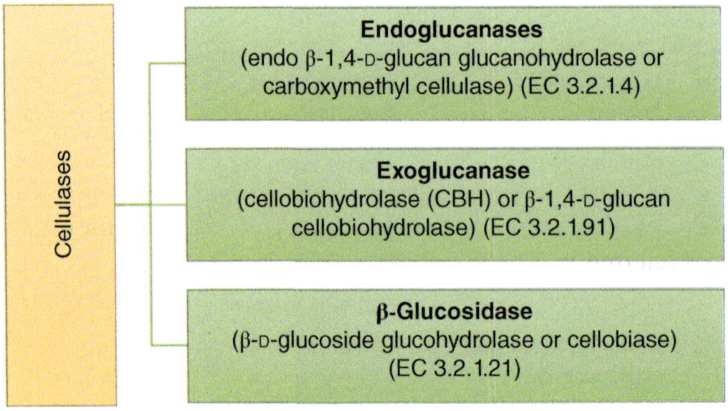

Figure 13.5 Cellulases according to their catalytic action mechanisms (*Source:* Designed by the author).

13.5.1 Endoglucanase (Endo β-1,4-D-Glucan Glucanohydrolase, EC 3.2.1.4)

Endoglucanases, a member of the glycosyl hydrolase (GH) family, initiate cellulose's degradation to reduce its polymerization degree. They randomly act on the β(1,4)glycosidic bonds of cellulose chains (internal amorphous sites) to form cello-oligosaccharides and glucose (Verma et al., 2021). They are produced by various organisms such as aerobic and anaerobic bacteria, archaea, and fungi (Nargotra et al., 2022).

13.5.2 Exoglucanase or Cellobiohydrolase (CBH) (β-1,4-D-Glucan Cellobiohydrolase, EC 3.2.1.91)

It is a group of enzymes that act on the reducing and non-reducing ends of cellulose microfibrils and produce cellotriose and cellobiose (Guo et al., 2023). It includes CBHI and CBHII. Cellobiose is located at the reducing end of the cellulose chain. Glucose (glucan hydrolases) and cellobiose (cellobiose hydrolase) are released through the hydrolytic activity of exoglucanases (Tiwari et al., 2018). About 70% of the cellulases produced by cellulolytic fungi are exoglucanases. Cellobiose hydrolases are also produced by various bacteria and fungi (Nargotra et al., 2022).

13.5.3 β-Glucosidase or Cellobiase (β-D-Glucoside Glucohydrolase, EC 3.2.1.21)

β-Glucosidases hydrolyze cellodextrin and cellobiose during the final stages of cellulose hydrolysis, aiding glucose release (Verma et al., 2021). They should be included in the cellulolytic enzyme mixture because they improve the conversion efficiency of cellulose into fermentable sugars. They are also vital in biofuel production since they prevent the catabolite repression of cellobiohydrolase and endoglucanase (Guo et al., 2023; Nargotra et al., 2022).

Bioethanol production from lignocellulosic biomass involves enzymatic hydrolysis, saccharification, and fermentation stages. The structure of lignocellulose, composed of resistant polymers such as cellulose, hemicellulose, and lignin, must undergo biological, physical, or chemical pretreatments. Biological pretreatments based on enzymatic hydrolysis are among the most effective methods for converting biomass into products (Le et al., 2024). The microbial enzymes responsible for hydrolysis are expected to be thermophilic and thermostable, which increases hydrolysis efficiency and saccharification rate while reducing contamination risk. Additionally, higher hydrolysis efficiency accelerates fermentation. Alongside cellulase, enzymes such as hemicellulase and laccase are also being researched (Dionisi et al., 2015). Strains of the *Bacillus* genus, known for producing these enzymes, are often preferred in bioethanol production from lignocellulose because of their industrial advantage in synthesizing large amounts of extracellular enzymes (Rajesh & Gummadi, 2023).

13.6 Conclusion

When fuels produced from renewable biomass replace fuels produced from non-renewable sources, the situation of resource depletion will be prevented, and the energy problems in the world will be solved. In this context, fuels produced using lignocellulose, a renewable biomass, as a raw material, are environmentally friendly and cost-effective. The most effective methods for processing lignocellulose are biological pretreatment methods. Of these, the use of microbial enzymes is of great importance in hydrolyzing the structural components of cellulose. Cellulase enzymes, in particular, hydrolyze the cellulose component of lignocellulose and provide a biological transformation of lignocellulose. Microbial cellulases, hemicellulases, and ligninolytic enzymes, as green catalysts, offer an environmentally friendly and cost-effective alternative for the biological pretreatment of lignocellulosic biomass. They also contribute to making bioethanol production processes more sustainable and economical. Enzymatic saccharification of lignocellulose has gained great importance in recent years as the most important step in synthesizing bioethanol, a value-added product in biorefineries. The food and energy problems experienced in the world will find solutions in light of these developments. Studies and research in this field are continuing.

References

Ahmed, A. A. Q., Babalola, O. O., & McKay, T. (2018). Cellulase-and xylanase-producing bacterial isolates with the ability to saccharify wheat straw and their potential use in the production of pharmaceuticals and chemicals from lignocellulosic materials. *Waste and Biomass Valorization*, *9*, 765–775. https://doi.org/10.1007/s12649-017-9849-5.

Areeshi, M. Y. (2022). Microbial cellulase production using fruit wastes and its applications in biofuels production. *International Journal of Food Microbiology*, *378*, 109814. https://doi.org/10.1016/j.ijfoodmicro.2022.109814.

Azadian, F., Badoei-Dalfard, A., Namaki-Shoushtari, A., & Hassanshahian, M. (2016). Purification and biochemical properties of a thermostable, haloalkaline cellulase from *Bacillus licheniformis* AMF-07 and its application for hydrolysis of different cellulosic substrates to bioethanol production. *Molecular Biology Research Communications*, *5*(3), 143.

Bagewadi, Z. K., Mulla, S. I., & Ninnekar, H. Z. (2017). Optimization of laccase production and its application in delignification of biomass. *International Journal of Recycling of Organic Waste in Agriculture*, *6*, 351–365. https://doi.org/10.1007/s40093-017-0184-4.

Bala, A., & Singh, B. (2019). Cellulolytic and xylanolytic enzymes of thermophiles for the production of renewable biofuels. *Renewable Energy*, *136*, 1231–1244. https://doi.org/10.1016/j.renene.2018.09.100.

Bhardwaj, N., Kumar, B., Agrawal, K., & Verma, P. (2021). Current perspective on production and applications of microbial cellulases: A review. *Bioresources and Bioprocessing*, *8*, 1–34. https://doi.org/10.1186/s40643-021-00447-6.

Broda, M., Yelle, D. J., & Serwańska, K. (2022). Bioethanol production from lignocellulosic biomass-challenges and solutions. *Molecules*, *27*(24), 8717. https://doi.org/10.3390/molecules27248717.

Chukwuma, O. B., Rafatullah, M., Tajarudin, H. A., & Ismail, N. (2020). Lignocellulolytic enzymes in biotechnological and industrial processes: A review. *Sustainability*, *12*(18), 7282. https://doi.org/10.3390/su12187282.

De Gonzalo, G., Colpa, D. I., Habib, M. H., & Fraaije, M. W. (2016). Bacterial enzymes involved in lignin degradation. *Journal of Biotechnology*, *236*, 110–119. https://doi.org/10.1016/j.jbiotec.2016.08.011.

De La Torre, M., Martín-Sampedro, R., Fillat, Ú., Eugenio, M. E., Blánquez, A., Hernández, M., Arias, M. E., & Ibarra, D. (2017). Comparison of the efficiency of bacterial and fungal laccases in delignification and detoxification of steam-pretreated lignocellulosic biomass for bioethanol production. *Journal of Industrial Microbiology and Biotechnology*, *44*(11), 1561–1573. https://doi.org/10.1007/s10295-017-1977-1.

Demirbas, A. (2010). Use of algae as biofuel sources. *Energy Conversion and Management*, *51*(12), 2738–2749. https://doi.org/10.1016/j.enconman.2010.06.010.

Dhillon, G. S., Oberoi, H. S., Kaur, S., Bansal, S., & Brar, S. K. (2011). Value-addition of agricultural wastes for augmented cellulase and xylanase production through solid-state tray fermentation employing mixed-culture of fungi. *Industrial Crops and Products*, *34*(1), 1160–1167. https://doi.org/10.1016/j.indcrop.2011.04.001.

Dionisi, D., Anderson, J. A., Aulenta, F., McCue, A., & Paton, G. (2015). The potential of microbial processes for lignocellulosic biomass conversion to ethanol: A review. *Journal of Chemical Technology & Biotechnology*, *90*(3), 366–383. https://doi.org/10.1002/jctb.4544.

Ejaz, U., Sohail, M., & Ghanemi, A. (2021). Cellulases: From bioactivity to a variety of industrial applications. *Biomimetics*, *6*(3), 44. https://doi.org/10.3390/biomimetics6030044.

Guerra, G., Casado, M. R. L. G., Arguelles, J., Sánchez, M. I., Manzano, A. M., & Guzman, T. (2006). Cellulase production with sugarcane straw by *Trichoderma citrinoviride* on solid bed. *Sugar Tech*, *8*, 30–35. https://doi.org/10.1007/BF02943738.

Guo, H., Zhao, Y., Chang, J. S., & Lee, D. J. (2023). Enzymes and enzymatic mechanisms in enzymatic degradation of lignocellulosic biomass: A mini-review. *Bioresource Technology*, *367*, 128252. https://doi.org/10.1016/j.biortech.2022.128252.

Hernández, C., Farnet Da Silva, A. M., Ziarelli, F., Perraud-Gaime, I., Gutiérrez-Rivera, B., García-Pérez, J. A., & Alarcón, E. (2017). Laccase induction by synthetic dyes in *Pycnoporus sanguineus* and their possible use for sugar cane bagasse delignification. *Applied Microbiology and Biotechnology*, *101*, 1189–1201. https://doi.org/10.1007/s00253-016-7890-0.

Ilić, N., Milić, M., Beluhan, S., & Dimitrijević-Branković, S. (2023). Cellulases: From lignocellulosic biomass to improved production. *Energies*, *16*(8), 3598. https://doi.org/10.3390/en16083598.

Infanzón-Rodríguez, M. I., Ragazzo-Sánchez, J. A., Del Moral, S., Calderón-Santoyo, M., & Aguilar-Uscanga, M. G. (2022). Enzymatic hydrolysis of lignocellulosic biomass using native cellulase produced by *Aspergillus niger* ITV02 under liquid state fermentation. *Biotechnology and Applied Biochemistry*, *69*(1), 198–208. https://doi.org/10.1002/bab.2097.

Julia, B. M., Belén, A. M., Georgina, B., & Beatriz, F. (2016). Potential use of soybean hulls and waste paper as supports in SSF for cellulase production by *Aspergillus niger*. *Biocatalysis and Agricultural Biotechnology*, *6*, 1–8. https://doi.org/10.1016/j.bcab.2016.02.003.

Kovács, E., Szűcs, C., Farkas, A., Szuhaj, M., Maróti, G., Bagi, Z., Rákhely, G., & Kovács, K. L. (2022). Pretreatment of lignocellulosic biogas substrates by filamentous fungi. *Journal of Biotechnology*, *360*, 160–170. https://doi.org/10.1016/j.jbiotec.2022.10.013.

Kumar, B., Bhardwaj, N., Alam, A., Agrawal, K., Prasad, H., & Verma, P. (2018). Production, purification and characterization of an acid/alkali and thermo tolerant cellulase from *Schizophyllum commune* NAIMCC-F-03379 and its application in hydrolysis of lignocellulosic wastes. *AMB Express*, *8*, 1–16. https://doi.org/10.1186/s13568-018-0696-y.

Le, Y., Zhang, M., Wu, P., Wang, H., & Ni, J. (2024). Biofuel production from lignocellulose via thermophile-based consolidated bioprocessing. *Engineering Microbiology*, *4*, 100174. https://doi.org/10.1016/j.engmic.2024.100174.

Lin, Y., & Tanaka, S. (2006). Ethanol fermentation from biomass resources: Current state and prospects. *Applied Microbiology and Biotechnology*, *69*, 627–642. https://doi.org/10.1007/s00253-005-0229-x.

Malik, W. A., & Javed, S. (2021). Biochemical characterization of cellulase from *Bacillus subtilis* strain and its effect on digestibility and structural modifications of lignocellulose rich biomass. *Frontiers in Bioengineering and Biotechnology*, *9*, 800265. https://doi.org/10.3389/fbioe.2021.800265.

Mihajlovski, K., Buntić, A., Milić, M., Rajilić-Stojanović, M., & Dimitrijević-Branković, S. (2021). From agricultural waste to biofuel: Enzymatic potential of a bacterial isolate *Streptomyces fulvissimus* CKS7 for bioethanol production. *Waste and Biomass Valorization*, *12*, 165–174. https://doi.org/10.1007/s12649-020-00960-3.

Millati, R., Syamsiah, S., Niklasson, C., Cahyanto, M. N., Ludquist, K., & Taherzadeh, M. J. (2011). Biological pretreatment of lignocelluloses with white-rot fungi and its applications: A review. *BioResources*, *6*(4), 5224–5259. https://doi.org/10.15376/BIORES.6.4.5224-5259.

Nargotra, P., Sharma, V., Lee, Y. C., Tsai, Y. H., Liu, Y. C., Shieh, C. J., Tsai, M.-L., Dong, C.-D., & Kuo, C. H. (2022). Microbial lignocellulolytic enzymes for the effective valorization of lignocellulosic biomass: A review. *Catalysts*, *13*(1), 83. https://doi.org/10.3390/catal13010083.

Nazar, M., Xu, L., Ullah, M. W., Moradian, J. M., Wang, Y., Sethupathy, S., Iqbal, B., Nawaz, M. Z., & Zhu, D. (2022). Biological delignification of rice straw using laccase from *Bacillus ligniniphilus* L1 for bioethanol production: A clean approach for agro-biomass utilization. *Journal of Cleaner Production*, *360*, 132171. https://doi.org/10.1016/j.jclepro.2022.132171.

Periyasamy, S., Isabel, J. B., Kavitha, S., Karthik, V., Mohamed, B. A., Gizaw, D. G., Sivashanmugam, P., & Aminabhavi, T. M. (2023). Recent advances in consolidated bioprocessing for conversion of lignocellulosic biomass into bioethanol—A review. *Chemical Engineering Journal*, *453*, 139783. https://doi.org/10.1016/j.cej.2022.139783.

Prasad, R. K., Chatterjee, S., Mazumder, P. B., Gupta, S. K., Sharma, S., Vairale, M. G., Datta, S., Dwivedi, S. K., & Gupta, D. K. (2019). Bioethanol production from waste lignocelluloses: A review on microbial degradation potential. *Chemosphere*, *231*, 588–606. https://doi.org/10.1016/j.chemosphere.2019.05.142.

Rai, R., Bibra, M., Chadha, B. S., & Sani, R. K. (2019). Enhanced hydrolysis of lignocellulosic biomass with doping of a highly thermostable recombinant laccase. *International Journal of Biological Macromolecules*, *137*, 232–237. https://doi.org/10.1016/j.ijbiomac.2019.06.221.

Rajesh, R., & Gummadi, S. N. (2023). Production of multienzymes, bioethanol, and acetic acid by novel *Bacillus* sp. PM06 from various lignocellulosic biomass. *Biomass Conversion and Biorefinery*, *13*(15), 13949–13961. https://doi.org/10.1007/s13399-022-02418-z.

Saini, J. K., Kaur, A., & Mathur, A. (2022). Strategies to enhance enzymatic hydrolysis of lignocellulosic biomass for biorefinery applications: A review. *Bioresource Technology*, *360*, 127517. https://doi.org/10.1016/j.biortech.2022.127517.

Samuel, M. S., Govarthanan, M., & Selvarajan, E. (2022). A comprehensive review on strategic study of cellulase producing marine actinobacteria for biofuel applications. *Environmental Research*, *214*, 114018. https://doi.org/10.1016/j.envres.2022.114018.

Saratale, G. D., Saratale, R. G., Ghodake, G. S., Jiang, Y. Y., Chang, J. S., Shin, H. S., & Kumar, G. (2017). Solid state fermentative lignocellulolytic enzymes production, characterization and its application in the saccharification of rice waste biomass for ethanol production: An integrated biotechnological approach. *Journal of the Taiwan Institute of Chemical Engineers*, *76*, 51–58. https://doi.org/10.1016/j.jtice.2017.03.027.

Shahzadi, T., Mehmood, S., Irshad, M., Anwar, Z., Afroz, A., Zeeshan, N., Rashid, U., & Sughra, K. (2014). Advances in lignocellulosic biotechnology: A brief review on lignocellulosic biomass and cellulases. *Advances in Bioscience and Biotechnology*, *5*, 246–251. https://doi.org/10.4236/abb.2014.53031.

Sharma, H. K., Xu, C., & Qin, W. (2019). Biological pretreatment of lignocellulosic biomass for biofuels and bioproducts: An overview. *Waste and Biomass Valorization*, *10*, 235–251. https://doi.org/10.1007/s12649-017-0059-y.

Sharma, V., Nargotra, P., & Bajaj, B. K. (2019). Ultrasound and surfactant assisted ionic liquid pretreatment of sugarcane bagasse for enhancing saccharification using enzymes from an ionic liquid tolerant *Aspergillus assiutensis* VS34. *Bioresource Technology*, *285*, 121319. https://doi.org/10.1016/j.biortech.2019.121319.

Sheng, T., Zhao, L., Gao, L. F., Liu, W. Z., Cui, M. H., Guo, Z. C., Ma, X.-D., Ho, S.-H., & Wang, A. J. (2016). Lignocellulosic saccharification by a newly isolated bacterium, *Ruminiclostridium thermocellum* M3 and cellular cellulase activities for high ratio of glucose to cellobiose. *Biotechnology for Biofuels*, *9*, 1–11. https://doi.org/10.1186/s13068-016-0585-z.

Siqueira, J. G. W., Rodrigues, C., de Souza Vandenberghe, L. P., Woiciechowski, A. L., & Soccol, C. R. (2020). Current advances in on-site cellulase production and application on lignocellulosic biomass conversion to biofuels: A review. *Biomass and Bioenergy*, *132*, 105419. https://doi.org/10.1016/j.biombioe.2019.105419.

Soccol, C. R., Faraco, V., Karp, S. G., Vandenberghe, L. P., Thomaz-Soccol, V., Woiciechowski, A. L., & Pandey, A. (2019). Lignocellulosic bioethanol: Current status and future perspectives. In A. Pandey, C. Larroche, C.-G. Dussap, E. Gnansounou, S. K. Khanal, & S. Ricke (Eds.), *Biofuels: Alternative feedstocks and conversion processes for the production of liquid and gaseous biofuels* (pp. 331–354). Academic Press. https://doi.org/10.1016/B978-0-12-816856-1.00014-2.

Tiwari, R., Nain, L., Labrou, N. E., & Shukla, P. (2018). Bioprospecting of functional cellulases from metagenome for second generation biofuel production: A review. *Critical Reviews in Microbiology, 44*(2), 244–257. https://doi.org/10.1080/1040841X.2017.1337713.

Vasić, K., Knez, Ž., & Leitgeb, M. (2021). Bioethanol production by enzymatic hydrolysis from different lignocellulosic sources. *Molecules, 26*(3), 753. https://doi.org/10.3390/molecules26030753.

Verma, N., Kumar, V., & Bansal, M. C. (2021). Valorization of waste biomass in fermentative production of cellulases: A review. *Waste and Biomass Valorization, 12*, 613–640. https://doi.org/10.1007/s12649-020-01048-8.

Xu, J., Gao, Z., Wu, B., & He, B. (2017). Lactose-inducted production of a complete lignocellulolytic enzyme system by a novel bacterium *Bacillus* sp. BS-5 and its application for saccharification of alkali-pretreated corn cob. *Cellulose, 24*, 2059–2070. https://doi.org/10.1007/s10570-017-1247-4.

Zhu, J. Y., & Pan, X. (2022). Efficient sugar production from plant biomass: Current status, challenges, and future directions. *Renewable and Sustainable Energy Reviews, 164*, 112583. https://doi.org/10.1016/j.rser.2022.112583.

Zoghlami, A., & Paës, G. (2019). Lignocellulosic biomass: Understanding recalcitrance and predicting hydrolysis. *Frontiers in Chemistry, 7*, 874. https://doi.org/10.3389/fchem.2019.00874.

Index

a

acetogenesis 35
acidification 153
acidogenisis 35
adoption drivers of CBP
  economic drivers 245–246
  policy and regulatory drivers 246
  technological and innovation drivers 246
advanced pretreatment techniques, CBP 131–133
agricultural waste 263
airlift bioreactors (ARLs) 43, 188–189
alcohol production feedstock 37–39
algal biomass 278
algal bioprocesses 41
anaerobic bioreactors 43
anaerobic digestion (AD) 35
artificial intelligence (AI) 156–157
  and digitalization 9–10
  driven process control 270
Aspen Plus 45
Aspergillus niger 34, 146

b

Bacillus subtilis 112, 146
baker's yeast for CBP 97–98
batch bioprocessing 145
batch bioreactor 42
β-glucosidase 286–287
bioartificial kidney (BAK) devices 200–202
bioartificial liver devices 197–199
Bio-Based Industries Joint Undertaking (BBI-JU) 158
BioBrick 149
biodiesel production 39–40
bioenergy bioprocess
  emerging trends 48
  policy, funding, and market considerations 47
  synthetic biology and metabolic engineering 47
  technical barriers 47
bioenergy production
  biochemical strategy 58
  biofuel production 67–68
  co-culture and microbial consortium strategies 59
    community stability and resilience 66
    metabolic labor division 66
    synergistic microbial interactions 65–66
  commercialization and scale-up challenges 61
    economic constraints 71
    infrastructure and logistics 71
    regulatory and policy challenges 71
    social acceptance and public perception 71
    technical barriers 70
  enzyme engineering
    for bioenergy 63–65
    and catalytic enhancement 59
  future prospects and emerging technologies 61
  microbial biofactory 61–63
  microbial strain development 57–59

bioenergy production (*contd.*)
  omics approaches in bioenergy 60–61, 68–70
  process intensification techniques 60
bioethanol feedstocks 277–278
bioethanol production process 276–277
  β-glucosidase or cellobiase 286–287
  bioethanol feedstocks 277–278
  endoglucanase 206
  exoglucanase/cellobiohydrolase 286
  lignocellulolytic enzymes 282–286
  lignocellulosic biomass in 278–282
biofuels
  blending goals and renewable fuel standards 87
  first-generation biofuels 11
  fourth-generation biofuels 12–13
  imports and exports policy 87–88
  monetary rewards 87
  production of 67–68
    capital expenditures 231
    feedstock costs 230–231
    operating expenditures 232
    process integration for 67–68
    product yields and conversion efficiency 232–233
  public education and awareness 88
  regulations for feedstock 87
  research and development 87
  rules pertaining to environment 88
  second-generation (advanced) biofuels 11–12
  support for infrastructure 88
  third-generation biofuels 12
  yields 78
biogas production
  diagram of 36–37
  stages 36
  via consolidated bioprocessing (CBP)
    importance in 126–127
    key performance metrics 130–131
    system design and operational parameters 129–130
biohydrogen production 40–42
biomass 167
  bioconversion 38
  feedstocks 38, 277–278
  microbial conversion of 113–114
Biomass Conversion Processes 31–32
biomass conversion strategy
  biochemical approach 78–79
  thermochemical approach 79–80
biopharmaceuticals 196–202
bioprocess
  bioenergy
    emerging trends 48
    policy, funding, and market considerations 47
    synthetic biology and metabolic engineering 47
    technical barriers 47
  biogas production 36–37
  bioreactors types 42–44
  defined 29
  engineering 30–32
  environmental and economic evaluations 45–46
  feedstock for 32–34
  fermentation
    biohydrogen production 40–42
    ethanol production 37–39
    transesterification (for biodiesel production) 39–40
  microorganisms and enzymes in 34–35
  process integration techniques 44–45
  types of 35–36
bioprocess technology
  advancements in 84–85
  assessment parameters for 151–154
  biofuels
    regulations and policies 87–88
    yields 78
  biomass conversion strategies 78–80
  challenges and bottlenecks 154–156
  classification of 145
  conversion efficiency 85–86
  feedstock diversity 85
  green biofuels 80–85
  for green biofuels productions 76

high production costs and initial investment 86
importance of technological assessment 142–143
key components and stages of 144
overview of 141–142
role of microorganisms and enzymes 145–146
scope and objectives 143–144
sustainability and availability of feedstock 85
technological advancements in 149–151
technology developments 85–86
bioreactors 167–168
applications
bioremediation 206–207
cell culture and tissue engineering 207–208
food production 209–210
production of biofuel 202–206
production of biopharmaceuticals 196–202
design 170–174
mass transfer 175–177
mixing 174–175
pressure 179
scale down 180
scale up 174, 179–180
shear 177–179
drug delivery 168
dynamic 169
future perspectives 210
general setup 168–169
perfusion 169
tissue engineering 168
types of 42–44, 180–184
airlift bioreactors 188–189
bubble column reactors 187–188
disposable bioreactors 194
drum bioreactor 192–193
fixed bed bioreactors 192–193
fluidized bed reactors 189–190
membrane bioreactors 191
packed bed bioreactors 188–190
perforated trays 191–192
photobioreactors 194–196
stirred tank reactors 184–187
wave bioreactors 193–194
bioremediation 206–207
BioSTEAM 152
blending goals and renewable fuel standards (RFS) 87
blood-brain barrier (BBB) models 199–200
bubble column reactors (BCRs) 187–188
bubble-free aeration system 178

c

Caldicellulosiruptor saccharolyticus 125, 147
capital expenditure (CAPEX) 152, 231
carbon capture and storage (CCS) technology 10
carbon capture, utilization, and storage (CCUS) technology 10
carbon intensity (CI) 235
carbon-neutral future 253
catalytic action mechanisms 286
CBP. *see* consolidated bioprocessing (CBP)
cell culture and tissue engineering 207–208
cellobiase 286–287
cellobiohydrolase 286
cellulolytic enzymes 113
cellulose 279, 281
cellulosic biofuel 236
circular bioeconomy 157, 233, 251
circular economy
bioprocess 46
green industrial energy 17
Clostridium thermocellum 125, 131, 147, 247
co-cultures consortia 112
commensalism 111
commercial viability 134
compound annual growth rate (CAGR) 3
Compounded Annual Growth Rate (CAGR) 267
consolidated bioprocessing (CBP) 45, 95
adoption challenges 248–249
adoption drivers 245–246
advanced pretreatment techniques 131–132
advantages and disadvantages of 102–103

consolidated bioprocessing (CBP) (*contd.*)
baker's yeast 97–98
bioethanol production employing consortium 229
for biofuel production cycle 235
biogas production via
importance in 126–127
key performance metrics 130–131
system design and operational parameters 129–130
biomass in 113–114
challenges and bottlenecks 154–156
circular bioeconomy
models 233
and sustainable development goals 157
commercial viability 134
conventional bioconversion *vs.* 228–229
core components 244
definition and core principles 146–147
emerging trends in 251
engineering strategies and synthetic biology 229–230
enzyme production, trade-offs 128
feedstock compatibility and pretreatment considerations 245
genetic and metabolic engineering of microbial consortia 131
geographic and sectoral trends 250–251
at high temperatures 101–102
historical development of 147–148
industrial actors leading implementation 249–250
innovation and strain development trajectory 236
integration with artificial intelligence and digital twins 156–157
integration with circular bioeconomy 251
lignocellulosic heterogeneity 128
low-carbon economy 234–236
metabolic limitations and genetic problems 128
microbial consortia 109–110
development 116–117
ecological interactions 110–111
integration of 114–116
types of 111–113
microbial strain limitations 127
microorganisms in 244–245
more than one microbe or consortium 100–101
overview of 125
pilot facilities 250
policy support and public–private partnerships 157–158
polymer structure 96
principles of 97
process efficiency and yield optimization 129
process integration and intensification 132–133
scalability and deployment potential 234
S. cerevisiae 244
SHF *vs.* SSF 148–149
single step technology and economical process 96–97
strain development/improvement 98–100
strategic recommendations 236–237
sustainability 134–135
sustainable biofuel production 251–253
technological advancements 133–134, 149–151
technological enhancements and digestibility improvements 246–248
thermophiles, consortia, and hybrid processes 245
tolerance problems and stability problems 128
traditional bioprocessing *vs.* 262
traditional multistep bioprocessing *vs.* 262
continuous bioprocessing 145
continuous bioreactor (chemostat) 43
continuous stirred tank reactors (CSTRs) 185, 234
conventional bioconversion *vs.* CBP 228–229
conversion efficiency 232–233
cost-benefit analysis (CBA) 46, 152, 268–269
cost of the capital (CoC) 20
CRISPR/Cas9 58, 62–63, 77, 100, 149

d

dark fermentation 40
degradation by-product inhibition 128
Design–Build–Test–Learn (DBTL) 247
digital twins 156–157
disposable bioreactors 194
downstream processing and product recovery 267
drum bioreactor 192–193
dynamic bioreactors 169

e

École Polytechnique Fédérale de Lausanne (EPFL) 129
ecological interactions, microbial consortia 110–111
economic drivers 245–246
economic feasibility and techno-economic assessment
  cost–benefit analysis of 268–269
  feedstock price and yield effect 269
  traditional bioprocessing *vs.* green technology 269–270
economic growth 14–15
electromagnetic generators (EMGs) 8
emerging trends in CBP 251
endoglucanase 286
energy balance 45
energy demand 153
energy return on investment (EROI) 251
energy security 15
Energy Storage System (ESS) 18
engineered CBP microorganisms 262–263
environmental sustainability impact, CBP
  cost components
    capital costs 267
    downstream processing and product recovery 267
    feedstock collection and pretreatment 265–266
    microbial strain development and inoculum preparation 266
    operating costs 267
  definition and evolution of 261
  economic feasibility and techno-economic assessment
    cost–benefit analysis of 268–269
    feedstock price and yield effect 269
    traditional bioprocessing *vs.* green technology 269–270
  lignocellulosic biomass and agricultural waste 263
  microorganisms used in 261–263
  opportunities 270
  process flow and system design
    enzyme production *in situ* 264
    integrated bioreactor design 264–265
    one-pot conversion 263–264
    process monitoring and control 265
  scope and objectives of 260–261
enzyme activity and product yield 249
enzyme engineering and cocktail optimization 149–150
enzyme engineering, for bioenergy
  cocktail optimization 65
  directed evolution 64
  enzyme immobilization 65
  rational design 64
enzymes 146
  in bioprocesses 34
  immobilization 65
  production *in situ* 264
Escherichia coli 146
ethanol production 37–39
European Climate Law 14
EU's Green Deal Industrial Plan 6
eutrophication 153
exoglucanase 286

f

fed-batch and continuous fermentation 68
fed-batch bioprocessing 145
fed-batch bioreactor 42
feedstock
  for bioprocess 32–34
  costs 230–231
  diversity of 85
  price and yield effect 269
  sustainability and availability of 85

fermentation bioprocess
 biohydrogen production 40–42
 ethanol production 37–39
 transesterification (for biodiesel production) 39–40
first-generation
 bioethanol 276
 biofuels 11, 80–81, 95
 feedstocks 33, 277
fixed bed bioreactors 192–193
flat-plate bioreactors 194–195
fluidized bed (FB) reactors 43, 189–190
food production 209–210
food waste (FW) 116
fossil fuels 275
fourth-generation
 biofuels 12–13, 82
 feedstock 33–34
free fatty acid (FFA) 40

g

gaseous biofuels 82–83
genetic instability 248–249
genome editing 62–63
genomics 68
global energy demand 227, 259, 275
global renewable energy market 3, 4
global warming potential (GWP) 153
Golden Gate 149
gradual depletion of energy sources 95
green biofuels
 gaseous biofuels 82–83
 liquid biofuels 83–85
 via bioprocess technology 80–82
green energy 1–2
green fuels
 biofuels 10–13
 green hydrogen 13
 market 21
greenhouse gas (GHG) 7, 167
 significant reduction in 13–14
green hydrogen production 8–9, 13
green industrial energy
 advantages of
  circular economy and resource efficiency 17
  energy security and independence 15
  greenhouse gas emission reduction 13–14
  grid stability and flexibility 16–17
  hard-to-abate sectors, decarbonization 15
  innovation, technological advancement, and competitiveness 15–16
  job creation and economic growth 14–15
  long-term cost efficiency and operational savings 14
  public health and environmental benefits 16
  regulatory and brand advantages 17–18
 challenges and market gaps
  energy storage and grid reliability 18
  geographic and temporal mismatches 20
  high initial capital cost 20
  infrastructure and supply chain constrains 19
  policy and regulation uncertainty 19–20
  supply–demand mismatch 18
  work force and skill shortage 19
 emerging technology 21
 features of 1–2
 green fuels
  biofuels 10–13
  green hydrogen 13
 market scenario of
  market size and growth 3–4
  policy and market drivers 6–7
  regional trends 4–6
 technology highlights of
  artificial intelligence and digitalization 9–10
  carbon capture and sustainable fuels 10
  green hydrogen production 8–9
  solar energy technology 7–8
  wind energy innovations 8
green industrial policy 6
green technology *vs.* traditional bioprocessing 269–270
Gross Domestic Product (GDP) 6

h

hard-to-abate (HtA) sectors 15
hemicellulose 281
high-solid fermentation (HSF) 60, 67
hydrogen production 247
hydrolysis 35
hydrolytic enzyme production 128

i

India's National Bio-Energy Mission 158
in situ product recovery (ISPR) 60, 68
integrated bioreactor design 264–265
International Labour Organization (ILO) 14

j

job creation 14–15

l

laboratory-scale bioreactors 170
Levelized Cost of Biogas (LCOB) 134
life cycle assessment (LCA) 46, 152–153, 270
lignin 281
lignocellulolytic enzymes
  with biological methods 283
  lignocellulose pretreatment 282
  microorganisms 285
lignocellulosic biomass (LCB) 38, 95–98, 227, 228, 259, 263, 276, 278–282
lignocellulosic heterogeneity 128
lignocellulosic materials 113
limited host diversity 248–249
liquid biofuels 83–85
low-carbon economy 234–236
Low Carbon Fuel Standard (LCFS) 235

m

Machine learning (ML) techniques 9–10
Mascoma Corporation pilot plant 250
membrane bioreactors (MBs) 191
metabolically cooperative consortia 112
metabolic engineering 47, 62
metabolomics 69–70
methanogens 35
microalga cultivation systems 41–42
microbial biofactory
  genome editing 62–63
  metabolic engineering 62
  synthetic biology 63
microbial consortia 109–110
  development 116–117
  ecological interactions 110–111
  genetic and metabolic engineering of 131
  integration of 114–116
  types of 111–113
Microbial Electrolysis Cells (MECs) 41
microbial reactions/processes 30
microbial strain development 57–59, 266
microbial strain limitations 127
microorganisms 146
  in bioprocesses 34
  engineered 262–263
  native 261–262
mixed-cultures consortia 112
multi-species biofilm membrane (MBM) 100
mutualism 111

n

native CBP microorganisms 261
natural consortia 111

o

omics approaches in bioenergy 60–61, 68–70
Ontario's Green Energy Act in Canada 14
open ponds 194
operating expenditures (OPEX) 232
operational elements (OPEX) 152

p

packed bed bioreactors (PBRs) 43, 188–190
perforated trays 191–192
perfusion bioreactors 169
Phenomena of World Resource and Ecology initiative 143
photo bioreactor (PBR) 43, 194–196
photofermentation 41
policy and regulatory drivers 245–246
process analytical technologies (PATs) 151
process integration and intensification 132–133
process intensification (PI) techniques 60, 67
proteomics 69

r

reactor design and process integration 150–151
recycling 155–156
Renewable Energy Act 20
Renewable Energy Directive (RED II) 236
renewable energy market 3
renewable fuel standards (RFS) 87, 235
Renewable Identification Number (RIN) 236
Renewable Natural Gas (RNG) 13

s

Saccharomyces cerevisiae 96–97, 100, 146–149, 244, 245
second-generation
  biofuels 11–12, 81–82, 95, 276
  biorefineries 114
  feedstocks 33, 277–278
semi-fed-batch strategies 247
separate hydrolysis and fermentation (SHF) 148–149, 229, 277
simultaneous hydrolysis and fermentation (SHF) 263
simultaneous saccharification and fermentation (SSF) 44, 60, 67, 148–149, 229, 277
solar energy technology 7–8
solid biofuels 77
solid-state fermentations (SSFs) 180
spatially structured consortia 112
stirred tank reactors (STRs) 150, 184–187, 264
strain development trajectory innovation 236
submerged fermentations (SmFs) 180
sugar-based feedstocks 37–39
SuperPro Designer 152
sustainable and circular CBP models 270
sustainable aviation fuel (SAF) 269
sustainable biofuel production 251–252
  carbon-neutral future 253
  strategic policy recommendations 252–253
  vision for future 252
sustainable development goals 157
Sustainable Development Goals (SDGs) 143
sustainable fuels 10
synthetic biology 47, 63, 247, 270
  in CBP 229–230
  and metabolic engineering 149
synthetic consortia 112
syntrophy 111

t

technical feasibility 151–152
technological and innovation drivers 245–246
temporally phased consortia 112–113
thermophile-based CBP 247
third-generation
  biofuels 12, 82
  feedstocks 33, 278
traditional bioprocessing
  vs. consolidated bioprocessing 262
  vs. green technology 269–270
traditional multistep bioprocessing *vs.* CBP 262
transcriptomics 68–69
transesterification 39–40
triboelectric nanogenerators (TENGs) 8
Trichoderma reesei 100–101, 125, 147
tubular photobioreactors 195

u

United Nations Framework Convention on Climate Change (UNFCCC) 17
US Inflation Reduction Act 6

v

vertical-column photobioreactors 195–196

w

water footprint 153
wave bioreactors 193–194
wind energy innovations 8

y

yeast strains 98, 99
yield, defined 152

z

Zymomonas mobilis 147

Printed and bound by CPI Group (UK) Ltd, Croydon, CR0 4YY
24/08/2026
14948392-0001